Food Security and Sustainability: Global Issues and Challenges

Food Security and Sustainability: Global Issues and Challenges

Editor: Ashley Ward

www.callistoreference.com

Callisto Reference,
118-35 Queens Blvd., Suite 400,
Forest Hills, NY 11375, USA

Visit us on the World Wide Web at:
www.callistoreference.com

ISBN: 978-1-64116-841-0 (Hardback)

Cataloging-in-Publication Data

Food security and sustainability : global issues and challenges / edited by Ashley Ward.
    p. cm.
Includes bibliographical references and index.
ISBN 978-1-64116-841-0
1. Farms, Small. 2. Food security. 3. Food supply. 4. Agriculture. I. Ward, Ashley.
HD1476.A3 F66 2023
338.1--dc23

Table of Contents

Preface

This book has been an outcome of determined endeavour from a group of educationists in the field. The primary objective was to involve a broad spectrum of professionals from diverse cultural background involved in the field for developing new researches. The book not only targets students but also scholars pursuing higher research for further enhancement of the theoretical and practical applications of the subject.

Sustainability refers to the use of goods and resources in such a way that meets the needs of the present generation without compromising the ability of the future generations to meet their needs. The ability of individuals within a geographical area to afford, access and source the required amount of food is known as food security. A sustainable food system refers to a type of food system that can fulfill the nutritional and food requirements of all people in a sustainable manner. Sustainability and food security are essential for reducing the effects of climate change while meeting the increasing demand for food in the world. The four pillars of food security are availability, access, usage and stability. Major issues affecting food security and sustainability include poverty, population, change in climatic conditions, inadequate distribution of food, and unmonitored nutrition programs. This book is a valuable compilation of topics, ranging from the basic to the most complex global issues and challenges related to food security and sustainability. It will help the readers in keeping pace with the rapid changes in this area of study.

It was an honour to edit such a profound book and also a challenging task to compile and examine all the relevant data for accuracy and originality. I wish to acknowledge the efforts of the contributors for submitting such brilliant and diverse chapters in the field and for endlessly working for the completion of the book. Last, but not the least; I thank my family for being a constant source of support in all my research endeavours.

Editor

Reducing Postharvest Losses during Storage of Grain Crops to Strengthen Food Security in Developing Countries

Deepak Kumar * and Prasanta Kalita

ADM Institute for the Prevention of Postharvest Loss, University of Illinois at Urbana-Champaign, Urbana, IL 61801, USA; pkalita@illinois.edu
* Correspondence: kumard@illinois.edu

Academic Editor: Christopher J. Smith

Abstract: While fulfilling the food demand of an increasing population remains a major global concern, more than one-third of food is lost or wasted in postharvest operations. Reducing the postharvest losses, especially in developing countries, could be a sustainable solution to increase food availability, reduce pressure on natural resources, eliminate hunger and improve farmers' livelihoods. Cereal grains are the basis of staple food in most of the developing nations, and account for the maximum postharvest losses on a calorific basis among all agricultural commodities. As much as 50%–60% cereal grains can be lost during the storage stage due only to the lack of technical inefficiency. Use of scientific storage methods can reduce these losses to as low as 1%–2%. This paper provides a comprehensive literature review of the grain postharvest losses in developing countries, the status and causes of storage losses and discusses the technological interventions to reduce these losses. The basics of hermetic storage, various technology options, and their effectiveness on several crops in different localities are discussed in detail.

Keywords: postharvest losses; food security; grain storage; smallholders; hermetic storage

---

1. Introduction

Meeting the food demand of a rapidly increasing global population is emerging as a big challenge to mankind. The population is expected to grow to 9.1 billion people by the year 2050, and about 70% extra food production will be required to feed them [1–3]. Most of this population rise is expected to be attributed to developing countries, several of which are already facing issues of hunger and food insecurity. Increasing urbanization, climate change and land use for non-food crop production, intensify these concerns of increasing food demands. In the last few decades, most of the countries have focused on improving their agricultural production, land use, and population control as their policies to cope with this increasing food demand. However, postharvest loss (PHL), a critical issue, does not receive the required attention and less than 5% research funding has been allocated for this issue in previous years [4–7]. Approximately one-third of the food produced (about 1.3 billion ton), worth about US $1 trillion, is lost globally during postharvest operations every year [8]. "Food loss" is defined as food that is available for human consumption but goes unconsumed [9,10]. The solutions to reduce postharvest losses require relatively modest investment and can result in high returns compared to increasing the crop production to meet the food demand.

Postharvest loss includes the food loss across the food supply chain from harvesting of crop until its consumption [9]. The losses can broadly be categorized as weight loss due to spoilage, quality loss, nutritional loss, seed viability loss, and commercial loss [11]. Magnitude of postharvest losses in the food supply chain vary greatly among different crops, areas, and economies. In developing countries, people try to make the best use of the food produced, however, a significant amount of produce is lost in postharvest operations due to a lack of knowledge, inadequate technology and/or poor storage infrastructure. On the contrary, in developed countries, food loss in the middle stages of the supply chain is relatively low due to availability of advanced technologies and efficient crop handling and storage systems. However, a large portion of food is lost at the end of the supply chain, known as food waste. "Food waste" can be defined as food discarded or alternatively the intentional non-food use of the food or due to spoilage/expiration of food [12]. In 2010, estimates suggested that about 133 billion pounds of food (31% of the total available food) was wasted at retail and consumer level in the United States. Among different agricultural commodities, the studies estimated that on a weight basis, cereal crops, roots crops, and fruit and vegetables account for about 19%, 20%, and 44% losses respectively [8,13]. On a calorific content basis, losses in cereal crops hold the largest share (53%). Cereal grains, such as wheat, rice, and maize are the most popular food crops in the world, and are the basis of staple food in most of the developing countries. Minimizing cereal losses in the supply chain could be one resource-efficient way that can help in strengthening food security, sustainably combating hunger, reducing the agricultural land needed for production, rural development, and improving farmers' livelihoods.

Postharvest loss accounts for direct physical losses and quality losses that reduce the economic value of crop, or may make it unsuitable for human consumption. In severe cases, these losses can be up to 80% of the total production [14]. In African countries, these losses have been estimated to range between 20% and 40%, which is highly significant considering the low agricultural productivity in several regions of Africa [15]. According to the World Bank report, sub-Saharan Africa (SSA) alone loses food grains worth about USD 4 billion every year [16]. These losses play a critical role in influencing the life of millions of smallholder farmers by impacting the available food volumes and trade-in values of the commodities. In addition to economic and social implications, postharvest losses also impact the environment, as the land, water and energy (agricultural inputs) used to produce the lost food are also wasted along with the food. Unutilized food also results in extra CO_2 emissions, eventually affecting the environment. A report from the Food and Agriculture Organization of the United Nations (FAO) using the life cycle perspective, estimated about 3.3 Gtonnes of CO_2 equivalent emissions due to food that was produced but not eaten, without even considering the land use change [17]. The blue water footprints (water use during life cycle of food) for the wasted food globally was estimated to be about 250 km^3 [14,17]. Similarly, the land used to grow the food is another valuable resource that goes to waste due to these losses. A study conducted on rice postharvest losses in Nigeria estimated that the lost paddy accounted for 19% of the total cultivated area [18]. On the global scale, about 1.4 billion hectares of land was wasted by growing food that was not consumed in the year 2007, an area larger than Canada and China [19].

Considering the criticality of PHL reduction in enhancing the food security, it becomes very important to know the pattern and scale of these losses across the world, especially in developing countries, and identify its causes and possible solutions. Although losses occur at each stage of the supply chain from production to consumer level, storage losses are considered most critical in developing countries. This paper provides a comprehensive review and discussion on the status of storage losses of major cereal crops, major factors that lead to these losses and possible solutions. Technology interventions play a critical role in addressing the issue of PHL, and several efforts have been made to develop and disseminate these technologies for smallholders in developing countries. However, there is a lack of compiled evidence-based information on the effectiveness of these technologies for various crops. This paper discusses in detail the technology interventions, especially the use and effectiveness of hermetic storage in reducing storage losses particularly for smallholders in developing countries.

2. Grain Supply Chain

During the crop transition from farm to consumer, it has to undergo several operations such as harvesting, threshing, cleaning, drying, storage, processing and transportation. During this movement, crop is lost due to several factors such as improper handling, inefficient processing facilities, biodegradation due to microorganisms and insects, etc. It is important to understand the supply chain and identify factors at various stages that cause food losses. The section below will discuss the various stages in grain supply chain and type of losses occurring at each stage.

2.1. Harvesting

Harvesting is considered as the first step in the grain supply chain and is a critical operation in deciding the overall crop quality. In the developing countries, crop harvesting is performed mainly manually using hand cutting tools such as sickle, knife, scythe, cutters. Almost all of the crop is harvested using combine harvesters in the developed countries.

Harvesting timing and method (mechanical vs. manual) are two critical factors dictating the losses during the harvesting operations. A large amount of losses occurs before or during the harvesting operations, if it is not performed at adequate crop maturity and moisture content. Too early harvesting of crop at high moisture content increases the drying cost, making it susceptible to mold growth, insect infestation, and resulting in a high amount of broken grains and low milling yields [20]. However, leaving the matured crop un-harvested results in high shattering losses, exposure to birds and rodents attack, and losses due to natural calamities (rain, hailstorms etc.) [21]. Most of the harvesting is performed manually in the developing countries, which is a highly labor intensive and slow process. During peak harvesting season, even the countries such as India and Bangladesh encounter labor shortages, which results in delays in the harvesting and subsequently large losses. According to a study conducted in Punjab, India, due to high shattering losses, the wheat harvesting losses were found increased by about 67% (2.5% from 1.5%) by delay in harvesting [22]. Another postharvest loss study in India estimated a 10.3% increase (1.74% to 1.92%) in paddy harvesting losses due to delayed harvesting because of a lack in adequate harvesting equipment [23]. The recommended optimum moisture content during harvesting of various crops is listed in Table 1.

Table 1. Maturity moisture content of various crops (Source: De Lucia and Assennato [24]).

Crop	Maturity Moisture Content	Crop	Maturity Moisture Content
Paddy	22–28	Beans	30–40
Maize	23–28	Groundnut	30–35
Sorghum	20–25	Sunflower	9–10

2.2. Threshing and Cleaning

The purpose of the threshing process is to detach the grain from the panicles. The process is achieved through rubbing, stripping, or impact action, or using a combination of these actions. The operation can be performed manually (trampling, beating), using animal power, or mechanical threshers. Manual threshing is the most common practice in the developing countries. Grain spillage, incomplete separation of the grain from chaff, grain breakage due of excessive striking, are some of the major reasons for losses during the threshing process [20,25]. Delay in threshing after harvesting of crop results in significant quantity and quality loss, as the crop is exposed to atmosphere, and is susceptible to rodents, birds, and insect attack [26]. As in the case of harvesting, lack of mechanization is the major reason for this delay that causes significant losses. High moisture accumulations in the crop lying in the field may even lead to start mold growth in the field.

The cleaning process is performed after the threshing to separate whole grains from broken grains and other foreign materials, such as straw, stones, sand, chaff, and weed seed. Winnowing is the most common method used for cleaning in the developing countries. Screening/sifting is another common method of cleaning, which can be performed either manually or mechanically. Inadequate cleaned grains can increase the insect infestation and mold growth during storage, add unwanted taste and color, and can damage the processing equipment. A large amount of grains are lost as spillage during this operation, and grain losses during winnowing can be as high as 4% of the total production [27].

2.3. Drying

As apparent from the Table 1, the grains are usually harvested at high moisture content to minimize the shattering losses in the field. However, the safe moisture content for long-term storage of most of the crops is considered below 13% [21]. Even for the short-term storage (less than 6 months), the moisture should be less than 15% for most of the crops. Inadequate drying can result in mold growth and significantly high losses during storage and milling. Therefore, drying is a critical step after harvesting to maintain the crop quality, minimize storage losses and reduce transportation cost.

Drying can be performed naturally (sun or shade drying) or using mechanical dryers. Natural drying or sun drying is the traditional and economical practice for drying the harvested crop, and is the most popular method in developing countries. Sometimes, whole crop without threshing is left in the field only for drying. For example, after wheat harvesting, stacks are made of 10–15 bundles of tied crop, and left in the field for drying. Sun drying is weather dependent, requires high labor, is slow, and causes large losses. Grains lying in the open for sun drying are eaten by birds and insects, and also get contaminated due to mixing of stones, dust, and other foreign materials. Unseasonal rains or cloudier weather may restrict the proper drying, and the crop is stored at high moisture, which leads to high losses due to mold growth. About 3.5% and 4.5% losses were reported during maize drying on raised platforms in Zambia and Zimbabwe respectively [15,28]. Some farmers use mats or plastic sheets for spreading the grains, which reduces the contamination with dust and makes the collection of grains easy. Mechanical drying addresses some of the limitations of natural drying, and offers advantages, such as reduction in handling losses, better control over the hot air temperature, and space utilization. However, they suffer with the limitations of high initial and maintenance cost, adequate size availability, and lack of knowledge to operate these dryers, especially with smallholders. Due to these limitations, these dryers are rarely used by smallholders in the developing countries [26].

2.4. Storage

Storage plays a vital role in the food supply chain, and several studies reported that maximum losses happen during this operation [9,29,30]. In most of the places, crops are grown seasonally and after harvesting, grains are stored for short or long periods as food reserves, and as seeds for next season. Studies report that in developing countries such as India, about 50%–60% of the grains are stored in the traditional structures (e.g., Kanaja, Kothi, Sanduka, earthen pots, Gummi and Kacheri) at the household and farm level for self-consumption and seed [22]. The indigenous storage structures are made of locally available materials (grass, wood, mud etc.) without any scientific design, and cannot guarantee to protect crops against pests for a long time. Costa [31] estimated losses as high as 59.48% in maize grains after storing them for 90 days in the traditional storage structures (Granary/Polypropylene bags). The causes of losses during grain storage will be discussed in detail in a later section.

2.5. Transportation

Transportation is an important operation of the grain value chain, as commodities need to be moved from one step to another, such as field to processing facilities, field to storage facilities, and processing facilities to market. The lack of adequate transportation infrastructure results in damage of food products through bruising and losses due to spillage. Transportation loses are relatively very low in the developed countries due to better road infrastructure and engineered facilities on the field and processing facilities to load and unload the vehicles rapidly with very little or no damage. At the field level, most of the crop is transported in bullock carts or open trollies in South Asian countries. Grains for self-usage are usually transported in bags from field storage to processing facilities in bullock carts, bicycles, small motor vehicles, or open trucks. Poor road infrastructure along with these improper and poorly maintained modes of transportation results in large spillage and high contamination. Multiple movements of crop is another major reason for high transportation losses. In countries such as India and Pakistan, sometimes bagged wheat is loaded and unloaded from vehicles up to ten times before it is milled [21]. During each movement some grains are lost as spillage. Unlike efficient bulk handling systems in developed countries, loading and unloading of grains from wagons, trucks, and rails at processing facilities is performed mostly manually in the developing nations, and results in high spillage. Low quality Jute bags are used commonly during transportation and even storage, which results in high spillage rates due to leakage from the sacks. Large quantities (usually 100 kg of grains) in each bag, and hooks used to lift these bags cause tear in these bags and results in high spillage [21]. Even the trucks used in the developing countries are not totally suitable to transport cereals and oil seed crops. Alavi et al. [26] reported 2%–10% losses during handling and transportation of rice in Southeast Asia.

2.6. Milling

The milling or processing operations vary for different grains. In the case of rice, the purposes of milling are to remove the husk and bran layers of paddy to provide cleaned and whole white rice kernels for human consumption. The operation can be performed manually or using milling machines. Traditionally, in rural areas, milling is performed manually by repeated pounding. Milling yields are highly dependent on the milling method, skills of the operator, and crop conditions before the milling process. Milling of paddy containing foreign materials results in a high amount of cracked and broken kernels and can also damage machines. Inadequately maintained milling machines result in a high amount of broken kernels and low milling yields. Alavi et al. [26] reported that milling losses are highest among the losses during postharvest operations of rice in five Southeast countries: China, Thailand, Indonesia, the Philippines, and Vietnam. Milling yields of rice in all these five countries were reported well below the theoretical yield of 71%–73%. The yields from village level small mills were as low as 57% due to small scale, poor calibration, and lack of maintenance. High moisture and an inadequately cleaned paddy aggravate the situation and reduce yields.

Figure 1 summarizes various losses that occur during the supply chain of cereal crops and major factors responsible for those losses in the developing countries.

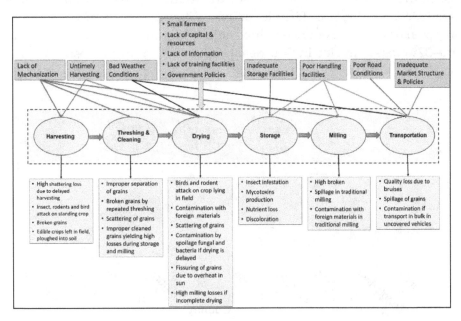

Figure 1.   Various factors and types of losses during the supply chain of cereal crops in developing countries.

3. Postharvest Losses of Cereal Crops in Developing Countries

Rice, wheat, and maize are major cereal grains in most of the developing countries. In countries such as Bangladesh, rice accounts for more than 90% of food produced and about 70% of calories intake [32]. In West Africa, Nigeria is currently the largest producer of rice with an annual production of about 3.3 million tonnes [18]. In spite of the large production and huge rice imports every year, a large number of people are undernourished in Nigeria. Similarly, Bangladesh is the fourth largest producer of rice worldwide, however, it is still food deficient and imports more than one million tons of rice every year. Saving the cereal crop lost during postharvest operations can help in meeting the food demand and reduce the load on the economy. A report from the World Bank, estimated 7%–10% of grain loss in postharvest operations at field level, and 4%–5% loss at the market and distribution stage in India for the year 1999 [25,33,34]. Estimates also suggest that these, approximately 12 to 16 million metric tons of grains wasted each year, could meet the food demand of about one-third of India's poor population [35]. However, despite the criticality of the issue, availability of consistent and reliable postharvest loss data is still a challenge. Very few loss assessment studies have been conducted in important developing economies such as India, China, and Brazil. After the FAO report in 2010, various institutes are making efforts to conduct comprehensive household surveys, interviews, and field measurements to determine the actual status of losses along food supply chains in various countries. This section discusses the status of losses in major staple crops based on the available literature data.

3.1. Rice

Being a high energy calorie food, rice accounts for one-fifth of the global calorie supply. The scale of postharvest losses in grain supply chains varies significantly, depending on the economy, agricultural conditions and practices, and climatic conditions of the region. An example of such variation of losses in the rice supply chain in different countries is illustrated in Figure 2. The rice losses have been reported as low as 3.51% in India to as high as 24.9% in Nigeria (Figure 2). Based on the 24.9% loss, the value of the total grain loss during the rice supply chain in Nigeria was estimated as 56.7 billion Nigerian Naira (NGN). According to Bala et al. [29], the losses in the rice value chain from producer to retailer were estimated as 10.74% to 11.71% (10.74% for Aman, 11.71% for Boro, and 11.59% for Aus) in Bangladesh. Most of the losses (85.28%–87.77% of the total) happened in the farm level operations, with storage losses (33.92%–40.99% of farm level losses) being the main contributor. Alavi et al. [26] compiled data on postharvest losses in rice value chains from different studies conducted by the FAO and reported 10%–37% losses in rice in Southeast Asia. In China, the losses were estimated in the range of 8%–26%.

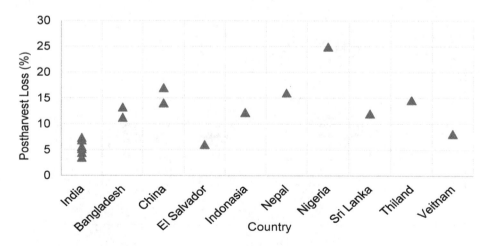

Figure 2. Postharvest losses in the rice value chain in various countries (in the case of a range of losses, an average of losses was used).

3.2. Wheat

Wheat is another major staple food of several countries in Europe, Asia, and North America. Similar to rice, significant losses happen during postharvest processing of wheat in developing countries. According to data compiled (before the year 1978) by US National Academy of Sciences, wheat losses in Sudan and Zimbabwe were estimated 6%–19% and 10% respectively [36]. Bala et al. [29] reported that the storage losses were maximum (41.7% of the total) among all the postharvest operation losses for wheat in Bangladesh, even considering the fact that the storage period of wheat is relatively small. Basavarja et al. [33] conducted a study to estimate losses in postharvest operations of wheat in the state of Karnataka, India. The estimations were based on a comprehensive survey from 100 farmers, 20 wholesalers, 20 processors and 20 retailers from the major producer district of each crop in the Karnataka state. The overall losses in the wheat supply chain from harvesting to retailer were estimated as 4.32%. Field level operations contributed 75.9% among the total postharvest losses (Table 2). The losses were maximum during storage operations due to a lack of availability, poor structures, presence of rodents, and improper drainage.

Table 2. Postharvest losses of wheat from various studies in different countries.

Country	Year	Losses (%)	Comments	Reference
Bangladesh	2010	3.62	- Maximum losses during storage	[29]
India	2013	1.84	- Maximum losses during harvesting - Punjab (study locality) is a developed state, so has better storage practices	[22]
	2013	2.74	- Study conducted in Uttar Pradesh - Maximum losses during harvesting (58.4% of the total)	[37]
	2004	4.32	- Study conducted in Karnataka - Maximum losses (0.95%) during storage at field level	[33]
	2012	4.32	- About 75% losses at the farm level - Maximum losses during storage at field level (28.9% of the total loss)	[38]
	2012	8.61	- Study conducted in Madhya Pardesh - Maximum losses during storage (56% of the total losses)	[39]
	2013	7.22	- Study conducted in West Bengal - Maximum losses during storage (54% of the total)	[37]
	2013	11.71	- Study conducted in Assam - Maximum losses during threshing (28.3%) and transportation (25.2% of the total)	[37]
Peru	2012	15–25	-	[40]
Sub-Saharan Africa	2013	15.2	-	[41]

3.3. Maize

Maize is an important part of staple food in Sub-Saharan Africa (SSA) and a major source (~36%) of daily calories intake. Pantenius [7] estimated 0.2%–11.8% weight loss due to insect infestation in maize after 6 months of storage in traditional granaries in Togo. Inter-American Institute for Cooperation on Agriculture (IICA) conducted a survey to estimate the postharvest losses in Latin America and Caribbean countries. The losses values estimated in various regions have been presented in the table below (Table 3). In almost all regions, most of the losses were observed occurring at the small and medium-scale farms due to a lack of adequate harvesting, drying, and storage technologies, along with a lack of information about the good agricultural practices. In Guatemala, due to a lack in storage structures along with the region's high humidity, storage losses were estimated between 40% and 45% [40]. Insect infestation was found as the major reason of storage losses in most of the cases. Kaminski and Christiansen [42] conducted a study to estimate the postharvest losses in maize crop in three SSA countries (Uganda, Tanzania, and Malawi) through comprehensive household surveys. The losses from the farm level activities were estimated in the range of 1.4% to 5.9%. Insects and pests were reported as the major reason of losses in maize during storage. Alavi et al. [26] reported an average of 23% losses in the maize value chain in ASEAN (the Association of Southeast Asian Nations) countries, with maximum losses happening during field drying (9%). Most of the maize is dried along the sides of road, especially in the Philippines. In Vietnam, major losses occur due to rodent attack and fungal disease during maize storage.

Table 3. Postharvest losses of maize from various studies in different countries.

Country	Year	Losses (%)	Comments	Reference
Bangladesh	2010	4.07	- Maximum losses during storage (60.4% of the total)	[29]
Ecuador	2012	10–30	- Major losses due to insect infestation during storage	[40]
Guatemala	2012	50	-	[40]
Malawai	2010	1.4	- These losses are only from farm level activities	[42]
Panama	2012	20	- Major losses at the small-scale level farm due to a lack of adequate technology	[40]
Peru	2012	15–25	-	[40]
Sub-Saharan Africa	2013	17.8	-	[41]
Tanzania	2008 2010	4.4 2.9	- These losses are only from farm level activities	[42]
Uganda	2009	5.9	- These losses are only from farm level activities	[42]

4. Storage Losses in Developing Countries

As discussed in the above section, the maximum amount of losses occurs during the storage of crops due to a lack of adequate infrastructure. Storage losses can be classified in two categories: direct losses, due to physical loss of commodities; and indirect losses, due to loss in quality and nutrition. It is important to consider both damage and losses by the insects during storage instead of just weight loss. "Damage" can refer to physical evidence of deterioration, for example, holes in the grains. It mainly affects the quality of grains. "Loss", on the other side, is the total disappearance of the food, which can be measured quantitatively [43]. The loss in quality results in value loss of the product, and sometimes leads to total rejection also. The rejection rate depends upon the individual's economic status and cultural background. For example, a subsistence farmer may consume the damaged food to some extent, whereas, affluent customers may reject even slightly damaged food. Some loss happens in the form of spillage from leaky sacks which can be identified when the store is emptied and the spilt grain remains on the floor.

The storage losses are affected by several factors, which can be classified into two main categories: biotic factors (insect, pest, rodents, fungi) and abiotic factors (temperature, humidity, rain) [32]. Moisture content and temperature are the most crucial factors affecting the storage life. Most of the storage molds grow rapidly at temperatures of 20–40 °C and relative humidity of more than 70% [32]. Low moisture keeps the relative humidity levels below 70% and limits the mold growth. In the traditional storage structure, temperature fluctuations due to weather changes cause moisture accumulation either at the top or bottom of the grains' bulk depending on the direction of air convection. This can be avoided by minimizing the temperature difference of inside and outside the storage structure. Grains should be dried to about 13% of the moisture content before storage to minimize the losses. At moisture contents of 16% or higher, the safe storage period of rice is only a few weeks [32]. Quality of grains before storage is another critical factor affecting the storage losses. Mechanical damage during harvesting and threshing can result in bruised areas on grains, which may serve as centers for infection and cause deterioration [25]. The criticality of a factor depends upon the storage conditions.

In most of the developing countries, especially in Africa and South Asia, grains are generally stored as bulk or bags in simple granaries constructed from locally available materials (straw, bamboo, mud, bricks). Mud bins and pots, bokharies (straw structure), kothis, and plastic containers are common storage structures in Asia [21]. Gunny bags and Plastic/Polythene bags are commonly used for the short duration storage, and Dole, Berh, Gola, Motka, Steel/Plastic drums are used for the long duration storage. Various types of granaries are used in African countries. Ebli-va, Kedelin, in-house smoked storage are some of the common maize storage structures in Togo [7]. In the in-house smoked storage methods, maize is stored within the dwelling in space between the ceiling and roof over the cooking spot to receive the heat. In West Africa, grains are commonly stored in the home or field in jute or polypropylene bags, raised platforms, conical structures, and baskets [15,44]. In East and Southern Africa, farmers use cow dung ash in small bags, wood cribs, pits, iron drums enclosed with mud, and metal bins for storing the grains [15,45]. "Nkokwe" is one of the most commonly used storage structure used in Malawi and Kenya. This is a kind of cylindrical basket made up of interwoven split bamboo and covered with a conical shaped roof of grass. The structure is raised off the ground on stilts [36]. Most of these structures are not scientifically designed and are made from locally available materials, and cause damage to stored grains due to biological, environmental and other factors.

4.1. Insect Infestation

Among all the biotic factors, insect pests are considered most important and cause huge losses in the grains (30%–40%) [15,43,46]. Some studies in Ghana reported that the maize losses due to insect infestation could be up to 50%, if all quantity losses, quality losses, and income loss due to early sale are considered in the estimation [43]. According to an economic model by Compton et al. [47], each percent of insect infestation results in 0.6%–1% depreciation in the value of maize [47,48]. From field studies in Togo, Pantenius [7] observed that insects and pests were responsible for 80%–90% of storage losses in grains. *Callosobruchus maculatus* (F.) alone, a common pulse weevil has been found responsible for up to 24% losses in stored pulses in Nigeria [46]. Losses due to insects in stored maize have been reported from 12% to 44% in the western highlands of Cameroon [46]. The maize weevil (*Sitophilus zeamais*), and larger grain borer (LGB) (*Prostephanus truncatus*) are the major pests in the maize. About 23% losses were observed in maize grains stored for six months, mainly due to infestation of maize weevil and LGB in Benin [49,50]. LGB originated in Central America and was accidently introduced in Africa in late 1970s [49]. Nowadays, it is found in most parts of Africa and is considered the most threatening pest, as it causes extensive damage in a very short time [43,51,52]. The sporadic nature of LGB makes even its control difficult: it does not infest all stores of the same area, and its reoccurrence in each year is not guaranteed. At farm level storage, more than 30% of weight loss have been observed in maize due to these pests [43,52]. Some studies on maize losses in Ghana estimated about 5% to 10% loss in market value due to infestation by only *Sitophuilus* spp., and 15% to 45% market value loss due to damage by LGB. Overall, these losses were equivalent to about 5% of the average household income in that area [43,53,54]. Abass et al. [15] reported that after six months of maize storage, LGB was responsible for more than half (56.7%) of the storage losses, followed by losses due to grain weevil and lesser grain borer. Patel et al. (1993) observed about 25% losses by *R. dominica* in wheat stored for 3 months under laboratory conditions [55].

4.2. Mycotoxins

Mycotoxin contamination is another big challenge, especially in the case of maize, which makes the food unsuitable for human consumption or animal feed. A large amount (25%–40%) of cereal grains are contaminated by the mycotoxins produced by storage fungi world-wide [56]. Molds and mycotoxins cause dry matter as well as quality loss, and are a hazard in the food value chain [57]. Aflatoxins, Fumonisins, Deoxynivalenol, and Ochratoxin are the most common and important mycotoxins, especially in maize [58–60]. Aflatoxins, produced as secondary metabolite by two fungi species *Aspergillus flavus* and *A. parasiticus*, are considered the most dangerous group of mycotoxins, as they increase the risk of liver cancer and affect growth in young children [59,61]. Because of food contamination, about 4.5 billion people are exposed to aflatoxins in developing countries [56,62]. High concentrations of aflatoxin can lead to aflatoxicosis, which can cause severe illness and even death [63]. *Penicillium verrucosum* (ochratoxin), a major mycotoxigenic mold is commonly found in damp cool climates (e.g., Northern Europe) and *Aspergillus flavus* is mostly observed in temperate and tropical conditions [57].

Mold during storage damages the grains as well as reducing grain germination. It also deteriorates the grain quality due to the must odour, increased fatty acid content, and reduced starch and sugar contents. Lipid peroxidation is another phenomenon that causes food deterioration and alters the taste and aroma, and may cause undesirable effects on human health [56]. High oil content varieties of oil seeds demand particular attention during storage, as the high level of moisture degrades the vegetable oil and produces high fatty acids, which sometimes also results in self-heating [14]. At farm level storage in developing countries, even rodents can damage a large portion of crop, whereas, fungi can be a major reason for spoilage at high relative humidity storage. Use of scientific storage structures and proper handling of grains can reduce storage losses to less than 1% [31,64].

Losses can be minimized by physically avoiding the entry of insects and rodents, and maintaining the environmental conditions that avoid growth of microorganisms. The knowledge of control points during harvesting and drying before storage can help in reducing losses during the storage of cereals. Taking the timely preventive actions for biotic and abiotic factors can be very effective in reducing the losses during storage.

5. Interventions to Reduce Storage Losses for Smallholders

Although a huge challenge, storage losses can be mitigated by use of efficient storage technology, upgrading infrastructure and storage practices. World Food Programme (WFP) with the help of the government and non-governmental organizations (NGOs) performed an Action Research Trial in Uganda and Burkina Faso to demonstrate the impact of improved postharvest management practices and using new storage technologies on the crop loss after harvesting [31]. The results concluded that irrespective of crop or storage periods, use of improved practices and new technologies resulted in about a 98% reduction in food loss [31]. It was also observed that losses in the traditional storage structures were much higher than those reported in the literature, because the storage period was longer than that commonly used by farmers in these countries. It is important to understand their usefulness, technical efficacy, and limitations to promote their adaptability among the consumers. This section of the paper will discuss various practices and technology interventions that can help in reducing storage losses for smallholders in developing nations. Other than saving losses, the availability of low cost and effective storage structures can motivate farmers to store their grains and obtain high prices instead of selling right after harvesting when there is an abundant supply of grains.

5.1. Chemical Fumigation

Synthetic insecticides are used in several countries and play an important role in controlling the pests and reducing losses during storage of grains. Methyl bromide (MB) and phosphine are the most commonly used chemicals in developing countries [65]. If the maize grains are sufficiently dry (moisture content less than 13%), use of phostoxin can control the LGB infestation in maize grains. However, phostoxin can be applied only by the licensed technicians in most parts of Africa, and farmers are allowed to use a mixture of pirimiphos-methyl (Actellic) and permethrin, commercially sold as Actellic Super [66]. Shelling the grains and storing them in polypropylene bags after proper application of Actellic Super can effectively avoid the pest infestation for a few months of storage [49,66]. This practice has been widely adopted by farmers in African countries, especially Kenya. More than 93% of farmers reported this as the most common method used for controlling pest during storage in Tanzania [49].

Irrespective of their effectiveness, the synthetic insecticides suffer from limitations such as high costs, development of genetic resistance in the treated pests, health hazards due to toxic residues, and environmental contamination [46,65]. Residuals from synthetic fumigants could cause considerable loss of seed viability [67]. Due to long use of phosphine, some insects have gained resistance to chemical fumigation in some countries [68,69]. Use of these chemical fumigation methods is even challenging in the traditional storage structures used in the developing countries, as most of them are open to reinfestation. Another challenge with these chemicals is knowledge and training to apply these pesticides at the correct time and at the correct dose. The delayed treatment, adulterated chemicals, and incorrect dosage can reduce the efficacy of the treatment and result in high storage losses.

5.2. Natural Insecticides

Several plant species and their extracts have been found with natural pesticide ability and are used very commonly as a traditional practice to protect the grains from insects in several African and Asian countries. Plant based chemicals and products would be biodegradable, environment friendly, and relatively safe for human health. The leaves and oil extract from leaves of *Chenopodium ambrosioides Linn.* (Chenopodiaceae) has been found to be very effective in controlling the damage of cereal grains by the insects during storage, in several studies. The plant is a branched herb and widely available in India [56]. Its natural pesticide abilities have been highlighted and investigated in several studies. Kumar et al. (2007) investigated the effectiveness of essential oil from wormseed against the fungal deterioration in stored wheat. The samples were analyzed for fungi after 12 months of storage at laboratory conditions. The oil was found to be significantly effective in controlling the *A. flavus* fungi in both inoculated (91.17% protection) and uninoculated (99.42% protection) wheat samples. The efficacy of oil was compared with synthetic fungicides (benzimidazole (Benomyl), diphenylamine (DPA), phenylmercuric acetate (Ceresan) and zinc dimethyl dithiocarbamate (Ziram)), and the oil was found to be relatively more effective, with minimum inhibitory concentration (MIC) (concentration at which the oil shows absolute fungitoxicity) lower than those of synthetic ones. Tapondjou et al. (2002) investigated the effectiveness of using leaves and extracted essential oil from wormseed (*Chenopodium ambrosioides* L.) against six common species of grain beetles in the western highlands of Cameroon: *Sitophilus zeamais* (maize weevil)., *S. granarius* (L.) (granary weevil), *Callosobruchus chinensis* (L.), *C. maculatus, Acanthocelides obtectus* (all Bruchidae), and *Prostephanus truncatus* (bruchids) (larger grain borer). The ground leaves were mixed with grains at the concentration of 0.05%–0.8% (w/w) for bruchids, and 0.8%–6.4% (w/w) for weevils and borers. The grains tested were maize for *S. zeamais* and *P. truncatus*, whole wheat for *S. granarius*, green peas for *C. chinensis*, mung bean for *C. maculatus* and white bean for *A. obtectus*. The oil effectiveness was checked on the filter paper discs by treating a Whatman No. 1 filter paper with the oil diluted in acetone at different concentrations of oil (0 to 1.6 μL/cm^2). The mortality rate was found to be relatively high for bruchids (100% for *C. chinensis* in 48 h) compared to that in the case of weevils and borers. The essential oil was observed to be more toxic than using leaves and resulted in high mortality. Other than mortality, the ground leaves were effective in inhibiting the F1 progeny production and adult emergence of the insects. Shaaya et al. [65] tested four edible oils: Pure soybean oil, pure and crude cottonseed oils, crude rice bran oil and crude palm kernel oil, as fumigants against common insects in beans and wheat. Crude palm kernel and crude rice bran oils were found to very effectively control *C. maculatus* in chickpea for the initial 4–5 months. Even at the end of 15 months, the insects in the fumigated samples were only 10% of those in the control sample. Similarly, crude cotton and soya bean oils were found to be effective against *S. oryzae* in wheat for the initial 4–5 months, and were effective later at high concentrations (10 g/kg). Only crude cotton oil was found to be significantly effective in controlling *S. zeamais* in maize, and that too mainly during the initial 4 months of storage period. After 8 months of storage, the oil was partially effective and had 20%–40% of the amount of insects of those in untreated maize samples. The major issue with these plant materials is that oil yields are low and might be expensive to use on a commercial scale, however, some plant leaves can be used as natural insecticides by smallholders.

5.3. Hermetic Storage

Hermetic storage (HS), also known called as "sealed storage" or "airtight storage", is gaining popularity as a storage method for cereal, pulses, coffee, and cocoa beans in developing countries, due to its effectiveness and avoidance of the use of chemicals and pesticides. The method creates an automatic modified atmosphere of high carbon dioxide concentration using sealed waterproof bags or structures. As the structures are airtight, the biotic portion of the grains (insects and aerobic microorganisms) creates a self-inhibitory atmosphere over time by increasing carbon dioxide concentration (oxygen decreases) due to its respiration metabolism. Some studies have reported that the aflatoxin production ability of *Aspergillus flavus* is also reduced at high concentrations of CO_2 [63,70]. Hermetic storage has been observed to be very effective in avoiding the losses (storage losses less than 1%) during long distance (international) shipments also [69]. Ease of installation, elimination of pesticide use, favorable costs, and modest infrastructure requirements are some of the additional advantages that make the hermetic storage options attractive [71].

The relationship of the main factors affecting the respiration of grain and microorganisms in hermetic storage has been described in the Figure 3 [72]. The CO_2 concentration inside the bags is usually used as an indicator of the biological activity of grains. [72,73]. Permeability of the bag and the gas partial pressure effect the movement of gases (O_2 and CO_2) in and out, whereas the concentration of these gases inside the bag depends on the balance between these exchanges and the respiration of the biotic portion of grains. Higher initial moisture content tends to increase the CO_2 concentration because of the increased respiration, however, the change was not found to be significant [72].

Another factor affecting the respiration rate is grain temperature. It has been observed in experimental studies that the temperature inside the bags follows the ambient temperature trend, and for every 10-degree increase in temperature, the CO_2 concentration increases by about 1.5% [72]. World Food Programme (WFP) in their Action Research Trial in Uganda and Burkina Faso, found out that if properly sealed, the hermetic storage units were themselves very efficient in killing the pests and insects without any use of phosphine fumigation [31]. Various hermetic storage options, such as Metallic silos, Purdue Improved Cowpea Storage (PICS) bags, SuperGrain bags, etc., have been developed and widely promoted in the last few years. These bags are being considered practical and cost-effective storage technology, and are becoming very popular in several countries [74].

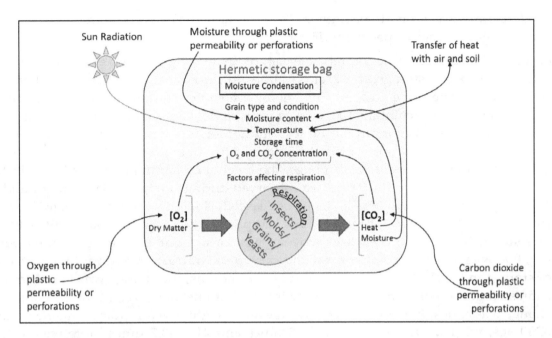

Figure 3. Illustration of the factors affecting the grain and microorganism respiration in the hermetic storage (Adapted from Cardoso et al. [72]).

A metal silo is a strong hermetically sealed structure (mostly cylindrical), built using a galvanized steel sheet, and has been found to be very effective for storing grains for long periods of time and avoiding insects and rodents [75,76]. In some locations, the siloes are made of painted aluminium sheeting which helps prevent corrosion and improves their appearance [76]. It is considered to be one of the key technologies which will be helpful in reducing postharvest losses and improving food security of smallholder farmers. PICS or Purdue Improved Crop Storage bags, originally developed for storage of cowpea, involve triple bagging the grains in hermetic conditions, and is widely used by farmers in sub-Saharan Africa. The grains are stored in double layer thick (80 μm) high density polyethylene (HDPE) bags and is held in a third woven nylon bag. After filling with the grains, the bags are sealed airtight. This will cut off the oxygen to the weevils and hinder their metabolic pathways preventing them from producing water, and killing them by desiccation (Murdock et al., 2012). More than 3 million PICS bags were sold in West and Central Africa during 2007–2013. SuperGrain, commercialized by GrainPro Inc. is another widely used water resistant and hermetic storage option. These bags are made up of a single thick layer of high density polypropylene with a thickness of about 78μm, and used as liner along with normal woven polypropylene bags [59]. ZeroFly bags are a product of Vestergaard, Switzerland. These are insecticide infused woven polypropylene bags designed to prevent damaging pest infestations. The bag is made with pyrethroid incorporated into polypropylene yarns.

Baoua et al. [77] conducted a study to compare the performance of SuperGrain bags and PICS bags, available in the West African market, to control pest infestation over the 4 months of storage. The change in temperature and relative humidity in both the bags were similar over time. The infestation level of *C. maculatus* eggs on the seeds after four months was found to be lower in SuperGrain bags (18.5 eggs per 100 seeds) in comparison to PICS bag (26.1 eggs per 100 seeds). Grain damage was observed relatively lower (18.5 grains with holes per 100 tested grains) for the PICS bag compared to that in the SuperGrain bags (29.5 grains with holes per 100 tested grains). Somavat et al. [78] compared the effectiveness of hermetic bin bags (GrainSafe IIITM, GrainPro Inc., Concord, MA, USA), metallic bins and gunny bags for storage of wheat under ambient conditions in India. There was no insect infestation found in clean grains stored in hermetic bags after 9 months of storage. For the artificially infested grains, bored grain percentage remained stable at 0.33% for hermetic bags in contrast to 2% and 8% for metallic bins and gunny bags respectively. At end of storage, seed viability was found to be higher (88%) for hermetic bags compared to 82% and 73% in metallic bins and gunny bags respectively (Figure 4) [78,79].

Baoua et al. [48] conducted a comprehensive study to investigate the effectiveness of PICS bags (50 kg capacity) for maize storage at eleven localities in Burkina Faso, Ghana, and Benin. Insect infestation in maize during storage varied from nil to highly infested. After about 196 days of storage, the PICS bags were able to maintain 100 seed grain weight, seed viability, and seed germination along with 95%–100% insect mortality at all localities. Moisture content was also observed to be unchanged during storage for most of the PICS bags. Although aflatoxin levels were observed in maize stored in both the PICS and woven bags, the level of contamination was lower in the PICs bags. Similar effectiveness of the PICS bags was observed during storage of Bambara groundnut [80], maize [81], mung bean and pigeonpeas [67], pigeonpeas [82] (Table 4). Mutungi et al. [67] investigated the effectiveness of the PICs bags for mung beans and pigeonpeas. Naturally and artificially infested grains were stored in the PICS bags and woven bags for 6 months. For both mung beans and pigeonpeas, the oxygen levels were reduced and carbon dioxide levels rose rapidly within the two months of storage in the PICS bags. For the initial two months, the change was found to be higher in highly infested grains compared to naturally infested grains, however, at the end of storage the average change was almost the same. Insect damage and weight loss for grains remained unchanged in the PICS bags, whereas, there was 24.2–27.5 times and 21.7–43.7 times more weight loss after 6 months of storage in the woven polypropylene bags for mung beans and pigeonpeas respectively. Treatment of grains with Actellic Super dust before storage in the woven bags did not help in reducing damage, and weight loss at the end of storage increased by factors of 20.8 and 22.5 for mung beans and pigeonpeas respectively.

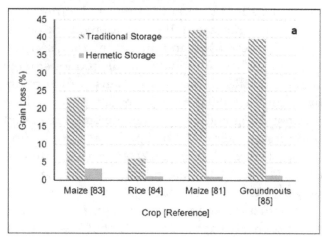 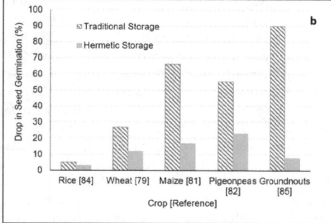

Figure 4. Amount of losses (a) weight loss; (b) seed germination losses, for various grains due to natural or artificial insect infestation during storage in traditional storage vs. hermetic storage (in the case of a range of losses, an average of the losses was used).

Figure 5 illustrates the effectiveness of various storage options compared to traditional polypropylene bags in reducing the losses of maize grains after 90 days of storage [31]. It can be clearly observed that losses in all new storage techniques were significantly lower than those in the traditional storage, with the minimum being in the case of metallic silos. A nationwide study on postharvest losses of rice in China reported 7%–13% grain losses at the rural household storage facilities, compared to only 0.2% losses at the national reserve level using scientific storage structures [64].

Table 4. List of the effective use of hermetic bags for various crops in developing countries

Type of Storage	Crop	Country	Duration of Storage	Investigations	Findings	Reference
	Maize	Kenya	6 months	Evaluated performance of hermetic storage (metal silos and super grain bags) and polypropylene bags to control infestation of pests.	Metal silo was the most effective option in controlling pest infestation. Metal silo was equally effective in controlling pest infestation even without any insecticide use. Supergrain bags were effective in controlling the infestation, however, the insect mortality was not complete. Bags were perforated by a larger grain borer.	[66]
SuperGrain Bags	Maize	Benin	150 days	Compared performance of hermetic bags and woven polypropylene bags for storage of maize infested with *Prostephanus truncatus* (Horn) and *Sitophilus zeamaisvas* (Motschulsky).	Moisture levels remained unchanged in hermetic bags. Growth of insects (*Prostephanus truncatus* and *Sitophilus zeamaisvas*) was significantly less in hermetic bags. There were **0.5%–6% losses** at end of storage compared to **19.2%–27.1% losses** in woven bags.	[83]
	Rice	Bangladesh	4 months	Compared performance of hermetic bags and traditional structures for storage of rice.	Moisture content of grains remained unchanged in hermetic bags. A total of **97%** seed germination in hermetic bags vs. **95%** in traditional storage. A total of **1%** damaged grains in hermetic bags in contrary to **6%** in traditional storage.	[84]

Table 4. *Cont.*

Type of Storage	Crop	Country	Duration of Storage	Investigations	Findings	Reference
	Mung bean, pigeonpea	Kenya	6 months	Evaluated performance of hermetic bags for naturally and artificially infested (*Callosobruchus maculatus* (F.)) grains.	One hundred grain weight, infestation, and grain damage remain unchanged in hermetic bags. There was **60.3 to 76.9%** damage in mung beans and **75.8%–95.7%** grain damage for pigeonpeas stored in woven polypropylene.	[67]
	Maize	Benin, Burkina Faso and Ghana	6.5 months	Evaluated performance of hermetic bags for preserving maize quality during storage.	There was **95%–100%** insect mortality in hermetic bags. PICS bags maintained the seed viability and germination.	[48]
	Maize	Kenya	6 months	Evaluated performance of hermetic bags for naturally and artificially infested (*Prostephanus truncates*) grains.	There was **0%–2%** weight loss in PICS bags compared to **36.3%–47.7%** weight loss in woven polypropylene bags. There was a **13%–20.1%** reduction in germination for grains stored in PICS bags compared to a **54.1%–78.4%** drop for grains stored in wove bags.	[81]
PICS bags	Bambara groundnut	Maradi, Niger	7 months	Evaluated performance of hermetic bags for preserving naturally infested Bambara groundnut quality during storage.	For highly infested grains, oxygen concentrations decreased significantly in hermetic bags contrary to unchanged in woven bags. Infestation level of *C. maculatus* in woven bags was **128 times higher** than that of hermetic bags. There was a **34.8%–89.3%** decrease in seed viability in woven bags, whereas, there was **no change** in grains stored in PICS bags. Abrasions were observed in inner HDPE bags.	[80]
	Pigeonpeas	India	8 months	Compared performance of hermetic bags vs. gunny bags for storage of pigeonpeas.	Germination of infested grains in gunny bags dropped to **44.5%** compared to high germination (**77%**) for grains stored for 8 months in hermetic bags.	[82]
	Groundnuts	India	4 months	Evaluated performance of hermetic bags for preserving the quality of natural and artificial infested groundnuts.	There was a **0.8%** decrease in seed weight for groundnut stored in hermetic bags compared to **7.2%** in cloth bags. Only **1.4%** weight loss for artificially infested groundnut stored in hermetic bags compared to **39.6%** in cloth bag. There was **92.3%** germination for artificially infested seeds stored in hermetic bags compared to only **10%** in the case of cloth bags.	[85]

PICS: Purdue Improved Cowpea Storage; HDPE: high density polyethylene.

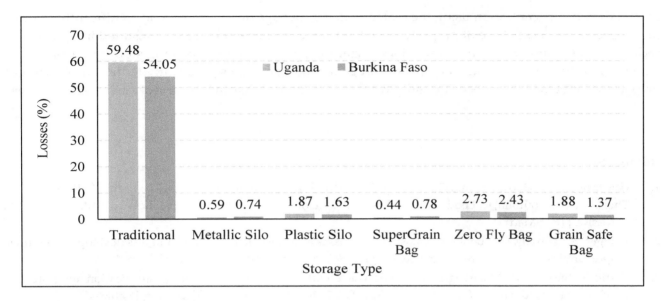

Figure 5. Losses in maize grain after 90 days of storage in various storage structures (data extracted from Figures 1 and 2 of Costa 2014 [31]).

Metal silos have been found to be effective in several other studies. However, their initial high cost is a major obstacle for their adoption by smallholders. Community level silos might be an economic alternative, as the cost per unit of grains decreases with increases in the size of silos. The maintenance cost is very low in the case of silos, which can compensate for the high initial cost to some extent. Kimenju and Hugo [49] conducted an economic analysis of using advanced storage structures, and reported that the economic gain (extra income by avoiding losses) using a metal silo compared to polypropylene bags could be up to USD 100 per ton of grains after 12 months of storage. However, farmers have to spend an extra USD 171 (1.8-ton capacity) to USD 316 (0.36-ton capacity) as the initial cost of silos over polypropylene bags. One of the main challenges of using hermetic bags is that the grain to be stored should be thoroughly dried to avoid mold and rotting of grains. Although these bags prevent the damage from insects, they do not provide an effective barrier from rodents. Specifically, SuperGrain bags are widely used in Russia and Latin America for storage of coffee. They are also popularized in Afghanistan for wheat storage, Nepal for corn, and in Vietnam for rice conservation [86]. These bags are being used successfully in several Latin American countries to store all major grain crops without application of pesticides [74]. Hermetic storage structures developed by GrainPro Inc. are used by several commercial companies in India to store high value spices and Basmati rice [71].

Technology interventions and improved storage structures can significantly reduce store losses. However, it is important to understand that training of smallholders is equally as necessary as the technology dissemination [87]. Along with making these technologies available at a reduced price, the government agencies and organizations have to ensure the development of facilities to provide information and training about the use and maintenance of these technologies in the local language, for successful adaptation and effective use of these technologies.

6. Conclusions

Postharvest loss is a complex problem and its scale varies for different crops, practices, climatic conditions, and country economics. Storage losses account for the maximum fraction of all postharvest losses for cereals in developing countries, and negatively affect the farmers' livelihoods. Most of the harvested grains are stored in the traditional storage structures, which are inadequate to avoid the insect infestation and mold growth during storage and lead to a high amount of losses. Technology interventions and improved storage structures can play a critical role in reducing postharvest losses and increasing farmers' revenues. Hermetic storage creates an automatic modified atmosphere of high carbon dioxide concentration using the sealed waterproof bags or structures, and significantly reduces insect infestation losses. Use of properly sealed hermetic storage structures has resulted in up to a 98% reduction in storage losses, maintained seed viability, and its quality for long storage times. Using better agricultural practices and adequate storage technologies can significantly reduce the losses and help in strengthening food security, and poverty alleviation, increasing returns of smallholder farmers.

References

1. Godfray, H.C.J.; Beddington, J.R.; Crute, I.R.; Haddad, L.; Lawrence, D.; Muir, J.F.; Pretty, J.; Robinson, S.; Thomas, S.M.; Toulmin, C. Food security: The challenge of feeding 9 billion people. *Science* **2010**, *327*, 812–818. [CrossRef] [PubMed]

2. Hodges, R.J.; Buzby, J.C.; Bennett, B. Postharvest losses and waste in developed and less developed countries: Opportunities to improve resource use. *J. Agric. Sci.* **2011**, *149*, 37–45. [CrossRef]

3. Parfitt, J.; Barthel, M.; Macnaughton, S. Food waste within food supply chains: Quantification and potential for change to 2050. *Philos. Trans. R. Soc. B Biol. Sci.* **2010**, *365*, 3065–3081. [CrossRef] [PubMed]

4. Bourne, M. *Post Harvest Food Losses—the Neglected Dimension in Increasing the World Food Supply*; Department of Food Science and Technology, Cornell University: Ithaca, NY, USA, 1977.

5. Greeley, M. Food, Technology and employment: The fram-level post-harvset system in developing countries. *J. Agric. Econ.* **1986**, *37*, 333–347. [CrossRef]

6. Kitinoja, L.; Saran, S.; Roy, S.K.; Kader, A.A. Postharvest technology for developing countries: Challenges and opportunities in research, outreach and advocacy. *J. Sci. Food Agric.* **2011**, *91*, 597–603. [CrossRef] [PubMed]

7. Pantenius, C. Storage losses in traditional maize granaries in Togo. *Int. J. Trop. Insect Sci.* **1988**, *9*, 725–735. [CrossRef]

8. Gustavsson, J.; Cederberg, C.; Sonesson, U.; van Otterdijk, R.; Meybeck, A. *Global Food Losses and Food Waste*; Food and Agriculture Organization of the United Nations: Rome, Italy, 2011.

9. Aulakh, J.; Regmi, A.; Fulton, J.R.; Alexander, C. Estimating post-harvest food losses: Developing a consistent global estimation framework. In Proceedings of the Agricultural & Applied Economics Association's 2013 AAEA & CAES Joint Annual Meeting, Washington, DC, USA, 4–6 August 2013.

10. Buzby, J.C.; Farah-Wells, H.; Hyman, J. The estimated amount, value, and calories of postharvest food losses at the retail and consumer levels in the United States. Available online: https://papers.ssrn.com/sol3/papers.cfm?abstract_id=2501659 (accessed on 31 May 2015).

11. Boxall, R.A. Post-harvest losses to insects—A world review. *Int. Biodeterior. Biodegrad.* **2001**, *48*, 137–152. [CrossRef]

12. Food and Agriculture Organization of United Nations. *Global Initiative on Food Losses and Waste Reduction*; FAO: Rome, Italy, 2014.

13. Lipinski, B.; Hanson, C.; Lomax, J.; Kitinoja, L.; Waite, R.; Searchinger, T. *Installment 2 of "Creating a Sustainable Food Future" Reducing Food Loss and Waste*; Working Paper; World Resource Institute: Washington, DC, USA, 2013.

14. Fox, T. *GlobaL Food: Waste Not, Want Not*; Institution of Mechanical Engineers: Westminster, London, UK, 2013.

15. Abass, A.B.; Ndunguru, G.; Mamiro, P.; Alenkhe, B.; Mlingi, N.; Bekunda, M. Post-harvest food losses in a maize-based farming system of semi-arid savannah area of Tanzania. *J. Stored Prod. Res.* **2014**, *57*, 49–57. [CrossRef]

16. Zorya, S.; Morgan, N.; Diaz Rios, L.; Hodges, R.; Bennett, B.; Stathers, T.; Mwebaze, P.; Lamb, J. *Missing Food: The Case of Postharvest Grain Losses in Sub-Saharan Africa*; The international bank for reconstruction and development/the world bank: Washington, DC, USA, 2011.

17. Food and Agriculture Organization of the United Nations. *Food Wastage Footprint—Impacts on Natural Resources*; FAO: Rome, Italy, 2013.

18. Gesellschaft für Internationale Zusammenarbeit. *Post-Harvest Losses of Rice in Nigeria and Their Ecological Footprint*; Deutsche Gesellschaft für Internationale Zusammenarbeit (GIZ) GmbH: Bonn/Eschborn, Germany, 2014.

19. Food and Agriculture Organization of the United Nations. *Toolkit: Reducing the Food Wastage Footprint*; FAO: Rome, Italy, 2013.

20. Khan, M.A. *Post Harvest Losses of RICE*; Khan, S.L., Ed.; Trade Development Authority of Pakistan: Karachi, Pakistan, 2010.

21. Baloch, U.K. *Wheat: Post-Harvest Operations*; Lewis, B., Mejia, D., Eds.; Pakistan Agricultural Research Council: Islamabad, Pakistan, 2010; pp. 1–21.

22. Grover, D.; Singh, J. Post-harvest losses in wheat crop in Punjab: Past and present. *Agric. Econ. Res. Rev.* **2013**, *26*, 293–297.

23. Kannan, E.; Kumar, P.; Vishnu, K.; Abraham, H. *Assessment of Pre and Post Harvest Losses of Rice and Red Gram in Karnataka*; Agricultural Development and Rural Transformation Centre, Institute for Social and Economic Change: Banglore, India, 2013.

24. De Lucia, M.; Assennato, D. *Agricultural Engineering in Development: Post-Harvest Operations and Management of Foodgrains*; FAO: Rome, Italy, 2006.

25. Shah, D. *Assessment of Pre and Post Harvest Losses in Tur and Soyabean Crops in Maharashtra*; Agro-Economic Research Centre Gokhale Institute of Politics and Economics: Pune, India, 2013.

26. Alavi, H.R.; Htenas, A.; Kopicki, R.; Shepherd, A.W.; Clarete, R. *Trusting Trade and the Private Sector for Food Security in Southeast Asia*; World Bank Publications: Washington, DC, USA, 2012.

27. Sarkar, D.; Datta, V.; Chattopadhyay, K.S. *Assessment of Pre and Post Harvest Losses in Rice and Wheat in West Bengal*; Agro-Economic Research Centre, Visva-Bharati, Santiniketan: Santiniketan, India, 2013.

28. Calverley, D. *A Study of Loss Assessment in Eleven Projects in Asia Concerned With Rice*; FAO: Rome, Italy, 1996.

29. Bala, B.K.; Haque, M.A.; Hossain, M.A.; Majumdar, S. *Post Harvest Loss and Technical Efficiency of Rice, Wheat and Maize Production System: Assessment and Measures for Strengthening Food Security*; Bangladesh Agricultural University: Mymensingh, Bangladesh, 2010.

30. Majumder, S.; Bala, B.; Arshad, F.M.; Haque, M.; Hossain, M. Food security through increasing technical efficiency and reducing postharvest losses of rice production systems in Bangladesh. *Food Secur.* **2016**, *8*, 361–374. [CrossRef]

31. Costa, S.J. *Reducing Food Losses in Sub-Saharan Africa (Improving Post-Harvest Management and Storage Technologies of Smallholder Farmers)*; UN World Food Programme: Kampala, Uganda, 2014.

32. Abedin, M.; Rahman, M.; Mia, M.; Rahman, K. In-store losses of rice and ways of reducing such losses at farmers' level: An assessment in selected regions of Bangladesh. *J. Bangladesh Agric. Univ.* **2012**, *10*, 133–144. [CrossRef]

33. Basavaraja, H.; Mahajanashetti, S.; Udagatti, N.C. Economic analysis of post-harvest losses in food grains in India: A case study of Karnataka. *Agric. Econ. Res. Rev.* **2007**, *20*, 117–126.

34. Shukla, B.D.; Patil, R.T. Post-harvest management, fight hunger with FAO. *India Grains* **2002**, *4*, 20–22.

35. Nagpal, M.; Kumar, A. Grain losses in India and government policies. *Qual. Assur. Saf. Crops Foods* **2012**, *4*, 143. [CrossRef]

36. Schulten, G. Post-harvest losses in tropical Africa and their prevention. *Food Nutr. Bull.* **1982**, *4*, 2–9.

37. Kannan, E. *Assessment of Pre and Post Harvest Losses of Important Crops in India*; Agricultural Development and Rural Transformation Centre, Institute for Social and Economic Change: Banglore, India, 2014.

38. Pathak, O.P.; Gupta, R.A. A study on post-harvest losses of food grains in Rajasthan. *Int. J. Appl. Res. Stud.* **2015**, *4*, 1–7.

39. Sharma, H.O.; Rathi, D. *Asseessment of Pre and Post Harvset Losses of Wheat and Soybean in Madhya Pradesh*; Jabalpur, M.P., Ed.; Agro-Economic Research Centre for Madhya Pradesh and Chhattisgarh, Jawaharlal Nehru Krishi Vishwa Vidyalaya: Jabalpur, India, 2013.

40. Inter-American Institute for Cooperation on Agriculture. *Post-Harvest Losses in Latin America and the Caribbean: Challenges and Opportunities for Collaboration*; IICA: Washington, DC, USA, 2013.

41. African postharvset loss information system. Available online: http://www.aphlis.net/ (accessed on 15 December 2015).

42. Kaminski, J.; Christiaensen, L. Post-harvest loss in sub-Saharan Africa—What do farmers say? *Glob. Food Secur.* **2014**, *3*, 149–158. [CrossRef]

43. Boxall, R. Damage and loss caused by the larger grain borer *Prostephanus truncatus*. *Integr. Pest Manag. Rev.* **2002**, *7*, 105–121. [CrossRef]

44. Hell, K.; Cardwell, K.; Setamou, M.; Poehling, H.M. The influence of storage practices on aflatoxin contamination in maize in four agroecological zones of Benin, West Africa. *J. Stored Prod. Res.* **2000**, *36*, 365–382. [CrossRef]

45. Wambugu, P.; Mathenge, P.; Auma, E.; van Rheenen, H. Efficacy of traditional maize (*Zea mays* L.) seed storage methods in western Kenya. *Afr. J. Food Agric. Nutr. Dev.* **2009**, *9*, 110–1128.

46. Tapondjou, L.; Adler, C.; Bouda, H.; Fontem, D. Efficacy of powder and essential oil from Chenopodium ambrosioides leaves as post-harvest grain protectants against six-stored product beetles. *J. Stored Prod. Res.* **2002**, *38*, 395–402. [CrossRef]

47. Compton, J.A.F.; Magrath, P.A.; Addo, S.; Gbedevi, S.R.; Amekupe, S.; Agbo, B.; Penni, H.; Kumi, S.P.; Bokor, G.; Awuku, M.; et al. The influence of insect damage on the market value of maize grain: A comparison of two research methods. In Proceedings of the Premier Colloque International: Lutte Contre les Déprédateurs des Denrées Stockées par les Agriculteurs en Afrique, Lome, Togo, 10–14 February 1997.

48. Baoua, I.; Amadou, L.; Ousmane, B.; Baributsa, D.; Murdock, L. PICS bags for post-harvest storage of maize grain in West Africa. *J. Stored Prod. Res.* **2014**, *58*, 20–28. [CrossRef]

49. Kimenju, S.C.; de Groote, H. Economic analysis of alternative maize storage technologies in Kenya. In Proceedings of the Joint 3rd African Association of Agricultural Economists (AAAE) and 48th Agricultural Economists Association of South Africa (AEASA) Conference, Cape Town, South Africa, 19–23 September 2010.

50. Meikle, W.; Markham, R.; Nansen, C.; Holst, N.; Degbey, P.; Azoma, K.; Korie, S. Pest management in traditional maize stores in West Africa: A farmer's perspective. *J. Econ. Entomol.* **2002**, *95*, 1079–1088. [PubMed]

51. Markham, R.; Bosque-Pérez, N.; Borgemeister, C.; Meikle, W. *Developing Pest Management Strategies for Sitophilus Zeamais and Prostephanus Truncatus in the Tropics*; FAO: Rome, Italy, 1994.

52. Tefera, T.; Mugo, S.; Likhayo, P. Effects of insect population density and storage time on grain damage and weight loss in maize due to the maize weevil *Sitophilus zeamais* and the larger grain borer *Prostephanus truncatus*. *Afr. J. Agric. Res.* **2011**, *6*, 2247–2254.

53. Magrath, P.; Compton, J.; Motte, F.; Awuku, M. *Coping With a New Storage Insect Pest: The Impact of the Larger Grain Borer in Eastern Ghana*; Natural Resources Institute: Chatham, UK, 1996.

54. Magrath, P.; Compton, J.; Ofosu, A.; Motte, F. *Cost-Benefit Analysis of Client Participation in Agricultural Research: A Case Study From Ghana*; Agricultural and Extension Network Paper No. 74b; Overseas Development Institute: London, UK, 1997; pp. 19–39.

55. Patel, K.; Valand, V.; Patel, S. Powder of neem-seed kernel for control of lesser grainborer (*Rhizopertha dominica*) in wheat (*Triticum aestivum*). *Indian J. Agric. Sci.* **1993**, *63*, 754–755.

56. Kumar, R.; Mishra, A.K.; Dubey, N.; Tripathi, Y. Evaluation of Chenopodium ambrosioides oil as a potential source of antifungal, antiaflatoxigenic and antioxidant activity. *Int. J. Food Microbiol.* **2007**, *115*, 159–164. [CrossRef] [PubMed]

57. Magan, N.; Aldred, D. Post-harvest control strategies: Minimizing mycotoxins in the food chain. *Int. J. Food Microbiol.* **2007**, *119*, 131–139. [CrossRef] [PubMed]

58. Kimanya, M.E.; Meulenaer, B.; Camp, J.; Baert, K.; Kolsteren, P. Strategies to reduce exposure of fumonisins from complementary foods in rural Tanzania. *Matern. Child Nutr.* **2012**, *8*, 503–511. [CrossRef] [PubMed]

59. Suleiman, R.A.; Kurt, R.A. Current maize production, postharvest losses and the risk of mycotoxins contamination in Tanzania. In Proceedings of the American Society of Agricultural and Biological Engineers Annual International Meeting, New Orleans, LA, USA, 26–29 July 2015.

60. Suleiman, R.A.; Rosentrater, K.A.; Bern, C.J. Effects of deterioration parameters on storage of maize: A review. *J. Nat. Sci. Res.* **2013**, *3*, 147.

61. Turner, P.; Sylla, A.; Gong, Y.; Diallo, M.; Sutcliffe, A.; Hall, A.; Wild, C. Reduction in exposure to carcinogenic aflatoxins by postharvest intervention measures in west Africa: A community-based intervention study. *Lancet* **2005**, *365*, 1950–1956. [CrossRef]

62. Williams, J.H.; Phillips, T.D.; Jolly, P.E.; Stiles, J.K.; Jolly, C.M.; Aggarwal, D. Human aflatoxicosis in developing countries: A review of toxicology, exposure, potential health consequences, and interventions. *Am. J. Clin. Nutr.* **2004**, *80*, 1106–1122. [PubMed]

63. Tefera, T.; Kanampiu, F.; de Groote, H.; Hellin, J.; Mugo, S.; Kimenju, S.; Beyene, Y.; Boddupalli, P.M.; Shiferaw, B.; Banziger, M. The metal silo: An effective grain storage technology for reducing post-harvest insect and pathogen losses in maize while improving smallholder farmers' food security in developing countries. *Crop Prot.* **2011**, *30*, 240–245. [CrossRef]

64. WeiFen, Q.; ZuXun, J. Paddy and rice storage in China. In *Advances in Stored Product Protection, Proceedings of the 8th International Working Conference on Stored Product Protection*; CAB International: Wallingford, UK, 2003; pp. 26–39.

65. Shaaya, E.; Kostjukovski, M.; Eilberg, J.; Sukprakarn, C. Plant oils as fumigants and contact insecticides for the control of stored-product insects. *J. Stored Prod. Res.* **1997**, *33*, 7–15. [CrossRef]

66. De Groote, H.; Kimenju, S.C.; Likhayo, P.; Kanampiu, F.; Tefera, T.; Hellin, J. Effectiveness of hermetic systems in controlling maize storage pests in Kenya. *J. Stored Prod. Res.* **2013**, *53*, 27–36. [CrossRef]

67. Mutungi, C.; Affognon, H.; Njoroge, A.; Baributsa, D.; Murdock, L. Storage of mung bean (*Vigna radiata* [L.] Wilczek) and pigeonpea grains (*Cajanus cajan* [L.] Millsp) in hermetic triple-layer bags stops losses caused by *Callosobruchus maculatus* (F.) (Coleoptera: Bruchidae). *J. Stored Prod. Res.* **2014**, *58*, 39–47.

68. Savvidou, N.; Mills, K.A.; Pennington, A. Phosphine resistance in Lasioderma serricorne (F.) (Coleoptera: Anobiidae). In *Advances in Stored Product Protection, Proceedings of the 8th International Working Conference on Stored Product Protection, York, UK, 22–26 July 2002*; Credland, P.F.A., Armitage, D.M., Bell, C.H., Cogan, P.M., Highley, E., Eds.; CAB International: Wallingford, UK, 2003; pp. 702–712.

69. Villers, P.; Navarro, S.; de Bruin, T. New applications of hermetic storage for grain storage and transport. In Proceedings of the 10th International Working Conference on Stored Product Protection, Estoril, Portugal, 27 June–2 July 2010.

70. Adler, C.; Corinth, H.G.; Reichmuth, C. Modified atmospheres. In *Alternatives to Pesticides in Stored-Product IPM*; Springer: New York, NK, USA, 2000; pp. 105–146.

71. Global Harvest Initiative. *Global Agricultural Productivity Report—Global Revolutions in Agriculture: The Challenges and Promise of 2050*; GHI: Washington, DC, USA, 2014.

72. Cardoso, M.L.; Bartosik, R.E.; Rodríguez, J.C.; Ochandio, D. Factors affecting carbon dioxide concentration in interstitial air of soybean stored in hermetic plastic bags (silo-bag). In Proceedings of the 8th International Conference on Controlled Atmosphere and Fumigation in Stored Products, Chengdu, China, 21–26 September 2008.

73. Bartosik, R.; Rodríguez, J.; Cardoso, L. Storage of corn, wheat soybean and sunflower in hermetic plastic bags. In Proceedings of the International Grain Quality and Technology Congress, Chicago, IL, USA, 15–18 July 2008.

74. Zeigler, M.; Truitt Nakata, G. *The Next Global Breadbasket: How Latin America Can Feed the World: A Call to Action for Addressing Challenges & Developing Solutions*; Inter-American Development Bank: Washington, DC, USA, 2014.

75. Food and Agriculture Organization of the United Nations. *Household MetalSilo: Key Allies in FAO's Fight against Hunger*; Agricultural and Food Engineering Technologies Service, FAO: Rome, Italy, 2008.

76. Yusuf, B.L.; He, Y. Design, development and techniques for controlling grains post-harvest losses with metal silo for small and medium scale farmers. *Afr. J. Biotechnol.* **2013**, *10*, 14552–14561.

77. Baoua, I.; Amadou, L.; Lowenberg-DeBoer, J.; Murdock, L. Side by side comparison of GrainPro and PICS bags for postharvest preservation of cowpea grain in Niger. *J. Stored Prod. Res.* **2013**, *54*, 13–16. [CrossRef]

78. Somavat, P.; Huang, H.; Kumar, S.; Garg, M.; Danao, M.; Singh, V.; Paulsen, M.; Rausch, K.D. Hermetic wheat storage for small holder farmers in India. In Proceedings of the First International Congress on Postharvest Loss Prevention, Rome, Italy, 4–7 October 2015.

79. Somavat, P.; Huang, H.; Kumar, S.; Garg, M.K.; Danao, M.G.C.; Singh, V.; Rausch, K.D.; Paulsen, M.R. Comparison of hermetic storage of wheat with traditional storage methods in India. In Proceedings of the American Society of Agricultural and Biological Engineers (ASABE) and the Canadian Society for Bioengineering (CSBE/SCGAB) Annual International Meeting, Montreal, QC, Canada, 13–16 July 2014.

80. Baoua, I.; Amadou, L.; Baributsa, D.; Murdock, L. Triple bag hermetic technology for post-harvest preservation of Bambara groundnut (*Vigna subterranea* (L.) Verdc.). *J. Stored Prod. Res.* **2014**, *58*, 48–52. [CrossRef]

81. Njoroge, A.; Affognon, H.; Mutungi, C.; Manono, J.; Lamuka, P.; Murdock, L. Triple bag hermetic storage delivers a lethal punch to *Prostephanus truncatus* (Horn) (Coleoptera: Bostrichidae) in stored maize. *J. Stored Prod. Res.* **2014**, *58*, 12–19. [CrossRef]

82. Vales, M.; Rao, C.R.; Sudini, H.; Patil, S.; Murdock, L. Effective and economic storage of pigeonpea seed in triple layer plastic bags. *J. Stored Prod. Res.* **2014**, *58*, 29–38. [CrossRef]

83. Ognakossan, K.E.; Tounou, A.K.; Lamboni, Y.; Hell, K. Post-harvest insect infestation in maize grain stored in woven polypropylene and in hermetic bags. *Int. J. Trop. Insect Sci.* **2013**, *33*, 71–81. [CrossRef]

84. Awal, M.A.; Hossain, M.A.; Ali, M.R.; Alam, M.M. Effective Rice Storage Technologies for Smallholding Farmers of Bangladesh. In Proceedings of the First International Congress on Postharvest Loss Prevention, Rome, Italy, 4–7 October 2015.

85. Sudini, H.; Rao, G.R.; Gowda, C.; Chandrika, R.; Margam, V.; Rathore, A.; Murdock, L. Purdue Improved Crop Storage (PICS) bags for safe storage of groundnuts. *J. Stored Prod. Res.* **2015**, *64*, 133–138. [CrossRef]

86. Ben, D.C.; van Liem, P.; Dao, N.T.; Gummert, M.; Rickman, J.F. Effect of Hermetic Storage in the Super Bag on Seed Quality and Milled Rice Quality of Different Varieties in Bac Lieu, Vietnam. *Int. Rice Res. Notes* **2009**. [CrossRef]

87. Kitinoja, L. Innovative small-scale postharvest technologies for reducing losses in horticultural crops. *Ethiop. J. Appl. Sci. Technol.* **2013**, *1*, 9–15.

2

Enrichment of Mango Fruit Leathers with Natal Plum (*Carissa macrocarpa*) Improves their Phytochemical Content and Antioxidant Properties

Tshudufhadzo Mphaphuli [1,2], Vimbainashe E. Manhivi [2], Retha Slabbert [1],
Yasmina Sultanbawa [3] and Dharini Sivakumar [2,*]

[1] Department of Horticulture, Tshwane University of Technology, Pretoria West 0001, South Africa;
chudufhadzo@gmail.com (T.M.); slabbertmm@tut.ac.za (R.S.)

[2] Phytochemical Food Network Group, Department of Crop Sciences, Tshwane University of Technology,
Pretoria West 0001, South Africa; vimbainashed@gmail.com

[3] Australian Research Council Industrial Transformation Training Centre for Uniquely Australian Foods,
Queensland Alliance for Agriculture and Food Innovation, Center for Food Science and Nutrition,
The University of Queensland, St Lucia, QLD 4069, Australia; y.sultanbawa@uq.edu.au

* Correspondence: SivakumarD@tut.ac.za

Abstract: Natal plum fruit (*Carissa macrocarpa*) is indigenous to South Africa and a rich source of cyanidin derivatives. Indigenous fruits play a major role in food diversification and sustaining food security in the Southern African region. Agro-processing of indigenous are practiced adopted by the rural African communities in order to reduce the postharvest wastage of fruit commodities. In the current study, Natal plum was added to mango pulp at different ratios (mango and Natal plum (5:1, 3:1, 2:1)) to develop a healthy-functional snack (fruit leather). The effects of added Natal plum on the availability of antioxidant constituents and in vitro antioxidant properties of a mango-based fruit leather were evaluated by comparing with mango fruit leather. Fruit leather containing mango and Natal plum (2:1) retained the highest content of cyanidin-3-O-glucoside chloride, cyanidin- 3-O-β-sambubioside, epicatechin, apigenin, kaempferol, luteolin, quercetin-3-O-rhamnosyl glucoside, catechin, quinic, and chlorogenic acids, and in vitro antioxidant activity. Proximate analysis showed that 100 g of fruit leather (2:1) contained 63.51 g carbohydrate, 40.85 g total sugar, 0.36 g fat, and 269.88 cal. Therefore, enrichment of mango fruit leather with Natal plum (2:1) increases its phytochemical content and dietary phytochemical intake, especially for school children and adolescents.

Keywords: cyanidin-3-O-glucoside; cyanidin-3-O-β-sambubioside; epicatechin; agro processing; phenols; food security; proximate composition

1. Introduction

Natal plum (*Carissa macrocarpa*) fruit is indigenous to South Africa, has an attractive red colour (Figure 1), and is a rich source of cyanidin derivatives (cyanidin-3-O-glucoside, cyanidin-3-O-β-sambubioside, and cyanidin-3-O-pyranoside) [1]. These cyanidin derivatives are associated with health benefits such as anti-inflammatory, antiviral, anti-proliferation, and anti-carcinogenic effects [2]. Natal plum is also rich in vitamin C, calcium, magnesium, and phosphorus [3]. Urbanization, change in food habits, and sustainable food systems contribute more to hidden hunger, and it is more prevalent among the urban population [4]. Therefore, diet diversification with traditional underutilized fruits can be one way to tackle hidden hunger [4]. Traditional underutilized fruits are readily available during different seasons; they are easily harvestable and cost much lower price than commercial fruits such as citrus or avocado. Therefore, Natal plum can be included in food diversification and traditionally rural communities make jams and jellies from this fruit. Thus, this fruit has the potential of improving rural industry and well-being of the communities. Based on the nutritional facts, Natal plum can be introduced in nutrition intervention programmes to sustain food and nutrition security.

Figure 1. Natal plum (*Carissa macrocarpa*) at maturity stage 4 (red).

Fruit leather is a dried-fruit or dehydrated fruit, high in fibre and carbohydrates, whilst low in fats. Fruit leathers are chewy, have a pleasant flavor, and are consumed as a sweet snack [5]. Due to its appealing nature, dried fruit leathers are another practical way to increase the consumption of solid fruit, especially for children and youth, or for communities affected by natural disasters or in war regions where food is in secure [5].

Since fruit leathers are concentrated, with a higher nutrient density compared to fresh fruit due to dehydration, this makes them a healthier and convenient alternative snack compared to candies and confections [6]. In addition, fruit leathers contain fewer calories per serving [7] and greater nutritional value in terms of antioxidants and minerals due to the dehydration process concentrating the nutrients [8]; therefore, they are suitably healthy food for heath food markets. The lower moisture content of fruit leathers also reduces the microbial infestation during storage and transportation.

Mangoes are widely produced in Venda Limpopo province at the village level. Each household has 3 to 4 mango trees in their home garden. The community cooperatives do not have infrastructure and a cold chain facility to store the mango fruits. Therefore, postharvest losses of mangoes during the mango season due to surplus supply are very high. Thus, using fresh mango pulp for fruit leather preparation is one way of preserving mangoes during off-season. It has been standard practice to add pectin, sugar, and sodium metabisulphite to fruit leathers; however, shelf-stable and acceptable mango (*Mangifera indica*) leather has been produced without any sugar and preservatives [9]. This makes mangoes ideal for making a leather, which suits consumer demands for natural and additive-free snacks. Apart from the aforementioned, mango fruit is well liked by consumers due to its attractive colour and flavour; it is also regarded as a good source of dietary antioxidants, such as ascorbic acid and carotenoids [10]. Mango pulp is rich in magnesium, potassium, and other essential nutrients [11]. Consumption of mango pulp effectively favorably affected postprandial glucose and insulin responses in individuals with Type 2 diabetes. The dietary source of carotenoids, flavonoids, and various phenolic acids and mangiferin present in mango pulp was proven to reduce blood glucose levels by inhibiting glucose absorption from the intestine, demonstrating antidiabetic properties [12]. Since mango pulp also contains some pectin, this makes it best suited for fruit leather application [13].

Inclusion of maqui berry extract to apple or quince leathers improves the functional properties and colour of the leathers from light brown to dark purple [14]. The enrichment of apple leathers with raspberry and blackcurrant improves their phenolic composition and antioxidant activity [15]. Adding Natal plum to mango pulp could increase the antioxidant properties of the mango–Natal plum fruit leather. In light of the aforementioned, the objective of the current study is to evaluate the incorporation of a Natal plum (berry) in different proportions to mango-based leather for the improvement of antioxidant constituents and to study in vitro antioxidant activity. Proximate analysis (total carbohydrate, sugar, fat, and energy) was performed to obtain information on the best Natal plum and mango pulp ratio that offers the highest antioxidant properties.

2. Materials and Methods

2.1. Sample Collection and Preparation

Natal plum (*Carissa macrocarpa*) fruits were harvested fresh from Tshwane University of Technology (TUT) main campus, Orion Residence, Pretoria, Gauteng Province. Mango fruits (*Mangifera indica* cv. Tommy Atkins) were selectively hand-harvested for this study, from home gardens in Venda (Vuwani), and transported to TUT at 10 °C within 6 h after harvest. Mango (brix° 20) and Natal plum (brix° 10) fruits were sanitised using free active chlorine solution (in a 50 mg L^{-1}), as recommended by Azeredo et al. [10]; thereafter, the mangoes were peeled, de-seeded, and cut into small cubes of 2 cm^3 each. Natal plums were cut into two halves and both types of fruit were weighed to determine the proper ratio required for the fruit leather preparation. In this study, the fruits were not cooked in order to prevent the loss of phytochemicals. Cut mango fruit was mixed with Natal plum fruit in three different ratios—5:1 Mango (M) Natal plum (N) MN5, 3:1 MN3, 2:1 MN2, and afterwards pureed using a food processor (Bosch, MCM3301BGB, Milton Keynes, UK) for 5 min. The resulting mixture was layered (9–10 mm thickness) on a non-stick sheet and dehydrated using a drying oven (Universal oven Memmert GmbH, Buechenbach, Germany) at 60 °C for 4 h until a moisture content of 15% was reached. Mango (M1) and Natal plum (N1) leathers were also made. The resulting fruit leather was cooled, cut into pieces (6 cm^2), rolled up, and packed in BOPP (water vapor transmission rate of 4 $\times 10^{-3}$ $kg/m^2/d$ at 90% RH, 38 °C and an OTR of 2.5 $L/m^2/d$ atmosphere at 25 °C to determine fruit leather colour properties. A set of 10 replicates of each fruit leather were snap-frozen in liquid nitrogen and stored at −80 °C for antioxidant and phytochemical analysis.

2.2. Fruit Leather Colour Properties

Colour properties of mango, Natal plum, and the mango-Nata plum mixed leathers were objectively measured at three points of the fruit leathers (two each on the opposite side and bottom) using a colour meter, Minolta CR-400 chromameter (Minolta, Osaka, Japan) [16]. A standard white tile was used to calibrate the chromameter. According to the colour system used, L^* measures lightness from black to white (0–100); a^* indicates red (+) to green (–), while b^* measures yellow (+) to blue (–). The intensity of the red colour is represented by positive a^* values, whereas the intensity of the yellow colour is represented by positive b^* values. Furthermore, the $h°$ values, known as a colour wheel, consist of red-purple angled at 0°, yellow at 90°, bluish-green at 180°, and blue colour at 270°. The total colour difference $\Delta E = [(\Delta L^*)^2 + (\Delta a^*)^2 + (\Delta b^*)^2]^{\frac{1}{2}}$ was calculated according to which ΔL = colour difference is calculated as the sample L^* value minus standard, Δa^* = colour difference is calculated as the sample a^* value minus standard, Δb^* = Colour difference is calculated as the sample b^* value minus standard [16].

2.3. Total Phenols

Total phenolics were determined using the Folin-Ciocalteu reagent [1,17]. Snap frozen fruit leather of 0.1 g was homogenised in 2 mL of 80% methanol containing 1% HCl, at room temperature using a BV1000 vortex mixer (Benchmark Scientific Inc., New Jersey, NJ, USA). The mixture was centrifuged at 10,000× g for 15 min (Model Hermle Z326k, Hermle Labortechnik GmbH, Wehingen, Germany); the supernatant 2 mL was used for the determination of total phenolic content. Briefly, 9 µL of extract (supernatant) was mixed with 109 µL of Folin-Ciocalteu reagent, followed by 180 µL of (7.5% *w/v*) Na_2CO_3. The solution was then mixed and incubated for 5 min at 50 °C, and after cooling to 25 °C, the absorbance was measured at 760 nm (BMG LABTECH SPECTROstar Nano microplate reader, Ortenberg, Germany). Total phenolic compounds were calculated using a standard curve of gallic acid and expressed as mg of gallic acid equivalents per 100 g fruit leather.

2.4. Antioxidant Scavenging Capacity Using Ferric Reducing Antioxidant Power (FRAP) Assay

Total antioxidant scavenging activity of fruit leathers was determined using ferric reducing ability (FRAP), according to a method described by Ndou et al. [1] and the authors of [18]. The analysis was conducted using aqueous stock solutions containing 0.1 mol/L acetate buffer (pH 3.6), 10 mmol/L TPTZ [2,4,6-tris(2-pyridyl)-1,3,5-triazine] acidified with concentrated hydrochloric acid, and 20 mmol/L ferric chloride. Thereafter, the stocks were combined (10:1:1, *v/v/v*, respectively) in order to form the FRAP reagent just prior to analysis. Fruit leather samples (0.2 g) were briefly snap-frozen and homogenised in 2 mL of acetate buffer. Thereafter, ferric reducing ability was quantified using 15 µL aliquot of fruit extract, mixed with 220 µL of FRAP reagent solution. The absorbance was determined at 593 nm using a spectrophotometer (BMG LABTECH GmbH, Spectro- Star Nano, Ortenberg, Germany). The antioxidant activity (FRAP assay) was expressed as ascorbic acid equivalent antioxidant activity.

2.5. Quantification of Different Phenolic Compounds

UPLC coupled to a quadrupole time-of-flight (QTOF) mass spectrometer (UPLCQTOF/MS) was used to detect and quantify various phenolic compounds, as previously described for Natal plum by Ndou et al. [1] (2019). Freeze-dried samples (50 mg per replicate) were mixed in 1 mL of ethanol/water solution (70:30, *v/v*), ultrasonicated for 30 min, then centrifuged (Hermle Z326k, Hermle Labortechnik GmbH, Wehingen, Germany) at 1000 g at 4 °C for 20 min. Using a 0.22-µm polytetrafluorethylene filter, the supernatants were subsequently decanted and filtered. The resulting filtrate was analysed using a Waters Acquity™ UPLC coupled to a SYNAPT G2 QTOFMS (Waters, Milford, MA, USA) using Masslynx (V4.1) software. The type of column used for separation, the mobile phase used, gradient concentrations, and the running time set and flow rate were similar to the methodology described by Ndou et al. [1]. The UPLC was connected directly to a QToF-MS equipped with a source of electrospray ions operating in negative ESI mode with 15 V cone voltage, 275 °C desolvation temperature, and 650 L/h desolvation gas flow. For best resolution and sensitivity, the rest of the MS settings was optimised. Data were obtained both in resolution mode and in MSE mode by scanning from *m/z* 150 to 1500 m/z. Two channels of MS data were obtained in MSE mode, one using low collision power (4 V) and the other using a collision power accelerator (40–100 V) to acquire fragmentation information. For accurate mass determination, leucine enkephaline was used as a reference mass and the tool was calibrated using sodium format. Progenesis QI (NonLinear Dynamics, NC, USA) was used to identify the compounds. Standards purchased from four distinct cocktails were produced at each stage to help identify isomers and compounds with comparable elemental formulas [18]. Since calibration standards are not available for all the compounds identified, the compounds were semi-quantitatively measured against calibration curves set up using chlorogenic acid, catechin, luteolin, epicatechin, and rutin (Four cocktails, Sigma-Aldrich (Johannesburg, South Africa). Methanol (50%) in water containing formic acid (1%) was used for the cocktail preparation. HyStar 3.2 and Data Analysis 4.2 (Bruker Daltonics) software were used to analyse and calculate information [1]. A TargetLynx processing method (part of MassLynx) was adopted to quantify the compounds responsible for the main peaks using the cocktail standards in each chromatogram, as described by Ndou et al. [1].

The targeted compounds 4-hydroxybenzoic acid, protocatechuic acid, gallic acid, caftaric *acid*, dicaffeoyltartaric acid, chlorogenic acid, caffeic acid, caffeic acid, apigenin, catechin, epicatechin, kaempferol, and luteolin were quantified using HPLC, with a photo diode array ultraviolet detector, Model Flexar ™ 89173-556 (PerkinElmer, Waltham, MA, USA), as described previously by Mpai et al. [18]. The column conditions and the mobile phase, flow rate, and gradient elution programme were according to Mpai et al. [18]. The chromatogram was read at 272, 280, 310, and 320 nm. The phenolic acids and flavonols were identified and quantified using pure external standards with 95% purity purchased from Sigma-Aldrich (Johannesburg, South Africa).

2.6. Proximate Analysis

Proximate analysis was performed on a selected fruit leather using AOAC standards methods [19].

2.7. Statistical Analysis

All experiments were carried out in triplicate. A complete randomised design was adopted; the data were analysed using analysis of variance (ANOVA) and means were compared using Fischer's Least Significant Differences Test at p values 0.01 and 0.001, using the Genstat for Windows 204 13th Edition (2010) (VSN International, Hempstead, UK).

3. Results and Discussion

3.1. Colour Properties of Mango-Natal Plum Fruit Leathers

Colour properties of the five different kinds of fruit leathers M1, N1, MN5, MN3, and MN2 are shown in Table 1. Fruit leather containing only mango (M1) demonstrated the highest brightness (lightness L^*), whilst fruit leather made of Natal plum (N1) was the darkest due to the lower luminosity (L^*); Table 1. Addition of Natal plum to reduced quantities of mango pulp in MN3 and MN2 significantly affected brightness (L^*) of the fruit leathers (Table 1). Similarly, adding maqui berry extract (*Aristotelia chilensis* Mol. Stuntz) to apple and quince formulations made the fruit leathers darker (brown) [13] due to its higher anthocyanin content [20].

Table 1. Colour properties of Natal plum-mango fruit leather at different ratios.

Fruit Leather Formulation	Colour Properties			
	L^*	a^*	b^*	ΔE
M1 (Mango)	57.2 ± 1.5 *a	5.8 ± 0.5 c	39.8 ± 1.7 a	Standard
N1(Natal plum)	36.4 ± 0.4 d	21.3 ± 0.9 b	7.7 ± 0.4 e	41.3 a
MN2 (2M:1N)	41.6 ± 0.4 c	27.7 ± 0.6 a	16.2 ± 0.7 d	35.8 b
MN3(3 M:1N)	43.5 ± 0.6 c	26.9 ± 0.6 a	19.4 ± 0.7 c	32.4 c
MN5(5 M:1N)	46.1 ± 0.5 b	27.4 ± 0.8 a	22.9 ± 0.8 b	29.6 d

Total color difference (ΔE). Values (Means ± S.D*) with different superscripts in a column are significantly different $p < 0.001$ by Fisher's protected least significant test. * standard deviation.

The colour coordinate positive a^* value relates to the red colour [10] (Ndou et al., 2019) of the fruit leather, which increased significantly when the Natal plum was added to the different ratios (5, 3 or 2) of mango fruit pulp (Table 1). A similar increase in a^* colour coordinates was noted when red acid calyx of roselle (*Hibiscus sabdariffa* L), rich in anthocyanin [21], was added to pineapple fruit leather [22]. In this study, although the redness in Natal plum dominates the yellow colour of the mango in the mango-Natal plum fruit leather, increasing Natal plum content did not significantly increase the redness of the mango-based fruit leather (Table 1). The yellowness (positive b^* values) of the mango pulp [23] in M1 and carotenoids are responsible for the yellowness in mango fruit leather [24]. Therefore, the addition of Natal plum to a mango pulp decreased the yellowness of the mango fruit leather by affecting the L^*, a^* and b^* colour coordinates (Table 1). Natal plum, at the mature stage (red color), contains cyanidin derivatives [1], which could be reasonable for the observed changes in colour properties of the mango-Natal plum leathers. It is also evident from the ΔE (the change in colour difference) with reference to mango fruit leather (M1) as standard, Natal plum (N1) showed the highest colour difference due to its pinkish red colour followed by the Natal plum-mango (MN2) fruit leather, which contained the highest amount of Natal plum (Table 1).

3.2. *Total Phenols and Antioxidant Property of Mango-Natal Plum Fruit Leathers*

Total phenols were approximately seven times higher in Natal plum leathers (N1) than in mango leathers (M1) (Figure 2). Addition of Natal plum at different ratios increased the total phenols in the Natal plum-mango fruit leathers MN5, MN3, and MN2 (Figure 2). Although the increasing concentrations of Natal plum in the mango fruit leather did not significantly increase the total phenol content in MN5, MN3, and MN2, in general, adding Natal plum to mango pulp improved the total phenols content of mango leathers by approximately 25–27% (Figure 2). The ferric-reducing antioxidant power (FRAP) of the Natal plum-mango fruit leathers MN5, MN3, and MN2 showed a similar trend as the total phenol (Figure 3). Previous research has showed that low FRAP activity relates to the low phenolic content in the fruit [25]. In this study, adding Natal plum to mango fruit pulp significantly increased the antioxidant activity by 33.3–41.6% (Figure 3). Similarly, Torres et al. [14] demonstrated the improvement of total phenolic content and antioxidant activity by 40% and 45%, respectively, in a functional snack developed by fortification of apple leather with maqui berry extract. Phenolics and carotenoids present in mango pulp were responsible for the antioxidant activity of mango pulp [26]. On the contrary, adding red acid calyx of roselle (*Hibiscus sabdariffa* L) to pineapple minimised the total phenol content and antioxidant property of roselle-pineapple fruit leather, and the higher heating process (100 °C) led to the decarboxylation of phenolic acids [8,22]. However, the temperature used in this study, which was less than 100 °C, could have prevented the decarboxylation of phenolic acids in Natal plum.

Previous research showed that Natal plum contains a wide range of phenolic compounds [1], whilst no cyanidin derivatives were previously observed in mango pulp [27]; this might be the reason for the higher total phenol content of Natal plum leather compared to mango leather.

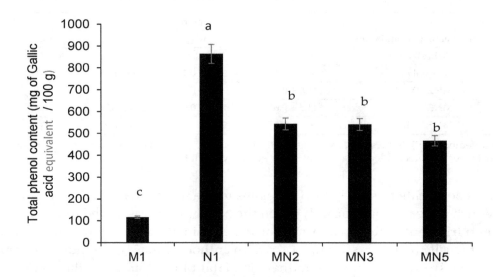

Figure 2. Total phenolic content of Natal plum-mango fruit leather at different ratios. Bars with different alphabets are significantly different $p < 0.001$ by Fisher's protected least significant test.

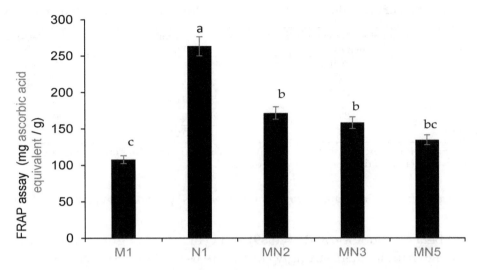

Figure 3. Antioxidant capacity (FRAP assay) of Natal plum-mango fruit leather at different ratios. Bars with different alphabets are significantly different $p < 0.001$ by Fisher's protected least significant test.

3.3. Cyanidin Derivatives in Mango-Natal Plum Fruit Leathers

The cyanidin derivative in the Natal plum is responsible for its red colour and is present in many berries [28]. Natal plum (N1) fruit leathers contained all three cyanidin derivatives, cyaniding-3-O-glucoside, cyanidin-3-O-β-sambubioside, and cyanidin-3-O-pyranoside (Table 2). These three cyanidin derivatives were reported in Natal plum at the red stage [1]; however, cyanidin-3-O-pyranoside was not detected in the mango-Natal plum leathers (MN5, MN3, MN2), which implied that the amount of Natal plum added did not significantly increase its concentration ($p < 0.01$) (Table 2). Although the concentrations of cyanidin-3-O-β-sambubioside and cyanidin-3-O-glucoside were significantly highest in Natal plum (N1), adding higher amounts of Natal plum (MN2) helped to retain these two compounds at higher levels (Table 2). The pH of the food matrix and temperature were reported to affect the stability of the cyanidin compounds [29]. Although the variability in the stability of the cyanidin compounds are dependent on the pH of the food matrix and the temperature, cyanidin-3-O-glucoside molecules were reported to be more stable than the cyanidin-3-O-pyranoside molecules [29]; thermal stability of cyanidin-3-O-β-sambubioside was also reported. However, formation of the quinonoid base and hemiketal (colourless) from the flavylium ion are favoured at higher temperatures [30]. Therefore, adding Natal plum, a rich source of cyanidin derivatives, to mango leathers, improved the functional compounds of mango fruit leather (Table 2). The drying temperature of the oven also affects the concentration of cyanidin derivatives; drying sour cherries at 50 °C reduced cyanidin-3-glucoside, compared to fresh ones [31]. In addition, extraction of the phenolic compounds affected the reduction in particle size, solvent types used for extraction, and the porosity of the dried samples [32].

Among the major anthocyanins, the bio availability of cyanidin-3-glucoside has been well documented [33]. Cyanidin-3-O-glucoside has been shown to exert insulin-like effects [34], anti-obesity effects [2], and anti-carcinogenic effects [35]; however, it has been shown that anthocyanins are bio-transformed in the gut before absorption. Many studies revealed that protocatechuic acid is one of the most likely major metabolites of anthocyanins [36,37]. It was demonstrated by Ormazabal et al. [38] that protocatechuic acid has the ability to regulate insulin responsiveness and inflammation during obesity. Findings of Ormazabal et al. [38] confirmed the advantageous effects of improving anthocyanin content in diets to combat against inflammation and insulin resistance in obesity.

Table 2. Concentration of cyanidin derivatives in Natal plum-mango fruit leather at different ratios.

Fruit Leather Formulation	Cyanidin-3-O-pyranoside (mg/kg)	Cyanidin-3-O-β_sambubioside (mg/kg)	Cyanidin-3-O-glucoside (mg/kg)
N1(Natal plum)	1.6 ± 0.01 a	46.9 ± 0.02 a*	52.5 ± 0.02 a
M1(Mango)	ND	8.6 ± 0.01 e	1.5 ± 0.04 e
MN2(2M:1N)	ND	29.2 ± 0.03 b	21.9 ± 0.01 b
MN3(3M:1N)	ND	20.9 ± 0.01 c	19.4 ± 0.02 c
MN5(5M:1N)	ND	18.9 ± 0.03 d	17.7 ± 0.01 d

Values (Means ± S.D) with different alphabets in a column are significantly different $p < 0.01$ by Fisher's protected least significant test. * Standard deviation. ND = not detected.

3.4. Correlation between the Predominant Cyanidin Components of Colour Properties and the Antioxidant Activity

Cyanidin-3-O-beta sambubioside ($R^2 = 0.89$) and cyanidin-3-O-glucoside chloride ($R^2 = 0.92$) in the Natal plum-mango fruit leathers strongly contribute to total phenolic content and also positively correlate to antioxidant properties (FRAP assay) (Table 3); this supports the observed higher antioxidant property in mango leather enriched with Natal plum, especially M2 and M3, compared to the mango fruit leather (M1) shown in Figure 3 This observation confirmed that the inclusion of Natal plum to mango-based leather improved antioxidant activity. Both cyanidin components negatively correlated to the brightness (L^*) ($R^2 = -0.83, -0.81$); similarly, the total phenols and antioxidant property correlated strongly and negatively with brightness (L^*) ($R^2 = 0.96$) (Table 3). In addition, the colour coordinate b^*, which that relates to the yellowness of the fruit leather, strongly and negatively correlated to the antioxidant activity ($R^2 = -0.78$) and total phenols ($R^2 = 0.97$). Therefore, the darker Natal plum-mango fruit leathers are rich in antioxidant properties and can be regarded as a functional snack.

Table 3. Pearson's correlation coefficients between the predominant cyanidin components of color properties and the antioxidant activity of Natal plum-mango fruit leather.

	Total Phenols	FRAP	Cyanidin-3-O-Beta Sambubioside	Cyanidin-3-O-Glucoside Chloride
		Correlation coefficients (R^2)		
Total phenols		0.869	0.90	0.92
FRAP			0.97	0.97
L^*(C)	−0.96	−0.76	−0.83	−0.81
a^*(C)	0.39	0.1	0.17	0.15
b^*(C)	0.97	−0.78	−0.85	−0.82

3.5. Flavonoids and Phenolic Acids in Mango-Natal Plum Fruit Leathers

Six dietary flavonoid aglycones and glycosides were quantified in the five types of leathers (N1, M1, MN2, MN3, MN5) namely, apigenin, catechin, epicatechin, kaempferol, luteolin, quercetin-3-O-rhamnosyl glucoside (rutin), quercetin-3-O-rhamnosyl galactoside, eriodictyol-7-O-glucoside, and naringenin-4-O-glucoside (Figure 4). Natal plum leather showed the highest concentration of epicatechin, rutin, quercetin-3-O-rhamnosyl galactoside, catechin, luteolin, naringenin-4-O-glucoside, and eriodictyol-7-O-glucoside (Figure 4). In mango fruit leather, the flavonoid concentrations were significantly lower and rutin, eriodictyol-7-O-glucoside, and quercetin-3-O-rhamnosyl galactoside, were not detected (Figure 4). Lower concentrations of flavonoids were detected in mango pulp [39]. However, the pulp of mango variety Tommy Aitkin, obtained from Ecuador, showed an absence of flavonoids [40]. The reduction in flavonoid concertation, or its disappearance, was associated with ripening [39]. Adding Natal plum to mango pulp in MN2 increased the concentration of epicatechin similar to the levels noted in Natal plum. Fruit leather MN2 retained the highest concentration of apigenin compared to the other fruit leathers, and the concentration of catechin content in fruit leather MN3 was similar to the levels noted in Natal plum. In general, fruit leathers MN2 showed higher levels of flavonoids than

MN3. Furthermore, epicatechin, catechins, kaempferol, apigenin, rutin, and eriodictyol were reported to demonstrate many health benefits [41].

The quantified phenolic acids were quinic acid and chlorogenic acid. Quinic acid was high in N1, and M1 had the least (Table 4). Natal plum was previously reported to contain less than 1 mg/kg chlorogenic acid in the red fruit [1]. The high amount of this phenolic in Natal plum leather may imply that phenolics were concentrated during dehydration. As the proportion of Natal plum to mango in the leather increased, the amount of quinic and chlorogenic acid in the leather increased ($p < 0.01$). Chlorogenic acid was not detected in mango fruit leather or reported in fresh mango pulp [40]. Quinic acid is also a major bioactive chemical found in Saskatoon berry genotypes [42]. Thus, the presence of these phenolic acids increases the potential of mango-Natal plum leather as a functional snack. Total ion chromatograms of metabolites of Natal plum–mango (MN2) fruit leather, in comparison to Natal plum fruit leather in ESI-mode by UPLC–QTOF/MS are shown in Supplementary Figure S1.

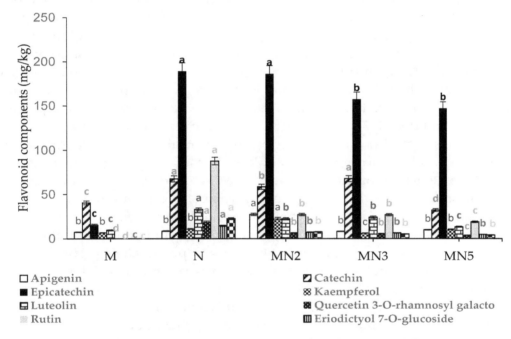

Figure 4. Flavonoid components of Natal plum-mango fruit leather at different ratios. Bars with different alphabets are significantly different $p < 0.001$ by Fisher's protected least significant test. In addition, bars with specific coloured alphabets relate to specific flavonoid compounds.

Table 4. Concentration of quinic and chlorogenic acids in the Natal plum-mango fruit leathers at different ratios.

Fruit Leather Formulation	Quinic Acid (mg/kg)	Chlorogenic Acid (mg/kg)
N1	1678.1 ± 1.30 [a*]	16.6 ± 0.03 [a]
M1	188.6 ± 0.6 [e]	Nd
MN5	565.8 ± 0.12 [d]	3.2 ± 1.00 [c]
MN3	646.7 ± 0.50 [c]	3.5 ± 0.06 [c]
MN2	708.0 ± 0.10 [b]	5.4 ± 0.09 [b]

Values (Means ± S.D) with different superscripts in a column are significantly different $p < 0.001$ by Fisher's protected least significant test. * Standard deviation, Nd—not detected.

3.6. Proximate Analysis of Mango-Natal Plum Fruit Leather (MN2)

Since the mango-Natal plum, fruit leather MN2 showed the highest concentration of functional compounds and antioxidant activity, it was further analysed for proximate composition. The proximate composition analysis included moisture, dry matter, sugars, total fat, carbohydrate, protein, total ash, energy, total dietary fibre, and sodium (Table 5). Moisture content in MN2 was observed to be quite high. Generally, a moisture content of 15–25% is preferred for fruit leathers; however, it is affected by sugar content, acidy of the fruit, drying process, temperature, and humidity [5,9]. Azeredo et al. [9] demonstrated six months' shelf life without preservatives for mango fruit leathers with low water activity (0.62), low pH (3.8), and a moisture content of 17.2%. Furthermore, although low moisture content can extend shelf life by preventing the fruit leathers from inhibiting microbial growth, the texture of the fruit leathers can be negatively affected due to extensive crispiness of the product [43,44]. The sugar content was 40.85 g/100 g—about 6.12 g of sugar for three servings of one fruit bar (5 g). According to the American Heart Association (AHA) [45], added sugar intake must be limited to less than 25 g sugar or 100 cal intake per day for children ranging from 2 to 18 years of age. Similarly, for men and women, 37.5 g (9 teaspoons) and 25 g of sugar intake per day, respectively, is adequate [45]. Therefore, three servings of one fruit bar (5 g) will contribute approximately towards $\frac{1}{4}$ portion of the 25 g that is allowable for children and women, and 1/6 portion of 37.5 g for men. In addition, this does not fall under the refined or processed sugar and is natural sugar found in the fruit. Refined sugar, which is an important contributor to dietary energy, has recently been a target of campaigns to reduce refined sugar intake [46]. Furthermore, better access to information on the amounts of sugar added to processed food is essential for appropriate monitoring of this important energy source [46]. The natural sugar present in Natal plum–mango fruit leather is fructose, with a number of micronutrients plus fiber. Natal plum–mango fruit leather contains 67.98 g/100g fiber. It is evident from previous research conducted by French and Read [47] that dietary fiber produces a viscous gel-like environment in the small intestine and this slows down gastric emptying, thereby facilitating a gradual release of sugar into the bloodstream. This avoids a glucose spike in the blood. Subsequently all the above-mentioned events lead to a reduced hunger sensation and eventually decrease food consumption. Digestion-resistant cellulose, hemicellulose, pectic substances, gums, mucilage, and lignin in dietary fibre [48] can be responsible for creating the viscous gel-like environment and have been proposed as important for a healthy gut microbiome [49]. Furthermore, the higher fibre content in this fruit leather may have also contributed to the high moisture content, since fibre is hydrophilic. Dietary fibre has been reported to lower the glycaemic indices of food. Supplementation of mango powder (10g) with carbohydrate, protein, fat, and fiber contents of 89.6%, 4.01%, 1.62%, and 13.4%, respectively, in obese individuals actually improved fasting glucose levels [50]. Thus, on this note, the sugar content in Natal-plum fruit leather snack may not offset any potential benefits of bioactive components.

Table 5. Proximate composition of Natal plum-mango fruit leather (MN2).

Proximate Composition	Results g/100 g
Moisture	32.02 ± 0.34
Dry matter	67.98 ± 0.82
Sugars sum	40.85 ± 2.86
Total fat	0.36 ± 0.03
Carbohydrates	63.51 ± 0.12
Protein	1.63 ± 0.08
Total ash	1.42 ± 0.07
Energy	1129.18 kJ
Total Dietary Fibre (TDF)	7.29 ± 0.87
Sodium (Na)	0.005 ± 0.01

In the presence of hydrophilic groups (such as those from pectin, fibre, and other soluble solids), drying can take longer, but longer drying times are associated with degradation of phytochemicals [32].

In this study, a lower temperature and shorter drying time were chosen in order to have a softer leather, as well as to preserve the phytochemicals.

The fruit leather MN2 was high in carbohydrates and low in proteins and fats (Table 5). The carbohydrate content in fruit leather MN2 was quite low, compared to blueberry fruit leather (89.00 mg/100g FW) and banana-pineapple-apple leather (80.00 ± 0.10 to 84.77 ± 0.06%) [51]. A lower carbohydrate content is associated with a lower calorific content of food. One hundred grams of fruit leather contains 1129.18 kJ (269.7 cal) (Table 5); therefore, three servings of one fruit bar (5 g) contains approximately 40 cal. The mango–Natal plum leather was also found to have low sodium content (Table 5). Strawberry leathers have been reported to contain 121.47 mg/kg of sodium [52], and this low sodium content makes it acceptable for people with high blood pressure and kidney problems. Furthermore, higher values of ash content relate to the higher mineral composition [53]. Mango-Natal plum fruit leather MN2 contained slightly higher ash content than that reported in the mixed fruit leather, which contained 40% banana, 40% pineapple, and 20% apple (1.20%). The visual appearance of mango fruit leather enriched with Natal plum (MN2) compared to the mango is illustrated in Figure 5.

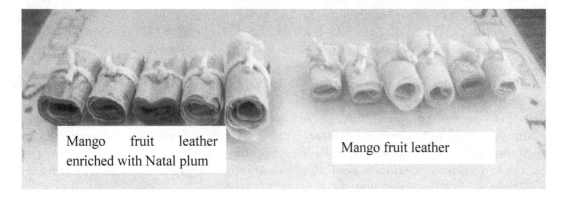

Figure 5. Mango fruit leather enriched with Natal plum.

4. Conclusions

Mango leather had a lower antioxidant activity and phenolic content compared to Natal plum leather. Increasing amounts of Natal plum in mango–Natal plum leather (MN2) significantly improved cyanidin components, flavonoids, and antioxidant activity. Mango–Natal plum leather (MN2) also contained carbohydrates, was high in fibre, and low in sodium, and can therefore be regarded as a potential functional snack.

Author Contributions: T.M. the first author, a master's degree student, performed the experiments, generated the data, and wrote the first draft of the manuscript. V.E.M., a postdoctoral researcher, validated and visualized the data, conducted certain statistical analysis, and rewrote the second draft of the manuscript. R.S. acted as the main supervisor of the postgraduate student, provided editorial assistance, and guidance in statistical analysis. Y.S. is a research collaborator and conducted some analyses at the Queensland Alliance for Agriculture and Food Innovation, Center for Food Science and Nutrition, The University of Queensland, Australia. D.S., the grant holder, conceptualized the research, was co-supervisor of the master's student, the first author; set out the methodologies, and substantially improved the second draft of this manuscript. All authors have read and agreed to the published version of the manuscript.

References

1.   Ndou, A.; Tinyani, P.P.; Slabbert, R.M.; Sultanbawa, Y.; Sivakumar, D. An integrated approach for harvesting Natal plum (*Carissa macrocarpa*) for quality and functional compounds related to maturity stages. *Food Chem.* **2019**, *293*, 499–510. [CrossRef] [PubMed]

2.   Kaume, L.; Gilbert, W.C.; Brownmiller, C.; Howard, L.R.; Devareddy, L. Cyanidin-3-O-β-D-glucoside-rich blackberries modulate hepatic gene expression, and anti-obesity effects in ovariectomized rats. *J. Funct. Foods* **2012**, *4*, 480–488. [CrossRef]
3.   Moodley, R.; Koorbanally, N.; Jonnalagadda, S.B. Elemental composition and fatty acid profile of the edible fruits of Amatungula (*Carissa macrocarpa*) and impact of soil quality on chemical characteristics. *Anal. Chim. Acta* **2012**, *730*, 33–41. [CrossRef] [PubMed]
4.   FAO Report Food in an Urbanised World, the Role of City Region Food Systems in Resilience and Sustainable Development 2015. Available online: http://www.fao.org/fileadmin/templates/agphome/documents/horticulture/crfs/foodurbanized.pdf (accessed on 23 March 2020).
5.   Ruiz, N.A.Q.; Demarchi, S.M.; Massolo, J.F.; Rodoni, L.M.; Giner, S.A. Evaluation of quality during storage of apple leather. *LWT J. Food Sci. Technol.* **2012**, *47*, 485–492. [CrossRef]
6.   Orrego, C.E.; Salgado, N.; Botero, C.A. Developments and trends in fruit bar production and characterization. *Crit. Rev. Food Sci. Nutr.* **2014**, *54*, 84–97. [CrossRef] [PubMed]
7.   Huang, X.; Hsieh, F.H. Physical Properties, Sensory Attributes, and Consumer Preference of Pear Fruit Leather. *J. Food Sci.* **2006**, *70*, E177–E186. [CrossRef]
8.   Sharma, P.; Ramchiary, M.; Samyor, D.; Baran Das, A. Study on the phytochemical properties of pineapple fruit leather processed by extrusion cooking. *LWT Food Sci. Technol.* **2016**, *72*, 534–543. [CrossRef]
9.   Hewavitharana, A.K.; Tan, Z.W.; Shimada, R.; Shaw, P.N.; Flanagan, B.M. Between fruit variability of the bioactive compounds, β-carotene and mangiferin, in mango (*Mangifera indica*). *Nutr. Diet.* **2013**, *70*, 158–163. [CrossRef]
10.   Azeredo, H.M.C.; Brito, E.S.; Moreira, G.G.E.; Farias, L.V. Effect of drying and storage time on the physico-chemical properties of mango leathers. *Int. J. Food Sci. Technol.* **2006**, *41*, 635–638. [CrossRef]
11.   United States Department of Agriculture (USDA). Nutritional Data Base for Standard Reference. Available online: http://www.nal.usda.gov/Fnic/Foodcomp/Data/Volume (accessed on 8 November 2011).
12.   Roongpisuthipong, C.; Banphotkasem, S.; Komindr, S.; Tanphaichitr, V. Postprandial glucose and insulin responses to various tropical fruits of equivalent carbohydrate content in non-insulin-dependent diabetes mellitus. *Diabetes Res. Clin. Pract.* **1991**, *14*, 123–131. [CrossRef]
13.   Diamante, L.M.; Bai, X.; Busch, J. Fruit Leathers: Method of Preparation and Effect of Different Conditions on Qualities. *Int. J. Food Sci.* **2014**, *2014*, 139890. [CrossRef] [PubMed]
14.   Torres, C.A.; Romero, L.A.; Diaz, R.I. Quality and sensory attributes of apple and quince leathers made without preservatives and with enhanced antioxidant activity. *LWT Food Sci. Technol.* **2015**, *62*, 996–1003. [CrossRef]
15.   Viskelis, J.; Rubinskiene, M.; Bobinas, C.; Bobinaite, R. Enrichment of Fruit Leathers with Berry Press Cake Powder Increase Product Functionality. In Proceedings of the 11th Baltic Conference on Food Science and Technology "Food Science and Technology in a Changing World", Jelgava, Latvia, 27–28 April 2017.
16.   Managa, M.G.; Remize, F.; Garcia, C.; Sivakumar, D. Effect of moist cooking blanching on colour, phenolic metabolites and glucosinolate content in chinese cabbage (*Brassica rapa* L. subsp. chinensis). *Foods* **2019**, *8*, 399. [CrossRef] [PubMed]
17.   Singleton, V.L.; Orthofer, R. Lamuela-Raventos, R.M. Analysis of Total Phenols and Other Oxidation Substrates and Antioxidants by Means of Folin-Ciocalteu Reagent. *Methods Enzymol.* **1999**, *299*, 152–178.
18.   Mpai, S.; du Preez, R.; Sultanbawa, Y.; Sivakumar, D. Phytochemicals and nutritional composition in accessions of Kei-apple (*Dovyalis caffra*): Southern African indigenous fruit. *Food Chem.* **2018**, *253*, 37–45. [CrossRef]
19.   Association of Official Analytical Chemists (AOAC). *Official Methods of Analysis*, 15th ed.; Association of Official Analytical Chemists: Washington, DC, USA, 2004.
20.   Escribano-Bailón, M.T.; Alcalde-Eon, C.; Muñoz, O.; Rivas-Gonzalo, J.C.; Santos-Buelga, C. Anthocyanins in berries of Maqui (*Aristotelia chilensis* Mol. Stuntz). *Phytochem. Anal.* **2006**, *17*, 8–14. [CrossRef]
21.   Castañeda-Ovando, A.; de Lourdes Pacheco-Hernández, M.; Elena Páez-Hernández, M.; Rodríguez, J.A.; Galán-Vidal, C.A. Chemical studies of anthocyanins: A review. *Food Chem.* **2009**, *113*, 859–871. [CrossRef]
22.   Ahmad, N.; Shafi'I, S.N.; Hassan, N.H.; Rajab, A.; Othman, A. Physicochemical and sensorial properties of optimised roselle-pineapple leather. *Malays. J. Anal. Sci.* **2018**, *22*, 35–44.
23.   Eyarkai Nambi, V.; Thangavel, K.; Shahir, S.; Chandrasekar, V. Color Kinetics During Ripening of Indian Mangoes. *Int. J. Food Prop.* **2016**, *19*, 2147–2155. [CrossRef]

24. Rumainum, I.M.; Worarad, K.; Srilaong, V.; Yamane, K. Fruit quality and antioxidant capacity of six Thai mango cultivars. *J. Agric. Nat. Resour.* **2018**, *52*, 208–214. [CrossRef]

25. Azizah, O.; Amin, I.; Nawalyah, A.G.; Ilham, A. Antioxidant capacity and phenolic content of cocoa beans. *Food Chem.* **2007**, *100*, 1523–1530.

26. Masibo, M.; He, Q.J. Major mango polyphenols and their potential significance to human health. *Compr. Rev. Food Sci. Food Saf.* **2008**, *7*, 309–319. [CrossRef]

27. Abbasi, A.M.; Guo, X.; Fu, X.; Zhou, L.; Chen, Y.; Zhu, Y.; Yan, H.; Liu, R.H. Comparative assessment of phenolic content and in vitro antioxidant capacity in the pulp and peel of mango cultivars. *Int. J. Mol. Sci.* **2015**, *16*, 13507–13527. [CrossRef] [PubMed]

28. Seeram, N.P.; Momin, R.A.; Nair, M.G.; Bourquin, L.D. Cyclooxygenase inhibitory and antioxidant cyanidin glycosides in cherries and berries. *Phytomedicine* **2001**, *8*, 362–369. [CrossRef]

29. Rakić, V.P.; Skrt, M.A.; Miljković, M.N.; Kostić, D.A.; Sokolović, D.T.; Poklar Ulrih, N.E. Effects of pH on the stability of cyanidin and cyanidin 3-O-β-glucopyranoside in aqueous solution. *Hem. Ind.* **2015**, *69*, 511–522. [CrossRef]

30. Vidot, K.; Achir, N.; Mertz, C.; Sinela, A.; Rawat, N.; Prades, A.; Dornier, M. Effect of temperature on acidity and hydration equilibrium constants of delphinidin-3-O- and cyanidin-3-O-sambubioside calculated from uni- and multi-wavelength spectroscopic data. *J. Agric. Food Chem.* **2016**, *64*, 4139–4145. [CrossRef]

31. Wojdyło, A.; Figiel, A.; Lech, K.; Nowicka, P.; Oszmiańsk, J. Effect of convective and vacuum–microwave drying on the bioactive compounds, color, and antioxidant capacity of sour cherries. *Food Bioproc. Technol.* **2014**, *7*, 829–841. [CrossRef]

32. Çoklar, H.; Akbulut, M. Effect of Sun, Oven and Freeze-Drying on Anthocyanins, Phenolic Compounds and Antioxidant Activity of Black Grape (Ekşikara) (*Vitis vinifera* L.). *S. Afr. J. Enol.* **2017**, *38*, 264–2732. [CrossRef]

33. Khoo, H.E.; Azlan, A.; Tang, S.T.; Lim, S.M. Anthocyanidins and anthocyanins: Colored pigments as food, pharmaceutical ingredients, and the potential health benefits. *J. Food Nutr. Res.* **2017**, *61*, 1361779. [CrossRef]

34. Scazzocchio, B.; Varì, R.; Filesi, C.; D'Archivio, M.; Santangelo, C.; Giovannini, C.; Galvano, F.J.D. Cyanidin-3-O-β-glucoside and protocatechuic acid exert insulin-like effects by upregulating PPARγ activity in human omental adipocytes. *Diabetes* **2011**, *60*, 2234–2244. [CrossRef]

35. Sun, C.; Zheng, Y.; Chen, Q.; Tang, X.; Jiang, M.; Zhang, J.; Chen, K.J.F.C. Purification and anti-tumour activity of cyanidin-3-O-glucoside from Chinese bayberry fruit. *Food Chem.* **2012**, *131*, 1287–1294. [CrossRef]

36. De Ferrars, R.M.; Czank, C.; Zhang, Q.; Botting, N.P.; Kroon, P.A.; Cassidy, A.; Kay, C.D. The pharmacokinetics of anthocyanins and their metabolites in humans. *Br. J. Pharmacol.* **2014**, *171*, 3268–3282. [CrossRef] [PubMed]

37. Ozdal, T.; Sela, D.A.; Xiao, J.; Boyacioglu, D.; Chen, F.; Capanoglu, E.J.N. The reciprocal interactions between polyphenols and gut microbiota and effects on bioaccessibility. *Nutrients* **2016**, *8*, 78. [CrossRef] [PubMed]

38. Ormazabal, P.; Scazzocchio, B.; Varì, R.; Santangelo, C.; D'Archivio, M.; Silecchia, G.; Lacovelli, A.; Giovannini, C.; Masella, R. Effect of protocatechuic acid on insulin responsiveness and inflammation in visceral adipose tissue from obese individuals: Possible role for PTP1B. *Int. J. Obes.* **2018**, *42*, 2012–2021. [CrossRef]

39. Abbasi, A.M.; Guo, X.; Fu, X.; Zhou, L.; Chen, Y.; Zhu, Y.; Yan, H.; Liu, R.H. Comparative Assessment of Phenolic Content and in Vitro Antioxidant Capacity in the Pulp and Peel of Mango Cultivars. *Int. J. Mol. Sci.* **2015**, *16*, 13507–13527. [CrossRef]

40. Ruales, J.; Baenas, N.; Moreno, D.; Stinco, C.; Meléndez-Martínez, A.; García-Ruiz, A.J.N. Biological active ecuadorian mango 'tommy atkins' ingredients—An opportunity to reduce agrowaste. *Nutrients* **2018**, *10*, 1138. [CrossRef]

41. Panche, A.N.; Diwan, A.D.; Chandra, S.R. Flavonoids: An overview. *J. Nutr. Sci.* **2016**, *5*, e47. [CrossRef]

42. Lachowicz, S.; Oszmiańsk, O.; Wiśniewski, R.; Seliga, T.; Pluta, S. Chemical parameters profile analysis by liquid chromatography and antioxidative activity of the Saskatoon berry fruits and their components. *Eur. Food Res. Technol.* **2019**, *245*, 2007–2015. [CrossRef]

43. Perera, C.O. Selected quality attributes of dried foods. *Dry Technol.* **2005** *23*, 717–730. [CrossRef]

44. Ciurzyńska, A.; Cieśluk, P.; Barwińska, M.; Marczak, W.; Ordyniak, A.; Lenart, A.; Janowicz, M. Eating Habits and Sustainable Food Production in the Development of Innovative "Healthy" Snacks. *Sustainability* **2019**, *11*, 2800. [CrossRef]

45. American Heart Association (AHA). Recommended Sugar Intake. Available online: https://www.healthline. com/nutrition/how-much-sugar-per-day#section (accessed on 3 March 2020).

46. Somerset, S.M. Refined sugar intake in Australian children. *Public Health Nutr.* **2003**, *6*, 809–813. [CrossRef] [PubMed]

47. French, S.J.; Read, N.W. Effect of guar gum on hunger and satiety after meals of differing fat content: Relationship with gastric emptying. *Am. J. Clin. Nutr.* **1994**, *59*, 87–91. [CrossRef] [PubMed]

48. Dhingra, D.; Michael, M.; Rajput, H.; Patil, R. Dietary fibre in foods: A review. *J. Food Sci. Technol.* **2012**, *49*, 255–266. [CrossRef] [PubMed]

49. Aziz, Q.; Doré, J.; Emmanuel, A.; Guarner, F.; Quigley, E.J.N. Gut microbiota and gastrointestinal health: Current concepts and future directions. *J. Neurogastroenterol.* **2013**, *25*, 4–15. [CrossRef] [PubMed]

50. Evans, S.F.; Meister, M.; Mahmood, M.; Eldoumi, H.; Peterson, S.; Perkins-Veazie, P.; Clarke, S.L.; Payton, M.; Smith, B.J.; Lucas, E.A. Mango supplementation improves blood glucose in obese individuals. *Nutr. Metabolic Insights* **2014**, *7*, NMI-S17028. [CrossRef] [PubMed]

51. Offia-Olua, B.I.; Ekwunife, O.A. Production and evaluation of the physico-chemical and sensory qualities of mixed fruit leather and cakes produced from apple (*Musa Pumila*), banana (*Musa Sapientum*), pineapple (*Ananas Comosus*). *Niger. Food J.* **2015**, *33*, 22–28. [CrossRef]

52. Concha-Meyer, A.A.; D'Ignoti, V.; Saez, B.; Diaz, R.I.; Torres, C.A. Effect of storage on the physico-chemical and antioxidant properties of strawberry and kiwi leathers. *J. Food Sci.* **2016**, *81*, C569–C577. [CrossRef]

53. Adedeji, A.A.; Gachovska, T.K.; Ngadi, M.O.; Raghavan, G.S.V. Effect of pretreatments on drying characteristics of Okra. *Dry Technol.* **2006**, *26*, 1251–1256. [CrossRef]

3

Food Origin Traceability from a Consumer's Perspective

Anna Walaszczyk * and Barbara Galińska

Faculty of Management and Production Engineering, Lodz University of Technology, 90-924 Lodz, Poland;
barbara.galinska@p.lodz.pl
* Correspondence: anna.walaszczyk@p.lodz.pl

Abstract: The awareness of food origin in the consumers' perspective has gradually become more significant not only in reference to consumers from highly developed countries but also from emerging ones, which are already on their way from a developing to developed economy. The purpose of the paper is to answer the research question by verifying four hypotheses formulated in the research process. The research question is: "Do the variables which characterize consumers of food products in Poland, including gender, age, education and financial status, affect the aspects related to food traceability, such as identification of the producer, importance of food product features when shopping, importance of the information given on food product packaging and influence of the shopping place and frequency on tracing the food origin?" The paper presents the results, analysis, and conclusions from the study in reference to the four assumed hypotheses related to the above-mentioned research question. The study was carried out on a group of 500 consumers of food products in Poland. The study topic selection is justified by the assumed significance of tracing back a food product's origin for a consumer who functions in a globalization-based economy; this was confirmed by the subject literature presented in the paper.

Keywords: traceability; food; consumer; consumption; behavior; security

1. Introduction

Food traceability, defined as the possibility to trace back the history of a food product, is a very important process from the point of view of ensuring food safety for consumers all over the world. The safety of produced food can be maintained only when full traceability of raw materials, semi-finished products, and processes are ensured on all stages of the food chain [1–5]. Adequate organization of the market of food products is difficult considering the size of the market and the strict legal requirements in force. Despite stringent control of the market by authorized control units, every week the Rapid Alert System for Food and Feed (RASFF) registers several dozen alerts in Europe, regarding food launched in the market that has been labeled dangerous for human life and health (RASFF was introduced by means of EC regulation No. 178/2002 of the European Parliament and the Council on 29 January 2002).

The literature on food safety presents different definitions of the term "traceability" [6]. The term originally appeared back in 1996 in the international ISO 8402 standard concerning quality management and quality assurance. According to the definition in the standard, traceability stands for the ability to trace the history, application, or location of a unit by way of an analysis of records allowing its identification [7]. The publication by Moe (1998) entitled "Perspectives on Traceability in Food Manufacture" contains a wider definition stating that traceability stands for the ability to trace a batch of products and its history throughout the entire production chain or part of it, from harvesting through to transport, storage, processing, distribution, and sales (this is known as the traceability chain) or internally—in a single step of the production chain (this is known as the internal traceability chain) [8]. According to the Efficient Consumer Response (ECR) strategy enforced in 2004, the essence of traceability lies in the possibility of monitoring the handling and origin of a reference product (production batch) in each stage of the supply chain, by all companies in the food sector (ECR Europe 2004). The ECR integrates producers, suppliers, and sellers in the supply chain to build a cost-effective system which responds to specific needs of a consumer; consequently, the total cost of the system and stock level are reduced with a simultaneous increase in value for the end customer [9]. Identification is an intrinsic element of traceability, determining its success in relation to tracing the history of products [10]. Schwagele defines traceability through the concept of identification and states that identification enables the acquisition of data from a previous stage of the chain (what was received and from whom) and the provision of information for the following stage (what was sent and to whom) [11]. In the opinion of Rabade and Alfaro, traceability stands for the registration and tracing of processes and materials used in production [12]. A definition of the concept of traceability can also be found in the first international standard—ISO 22005 of 2007—applying to food safety management in the context of tracing its history. The standard contains the following definition: "traceability stands for the ability to trace the flow (movement) of feed or food through specific production stage(s), processing and distribution; movement can refer to the origin of materials (raw materials), history of processing and distribution of feed or food" [13,14].

Gradually more attention was paid to scientific research carried out not only in Europe but also globally on the consumers' perception of food origin tracing. Recently such research was conducted in China [15], Japan [16], and Brazil [17]. The general conclusion from the research, which either applies to a particular country or compares the issue of consumers' approach to food traceability in different countries [18], is that consumers want information regarding where a product comes from, what production methods were used, whether production was certified, and if information on the packaging is complete and reliable [19,20]. Consumers have become highly concerned about the safety of their food and so the speed of obtaining information about contamination and diseases transmitted by food, effective risk management, and efficient management of non-compliant products' withdrawal from the market, have become more important [21]. Consumers' concerns about food safety are also related to genetic modification of food [22]. In order to face the challenge of an increasingly demanding process of food product traceability, China uses extensive and advanced food origin tracing methods based on DNA [23], whereas in reference to food origin tracing management, an additional consumer fee, called WTP (Willingness To Pay), is used for information about the food [24].

Regulation No. 178/2002 issued in January 2002, sets out the general rules and requirements of food law; the establishment of the European Food Safety Authority and procedures concerning food safety are fundamental for the food laws of communities (including Poland, on which the research was based). Article 18 of the Regulation obliges all EU member states to establish and implement traceability procedures as of 1 January 2005. The core of traceability is in tracing the path of a finished product "forwards" and "backwards" [25]. Backward traceability helps to identify the cause and sources of a hazard (e.g., via data by the producer or number of product batches). Forward traceability involves withdrawing a finished product from the market owing to identification of the location where a non-compliant batch was delivered. The trade quality of products in Poland is supervised by two bodies: the Trade Inspectorate (retail) and the Inspectorate for Trade Quality of Agricultural and Food Products (producers). Moreover, in the food supply chain, there is no continuous exchange of information about the course of each measure taken by subsequent actors, which impairs the traceability of products. The growth of a company in the food sector and the adaptation of the internal and external traceability system to ensure food safety, are possible if the traceability system is treated as a subsystem whose presence is necessary to manage product quality [26,27]. This is because traceability is indispensable to ensuring the quality of production and the product itself [28].

This study's research problem, formulated based on the subject literature review, is expressed as a research question, the answers to which are the paper's objective. The research question is as follows: "Do the variables which characterize consumers of food products in Poland, including gender, age, education and financial status, affect the aspects related to food traceability, such as identification of the producer, importance of food product features when shopping, importance of the information given on food product packaging and influence of the shopping place and frequency on tracing the food origin?".

The validity of the research issue to be verified is justified by the fact that under current globalization-based economic conditions, food origin is of paramount importance. Consumers in highly developed and developing countries have a wide variety of food product criteria (including origin) to choose from [29]. Observations of consumer behavior revealed that people want to know more about the food they buy. Knowledge of the food origin, and as a conscious choice in this respect, contributes to the consumer's self-assurance about food safety [30–33]. The issue of the requirements that particular consumer groups establish for particular food products is also very important for food producers and developers of production-related processes. Clients' requirements are the key requirements to be faced by producers because a consumer is the key stakeholder for a conscious producer [34].

In order to assign a quantitative dimension to the research issue, which can be confirmed by statistical indicators, four research hypotheses were formulated and verified in the study:

1. **H1.** *There is a relationship between the identified groups of consumers and the kind of food producer they choose, as regards market coverage.*

2. **H2.** *There are significant differences in the ranking of the validity of aspects taken into account when people from the identified consumer groups buy food.*

3. **H3.** *There are significant differences in the way that people from the identified consumer groups rank the validity of information on food packaging.*

4. **H4.** *There are significant differences in the way that people from the identified consumer groups rank the significance of a food product's origin.*

The presented hypotheses determined the use of the following study methods.

2. Materials and Methods

The study was carried out in the first half of 2017 in Poland on a population of 500 consumers. It was a pilot study in which 50% of the studied group of respondents were men and 50% were women. The integrity of the study sample was the only parameter of purposeful selection of the study subject matter. Other selection conditions of consumers participating in the study were randomized and comprised of:

- respondents in four age brackets (20–30 years, 31–40 years, 41–50 years, and over 60 years);
- four levels of education (primary or lower secondary education, vocational education, secondary education, and university education);
- size of private household in terms of number of persons included (1 person, 2 people, 3–4 people, more than 5 people);
- place of residence presented in four categories (countryside, city up to 50,000 people, city with 50,000–100,000 inhabitants, and city with a population above 100,000);
- per capita gross income in the household (up to 400 EUR, 401–600 EUR, 601–950 EUR, over 950 EUR).

Following a demographic analysis of the respondents taking part in the study, regarding their sex, age, education and income, some similarities and relationships were observed between the characteristics. An analysis of concentrations was performed in order to group the respondents according to uniform or highly similar characteristics [35]. Grouping was intended to focus attention on the groups of consumers participating in the study, which could become representative groups from the point of view of the conclusions drawn from the study. A two-step concentration analysis was applied to divide the study participants into four groups. The Bayesian information criterion (BIC) was used to separate the subgroups, while the distance between the study participants was measured with a credibility logarithm. The two-step concentration analysis facilitated the identification of the following four groups; most homogeneous within a specific group and also most diversified as compared to other groups, based on such variables as sex, age, education, and income:

- men of all ages, with a secondary degree, medium income
- people aged 31–60, with a university degree, medium income
- women of all ages, with a secondary degree, medium income
- younger people (20–30 years old), with a university degree, medium and higher income.

Table 1 presents the quantitative characteristics of the identified groups.

Table 1. Profiles of the four identified groups of respondents.

		Concentration Number in Two-Step Grouping							
		1		2		3		4	
		n	%	n	%	n	%	n	%
Age	20–30	39	32.5			10	6.0	50	47.2
	31–40	11	9.2	3	2.9	45	26.8	40	37.7
	41–50	2	1.7	52	51.0	34	20.2	11	10.4
	51–60	24	20.0	21	20.6	49	29.2	5	4.7
	60+	44	36.7	26	25.5	30	17.9		
Sex	Male	120	100.0	85	83.3			43	40.6
	Female			17	16.7	168	100.0	63	59.4
Education	Primary/lower secondary			12	11.8				
	Vocational			36	35.3	39	23.2		
	Secondary	120	100.0	10	9.8	128	76.2		
	Tertiary			44	43.1	1	0.6	106	100.0
Income	401–600 EUR	6	5.0	2	2.0	6	3.6		
	601–950 EUR	83	69.2	98	96.1	136	81.0	54	50.9
	over 950 EUR	31	25.8	2	2.0	26	15.5	52	49.1

Source: data developed based on our own study.

A proprietary questionnaire consisting of eight main questions and six identification questions concerning sex, age, education, household size, and income of the respondents (based on the analysis of concentrations, household size and place of residence did not matter from the point of view of respondent grouping) was the study tool. Among the main questions there were six multiple-choice questions and two questions where one needed to assign rank on a 1–6 and 1–5 scale. "1" meant most important, and "5" or "6" least important. The ranked questions concerned those factors most important for consumers when they buy food. They involved such aspects as knowledge of the brands, guarantee of product quality and freshness, low price, best before date, product origin, and ingredients. Moreover, the questions touched upon information that in the consumers' opinions should be placed on food packaging (i.e., information about the producer, the product's energy value, best before date, detailed information about raw materials, and product ingredients).

Collection of empirical materials by means of an interview questionnaire developed by the authors was justified by the specificity of the research and the necessity to reach a wide range of respondents.

A focused group interview (i.e., a focus group), composed of representatives of academic staff and entrepreneurs was involved in the study in order to:

- use the opinions of the focus group members (as one of the factors) to formulate four study hypotheses (presented in the introduction of the paper) and to develop an interview questionnaire;
- extend the interpretation of the obtained results of the study.

The study used both direct tools (proprietary interviews) and indirect tools (e-mails and Google surveys). The data from the study were collected in an Excel spreadsheet and then summarized and analyzed quantitatively using Statistica (a statistical data-processing software). The following statistical measures were used to analyze the results of the study:

- Chi-squared test of cross tabulation;
- Cramer's V test, effect size;
- non-parametric Mann–Whitney U test;
- non-parametric Kruskal–Wallis test.

The majority of analyses in the study were based on data in the form of ranks or ordinal variables. Therefore, referring to the characteristics of each statistical measure, when testing hypotheses mentioning different distributions of ranks in different groups, the Kruskal–Wallis test was used. Pearson's Chi-squared test was used for verification of the presence of relationships between the variables according to the hypotheses. The following symbols were used for presenting the study results:

- M—arithmetic mean
- SD—standard deviation (the lower the value, the more focused the observations around M)
- MR—mean range (non-parametric equivalent of M)
- Mdn—median (value of a characteristic above and below which the number of observations N is the same)
- N—number of observations
- "p"—relevance, for the value <0.05 we have over 95% certainty that there is a relationship between classifications and the selected variable (i.e., that the observed relationship is not incidental)
- χ^2 —general statistics for the reference analysis, based on which "p" is calculated
- Cramer's r, V—effect size: it gives information about the size of the relationship between two variables (up to 0.30 the effect is assumed to be weak; between 0.30 and 0.50 the effect is assumed to be medium, while over 0.50 the effect is assumed to be strong)
- Z—general statistics for the reference analysis, based on which "p" and "r" are calculated.

3. Results of the Study

The first hypothesis assumed there would be a relationship between the identified consumer groups and the food producers they chose for their market coverage. An analysis performed with a Chi-squared test of cross tabulations revealed the presence of a strong relationship between the identified groups and their preferred producers. Details of the study results in this respect are presented in Table 2.

The second hypothesis assumed there would be significant differences in the ranking of the validity of aspects, which the representatives of the four identified groups of consumers took into account when buying food. Analysis using the Kruskal–Wallis test revealed that differences in the ranking of validity occurred in three (brand, best before date, and product ingredients) out of six analyzed aspects, which included brand, quality guarantee, price, best before date, origin, and ingredients. Details of the study results in this respect are presented in Table 3.

Table 2. First hypothesis results.

			Most often You Choose Producers among:				χ^2	p	Cramer's V
			Global Market Leaders	Domestic Producers	Regional Producers	It does not Matter to Me			
Group number	1	N	51	52	10	7	20.78	<0.05	0.12
		%	42.5%	43.3%	8.3%	5.8%			
	2	N	49	50	1	2			
		%	48.0%	49.0%	1.0%	2.0%			
	3	N	65	82	19	2			
		%	38.7%	48.8%	11.3%	1.2%			
	4	N	51	42	12	1			
		%	48.1%	39.6%	11.3%	0.9%			

Source: data developed based on our own study.

Table 3. Second hypothesis results.

Question	Group Number	N	MR	M	Mdn	SD	χ^2	p	Significant Differences
How important is the brand?	1	120	221.52	2.50	3.00	1.34	12.56	<0.01	1 < 4
	2	102	245.48	2.72	3.00	1.22			
	3	168	248.51	2.76	3.00	1.26			
	4	106	281.93	3.07	3.00	1.17			
How important is the best before date?	1	120	282.14	2.47	2.00	1.66	11.74	<0.01	1 > 3; 1 > 4
	2	102	247.41	1.94	1.50	1.24			
	3	168	239.65	1.93	1.00	1.36			
	4	106	225.50	1.80	1.00	1.29			
How important are the ingredients?	1	120	208.39	3.88	4.00	1.01	13.47	<0.01	1 < 2; 1 < 3; 1 < 4
	2	102	260.32	4.34	4.00	1.25			
	3	168	262.68	4.38	4.00	1.30			
	4	106	260.06	4.38	4.00	1.32			

Source: data developed based on our own study. N, number of observations; MR, mean range; M, arithmetic range; Mdn, mean; SD, standard deviation.

The third hypothesis assumed there would be significant differences in the ranks assigned by the representatives of the identified consumer groups to the validity of information given on food packaging. A quantitative analysis revealed that the assumed differences did not actually occur. Details of the study results in this respect are presented in Table 4.

In the last (fourth) research hypothesis, it was assumed that significant differences existed in the ranks that the identified groups of consumers assigned to food product origin. The variables included the place of food purchase (Table 5) and the frequency of shopping (Table 6). The results of the performed analysis did not confirm the hypothesis.

Table 4. Third hypothesis results.

Questions	Group No.	N	MR	M	Mdn	SD	χ^2	p	Significant Differences
How important is information about the producer?	1	120	253.81	5.13	5.00	0.84	1.72	0.63	–
	2	102	243.72	5.06	5.00	0.87			
	3	168	240.32	5.04	5.00	0.83			
	4	106	260.06	5.16	5.00	0.86			
How important is information about the energy value?	1	120	251.33	2.29	2.00	0.46	2.47	0.48	–
	2	102	242.22	2.25	2.00	0.44			
	3	168	257.24	2.32	2.00	0.47			
	4	106	237.49	2.24	2.00	0.43			
How important is information about the best before date?	1	120	245.67	2.71	3.00	0.46	2.47	0.48	–
	2	102	254.78	2.75	3.00	0.44			
	3	168	239.76	2.68	3.00	0.47			
	4	106	259.51	2.76	3.00	0.43			
How important is information about the raw materials the product is made of?	1	120	242.48	5.00	5.00	0.77	1.11	0.77	–
	2	102	257.92	5.09	5.00	0.77			
	3	168	243.74	5.01	5.00	0.80			
	4	106	253.79	5.07	5.00	0.73			
How important is information about ingredients?	1	120	248.15	4.88	5.00	0.84	3.20	0.36	–
	2	102	245.35	4.85	5.00	0.80			
	3	168	261.22	4.95	5.00	0.82			
	4	106	231.77	4.77	5.00	0.81			

Source: data developed based on our own study.

Table 5. Product origin vs. place of purchase.

	Where do You Buy Food Most Often?	N	MR	M	Mdn	SD	χ^2	p
How important is product origin?	Supermarket	103	237.03	5.09	5.00	0.84	1.22	0.75
	Mini market	279	253.37	5.18	5.00	0.83		
	Grocer's	89	251.61	5.19	5.00	0.78		
	Marketplace	27	250.17	5.07	5.00	1.07		

Source: data developed based on our own study.

Table 6. Product origin vs. frequency of doing shopping.

	Frequency of Doing Shopping	N	MR	M	Mdn	SD	χ^2	p
How important is product origin?	Every day	55	251.27	5.16	5.00	0.83	1.37	0.71
	Several times a week	107	237.54	5.12	5.00	0.75		
	Once a week	318	254.39	5.18	5.00	0.84		
	Once or twice a month	20	255.83	5.05	5.00	1.19		

Source: data developed based on our own study.

4. Analysis of Results and Discussion

In reference to the four hypotheses made at the beginning of the study, only two were confirmed; there are relationships between the identified groups of consumers and the kind of producer they chose as regards market coverage (H1), and there are significant differences in the ranking of the validity of aspects taken into account when representatives of the four identified groups of consumers buy food (H2).

As for the first hypothesis, women of all ages (no age diversification), with secondary education and medium income, were the ones who chose products from "global market leaders" least often (group 3). Global brand products were most often chosen by people aged 31–60, with higher education and medium income (group 2), and by younger people (aged 20–30) with higher education and medium or higher income (group 4). Consumers from group 2 tended to buy products from domestic producers as often as products from global market leaders, while regional producers were the least popular in the group. This can be justified by the maturity of this group of consumers, as exemplified by their age and education, and hence greater confidence in a producer operating on a global market and having a more extended control system of the food produced.

Confirmation of the second hypothesis reveals that males of any age, medium income, and secondary education (group 1) ranked the significance of a brand much lower than younger people (aged 20–30) with higher education, and medium and higher income (group 4). Additionally, the same men ranked the significance of the product ingredients lower than any other identified group. They considered the best before date as a much more important factor than people from group 3 (women of any age, secondary education, and medium income) and group 4 (younger people aged 20–30, higher education, of medium and higher income). From the point of view of the aim of the study, product ingredients form the most important aspect for food traceability. The study revealed that this mattered to all consumers participating in the study, but it was least important for men with secondary education and medium income. Such a result can be justified by the fact that medium income does not always give consumers the freedom to choose food with the ingredients they would prefer, and secondary education may be indicative of a lack of knowledge on how to analyze the ingredients of a food product correctly. In addition to the above, men typically pay less attention to more pragmatic issues, such as best before dates or prices, than the product's composition/ingredients. The reasons for this behavior could form the subject of a sociological or psychological enquiry.

Two hypotheses were not confirmed in the course of the study. They included H3: there are important differences in the rank assigned by representatives of the identified consumer groups to the validity of information given on food packaging, and H4: there are significant differences in the rank assigned by the identified groups of consumers to the aspect of food product origin. Referring to H3, it should be pointed out that the identified groups of consumers were quite unanimous about the significance of information on the raw materials that the product is made of (which is a positive result from the point of view of the importance of traceability for consumers). The greatest discrepancies in answers were observed in relation to information about the product composition/ingredients (the greatest differences were observed between groups 3 and 4). It can be concluded that all studied information on food packaging, including identification of the producer, energy value of the product, best before date, raw materials used for production, and composition/ingredients are of equal importance for men and women, regardless of age, education, or income.

In reference to H4, assuming significant differences in the assessment of the validity of food origin, it turned out that differences in the answers given by different consumers were so small that they could be considered unanimous as for the significance of product origin vs. the following two variables: place of doing shopping and frequency of doing shopping. This suggests that food traceability is important for all study participants, regardless of the frequency or location of doing shopping.

Referring to the research question posed, the results of the study showed that the origin of food products mattered to consumers, regardless of the analyzed conditions. Differences in relation to the identified groups of consumers were noticeable in the following respect: some aspects of traceability were approached in a different way and have a different significance priority assigned to them. Perhaps it would be worth analyzing the issue of specific information that should—in the consumers' opinion—be placed on food products and reviewing if the scope of data complies with the current legal requirements. Identification of how far back the process of backward traceability of food should reach from the point of view of consumers could form an interesting focus for further studies.

5. Conclusions

The results of the study on food traceability from a consumer's perspective, carried out on a group of 500 randomly selected representatives in Poland and presented here, revealed that the issue of food traceability in Poland is important. This is evidenced by the fact that of all the aspects available in the study for selection, the one that consumers paid most attention to was food origin, while the presence of information about producers was considered necessary for food packaging.

Food traceability by consumers is an interesting and diversified phenomenon, which strongly affects the organization of a food chain. Consumer requirements are conditioned by organizational and management decisions by food suppliers on different levels of production, especially when their decisions apply to food safety. The research constraints, which could be used as an idea for subsequent consumer research related to food traceability include:

- a lack of comparison with other research results conducted worldwide, pertaining to similar research topics;
- a lack of initial in-depth verification of the knowledge of the traceability notion among the responders.

Author Contributions: Conceptualization, A.W. and B.G.; methodology, A.W.; software, A.W.; validation, A.W. and B.G.; formal analysis, A.W.; investigation, A.W.; resources, A.W. and B.G.; data curation, A.W.; writing—original draft preparation, A.W.; writing—review and editing, B.G.; visualization, A.W.; supervision, A.W. and B.G.; project administration, B.G.; funding acquisition, B.G.. All authors have read and agreed to the published version of the manuscript.

References

1. Skilton, P.F.; Robinson, L.J. Traceability and normal accident theory: How does supply network complexity influence the traceability of adverse events? *J. Supply Chain Manag.* **2009**, *45*, 40–53. [CrossRef]
2. Adrian, A.M.; Norwood, S.H.; Mask, P.L. Producers' perceptions and attitudes toward precision agriculture technologies. *Comput. Electron. Agric.* **2005**, *48*, 256–271. [CrossRef]
3. Hoorfar, J.; Jordan, K.; Butler, F.; Prugger, R. *Food Chain Integrity: A Holistic Approach to Food Traceability, Safety, Quality and Authenticity*; Woodhead Publishing Limited: Sawston, UK; Cambridge, UK, 2011.
4. Bennet, G.S. *Food Identity Preservation and Traceability: Safer Grains*; CRC Press: Boca Raton, FL, USA, 2009.
5. Bosona, T.; Gebresenbet, G. Food traceability as an integral part of logistics management in food and agricultural supply chain. *Food Control* **2013**, *33*, 32–48. [CrossRef]
6. Olsen, P.; Borit, M. How to define traceability. *Trends Food Sci. Technol.* **2013**, *29*, 142–150. [CrossRef]
7. ISO 8402:1994—Quality Management and Quality Assurance—Vocabulary. British Standards Institution, London, 1994. Available online: https://www.iso.org/standard/20115.html (accessed on 20 March 2019).
8. Moe, T. Perspectives on tracebility in food manufacture. *Trends Food Sci. Technol.* **1998**, *9*, 211–214. [CrossRef]
9. ECR Europe (Efficient Consumer Response). *Using Traceability in the Supply Chain to Meet Consumer Safety Expectations*. ECR Europe, 2004. Available online: https://www.ecr-community.org/publications/ (accessed on 15 June 2015).
10. Beulens, A.J.M.; Broens, D.F.; Folstar, P.; Hofstede, G.J. Food safety and transparency in food chains and networks: Relationships and challenges. *Food Control* **2005**, *16*, 481–486. [CrossRef]
11. Schwagele, F. Traceability from a European Perspective. *Meat Sci.* **2005**, *71*, 164–173. [CrossRef] [PubMed]
12. Rabade, L.A.; Alfaro, J.A. Buyer-supplier relationship's influence on traceability implementation in the vegetable industry. *J. Purch. Supply Manag.* **2006**, *12*, 39–50. [CrossRef]
13. ISO 22005:2007—Traceability in the Feed and Food Chain—General Principles and Basic Requirements for System Design and Implementation. European Committee for Standardization, CEN 2007. Available online: https://www.iso.org/standard/36297.html (accessed on 29 February 2020).
14. Shengnan, S.; Xinping, W. Promoting traceability for food supply chain with certification. *J. Clean. Prod.* **2019**, *217*, 658–665.

15. Kendall, H.; Kuznesof, S.; Dean, M.; Chan, M.-Y.; Frewer, L. Chinese consumer's attitudes, perceptions and behavioural responses towards food fraud. *Food Control* **2019**, *95*, 339–351. [CrossRef]

16. Jin, S.; Zhou, L. Consumer interest in information provided by food traceability systems in Japan. *Food Quality and Preference* **2014**, *36*, 144–152. [CrossRef]

17. Matzembacher, D.E.; Stangherlin, I.; Slongo, L.A.; Cataldi, R. An integration of traceability elements and their impact in consumer's trust. *Food Control* **2018**, *92*, 420–429. [CrossRef]

18. Rijswijk, W.; Frewer, L.J.; Menozzi, D.; Faioli, G. Consumer perceptions of traceability: A cross-national comparison of the associated benefits. *Food Qual. Prefer.* **2008**, *19*, 452–464. [CrossRef]

19. Smith, I. *Meeting Customer Needs*; Routledge: London, UK, 2012.

20. Smith, I.; Furness, A. *Improving Traceability in Food Processing and Distribution*; Woodhead Publishing: Sawston, UK; Cambridge, UK, 2006.

21. Astill, J.; Dara, R.A.; Campbell, M.; Farber, J.M.; Yada, R.Y. Transparency in food supply chains: A review of enabling technology solutions. *Trends Food Sci. Technol.* **2019**, *91*, 240–247. [CrossRef]

22. Boccia, F.; Covino, D.; Sarnacchiaro, P. Genetically modified food versus knowledge and fear: A Noumenic approach for consumer behaviour. *Food Res. Int.* **2018**, *111*, 682–688. [CrossRef] [PubMed]

23. Zhao, J.; Xu, Z.; You, X.; Zhao, Y.; Yang, S. Genetic traceability practices in a large-size beef company in China. *Food Chem.* **2019**, *277*, 222–228. [CrossRef] [PubMed]

24. Liu, R.; Gao, Z.; Nayga, R.M.; Snell, H.A.; Ma, H. Consumers'valuation for food traceability in China: Does trust matter? *Food Policy* **2019**, *88*, 101768. [CrossRef]

25. Council of the European Union. *Regulation (EC) No. 178/2002 of the European Parliament and of the Council*; Council of the European Union: Brussels, Belgium, 2002.

26. Coff, C.; Barling, D.; Korthals, M.; Nielsen, T. *Ethical Traceability and Communicating Food*; Springer Science & Business Media: Dordrecht, The Netherlands, 2008; Volume 15, pp. 1–18.

27. Eckschmidt, T. *The Little Green Book of Food Traceability: Concepts and Challenges*; Booksurge Publishing: Charleston, SC, USA, 2009.

28. Bertolini, M.; Bevilacqua, M.; Massini, R. FMECA approach to product traceability in the food industry. *Food Control* **2006**, *17*, 137–145. [CrossRef]

29. Galińska, B. Changes in Supply Chains in the Light of Emerging Market Procurement. In Proceedings of the 19th International Scientific Conference Business Logistics in Modern Management (BLMM2019), Osijek, Croatia, 10–11 October 2019; pp. 533–545.

30. Wales, C.; Harvey, M.; Warde, A. Recuperating from BSE: The shifting UK institutional basis for trust in food. *Appetite* **2006**, *47*, 187–195. [CrossRef] [PubMed]

31. Montserrat, E.; Santaclara, F.J. *Advances in Food Traceability Techniques and Technologies: Improving Quality Throughout the Food Chain*; Woodhead Publishing Limited: Sawston, UK; Cambridge, UK, 2016.

32. Resende-Filho, M.A.; Terrance, M. Hurley, Information asymmetry and traceability incentives for food safety. *Int. J. Prod. Econ.* **2012**, *139*, 596–603.

33. Lees, M. *Food Authenticity and Traceability*; Woodhead Published Limited: Sawston, UK; Cambridge, UK, 2003.

34. ISO 9001:2015—Quality Management System. Requirements. ISO 2015. Available online: https://www.iso.org/standard/62085.html (accessed on 29 February 2020).

35. Everitt, B.S.; Landau, S.; Leese, M. *Cluster Analysis*; Oxford University Press: New York, NY, USA, 2001.

4

Salmon Intake Intervention in the Vulnerable Group of Young Polish Women to Maintain Vitamin D Status during the Autumn Season

Zofia Utri and Dominika Głąbska *

Department of Dietetics, Institute of Human Nutrition Sciences, Warsaw University of Life
Sciences (SGGW-WULS), 159c Nowoursynowska Street, 02-776 Warsaw, Poland; zofia_utri@mail.sggw.pl
* Correspondence: dominika_glabska@sggw.pl

Abstract: Fish products are the main dietary source of vitamin D, but due to a low fish intake in the majority of European countries, an inadequate vitamin D intake is common, especially in the vulnerable group of young women for whom it is essential for the osteoporosis prevention. The aim of the presented study was to assess the possibility of applying salmon intake intervention for maintaining vitamin D status in young Polish women during the autumn season, in which in Poland there is not enough sunshine exposure to generate skin synthesis. The dietary intervention within VISA Study (Vitamin D In Salmon) comprised eight weeks of daily consumption of 50 g of Atlantic salmon and was conducted in a group of 47 women aged 20–30 years. Within the study, their changes of total serum 25-hydroxyvitamin D (25(OH)D) levels were analyzed and the effectiveness of the intervention depending on age, body mass index (BMI), and baseline 25(OH)D were assessed. Until the 4th week, 25(OH)D in the studied group decreased from 57.1 nmol/L to 39.9 nmol/L ($p < 0.0001$), but afterward it increased until the 8th week to 54.1 nmol/L ($p = 0.0005$), contributing to results not differing from the baseline ($p = 0.7964$). At the same time, the share of respondents characterized by an inadequate vitamin D status increased until the 4th week, but afterward, it decreased until the 8th week ($p = 0.0002$). Neither the age (in the assessed range), nor the BMI influenced 25(OH)D during the study, but only the baseline 25(OH)D was correlated with the BMI ($p = 0.0419$; R = −0.2980). The baseline 25(OH)D was associated with its levels during the intervention, as well as with 25(OH)D change from the baseline values ($p < 0.0001$). It may be concluded that, in spite of the initial decline of the 25(OH)D observed (probably connected to the starting time of the study), afterward the salmon intake intervention contributed to its increase, while the baseline 25(OH)D status was an important determinant of the intervention effectiveness during the autumn season.

Keywords: vitamin D; dietary intake; dietary intervention; fish intake; salmon; 25-OH-cholecalciferol; 25-hydroxyvitamin D; 25(OH)D; young women

1. Introduction

Vitamin D is of the highest importance for the skeletal system, promoting calcium and phosphorus absorption in the gut, maintaining their adequate concentrations, and for bone growth and remodeling by osteoblasts and osteoclasts [1]. Studies also suggest an association between vitamin D status or intake and various diseases and disorders. Recent meta-analyses show that vitamin D deficiency is associated with a higher risk of sleep disorders [2] and may be related to autoimmune thyroid disease [3], whereas high vitamin D serum levels have a protective effect on breast cancer in premenopausal women [4]. They also indicate that vitamin D supplementation significantly reduces cancer mortality [5], as well as the rate of asthma exacerbations [6]. Taking it into account, scientists agree that vitamin D goes beyond influencing bone health only [7,8].

However, vitamin D status is a global problem. Based on protocols of the Vitamin D Standardization Program (VDSP) by National Institutes of Health (NIH) [9], the analysis of data for 14 European countries indicated the prevalence of vitamin D deficiency of 13% [10], while defined by the United States (US) Institute of Medicine (IoM) and the United Kingdom National Osteoporosis Society (NOS) as serum 25-hydroxyvitamin D (25(OH)D) level lower than 30 nmol/L (12 ng/mL, 1 nmol/L = 0.4 ng/mL) [11,12]. At the same time, for the alternate threshold for vitamin D deficiency of serum 25(OH)D level lower than 50 nmol/L, which is suggested by the US Endocrine Society [13], European Food Safety Authority (EFSA) [14] and Polish recommendations [15], the prevalence was 40.4% [10]. Among healthy adolescents, the prevalence of vitamin D deficiency (25(OH)D level lower than 50 nmol/L) sometimes reaches even 42% [16] and in teenage girls living in northern Europe, in the winter season, even 92% [17]. In young women aged 18–29 years living in France, the prevalence of vitamin D deficiency (25(OH)D level lower than 50 nmol/L) amounts to 48.4% and is the highest among female groups [18].

The main source of vitamin D is the skin synthesis of cholecalciferol from 7-dehydrocholesterol [19] as a result of UVB radiation (290–320 nm) from sunshine exposure, which depends on the latitude, the season, and the time of the day [7]. In countries such as Poland, which are situated in a moderate climate, endogenous vitamin D synthesis is possible from April to October only [20]. The other source of vitamin D is dietary intake—mainly from animal products, especially from fatty fish [21]. However, the vitamin D content in fish species differs significantly—in Poland, from 30 µg/100 g for eel, 19 µg/100 g for herring, and 13 µg/100 g for salmon to 1 µg/100 g for cod and 0.8 µg/100 g for flounder [22].

Due to a low fish intake in most European countries [23], inadequate vitamin D intake has been observed for years [19]. In Poland, it has been reported that for women the average vitamin D intake is 3.3 µg [24], whereas the most prominent recommendations indicate at least 10 µg as the reference intake of vitamin D for adults [11,25,26]. The problem of inadequate vitamin D intake is most serious when it comes to women under 30 years of age, since this is the group in which the intake of vitamin D is the lowest according to the 2003–2006 National Health and Nutrition Examination Survey (NHANES) [27]. Moreover, until the age of 30, the maximum bone density and peak bone mass are reached [28], which is influenced by vitamin D, so obtaining adequate vitamin D intake and status would be especially important in that vulnerable group of young women.

In recent years, intervention studies with the aim of improving vitamin D status through fish intake increase have been conducted in various European countries [29]. Their efficacy differed depending on the studied group, the type of intervention (fish species, dose, frequency), and intervention duration. However, no effective and recommended intervention has been defined in order to improve vitamin D intake and status in young women. Taking this into account, the aim of the presented study was to assess the efficacy of salmon intake intervention on vitamin D status in young Polish women during autumn.

2. Materials and Methods

2.1. Ethical Statement

The VISA Study (Vitamin D In Salmon) was conducted according to the guidelines of the Declaration of Helsinki and it was approved by the Ethics Committee of the Faculty of Human Nutrition and Consumer Sciences of the Warsaw University of Life Sciences (No 27/2018).

2.2. Studied Group

The studied group of young women was recruited in a procedure of convenience sampling, with the snowball effect, while announced in university social media. The inclusion criteria were as follows:

–   females,
–   Caucasian,

- aged 20–30 years,
- living in Warsaw or its surroundings (necessary to visit Dietetic Outpatient Clinic of the Department of Dietetics, Warsaw University of Life Sciences (WULS-SGGW) once a week for 8 weeks of the study duration),
- providing written informed consent to participate in the study.

The exclusion criteria were as follows:

- pregnancy,
- lactation,
- obesity, defined based on the criteria of the World Health Organization [30] as body mass index (BMI) ≥ 30 kg/m^2,
- fish and/or seafood allergy,
- following any diet with fish consumption restriction (e.g., vegetarian diet),
- vitamin D supplementation use up to 3 weeks before beginning of study and/or planned during study time,
- diseases and/or use of medicines changing vitamin D metabolism,
- planned travels to countries below the 40th parallel (countries with adequate sun exposure to obtain cutaneous vitamin D synthesis during study time),
- planned solarium use during study time.

In total, 51 individuals met the criteria and were included in the study and in this group the salmon intake intervention was conducted.

2.3. Dietary Intervention

The intervention study was designed to assess the efficacy of daily Atlantic farmed smoked salmon intake intervention on serum 25(OH)D level. All participants were enrolled in and finished the study at the same time. The fish intake intervention lasted from October 24, 2018 to December 18, 2018. The period of intervention was planned because from other studies it is known that in countries such as Poland, skin synthesis of vitamin D, which is the major source of the vitamin, is only possible from April to October [20], so it was decided to conduct the study directly after this period.

Atlantic farmed smoked salmon was chosen in the intervention as a source of vitamin D that is possible to be applied in practice, for numerous reasons. Salmon contains significant amounts of vitamin D (17.1 µg/100 g for smoked salmon, based on United States Department of Agriculture—USDA food composition tables [31]) and is widely available on the Polish market, independently from the region of the country [32]. Moreover, Atlantic farmed salmon is a species that contains very little mercury (≤ 0.1 µg/g) and dioxins (0.5–4 pg Toxic Equivalent (TEQ)/g), so it is classified to the first and second group respectively when it comes to the smallest content of those contaminants [33]. The very high content of eicosapentaenoic acid (EPA) and docosahexaenoic acid (DHA) (> 15 mg/g) also contributes to the Food and Agriculture Organization (FAO) of the United Nations and the World Health Organization (WHO) [33] recommending salmon as a fish species having much more benefits than risks.

The Atlantic farmed smoked salmon that was used for the study was obtained from one producer, one of the leading salmon sellers in Poland (Suempol Polska Ltd.), and for all the participants, the provided salmon was obtained from the same batch, in identical sliced tray modified atmosphere packaging.

Studies show a large variation in the content of vitamin D in salmon [34], therefore its content was measured in the smoked Atlantic salmon used in the study by a leading vitamin laboratory in Europe – Eurofins Vitamin Testing Denmark (EN 12821: 2009-08, LC-DAD, accredited methodology no. 581). The measured content of vitamin D was 21.3 ± 5.55 µg/100 g, which is higher than the value typical for smoked salmon, like the one shown in the Polish Food Composition Tables [22].

The daily intake was planned as 50 g, so it was attributed to around 10.65 μg of vitamin D, which covers the recommended 10 μg [11,25,26]. Atlantic farmed smoked salmon was packed in 50 g portions to facilitate consuming 1 portion (50 g) each day. Participants were given 7 packages (7 × 50 g) of smoked salmon once a week, for 8 weeks of the study duration. To increase adherence to intervention, participating women were asked to report their intake daily. Participants were asked to incorporate the given salmon in their daily food intake substituting it for other products such as meat, cheese, or eggs. They were not recommended to exclude other fish products from their diet, so if they had previously consumed them, they were recommended not to change this habit.

Participants had their serum 25(OH)D level measured three times: before the dietary intervention (at baseline (t0)), after 4 weeks of dietary intervention (t4), and after 8 weeks of intervention (after the study (t8)). The study course is shown in Figure 1.

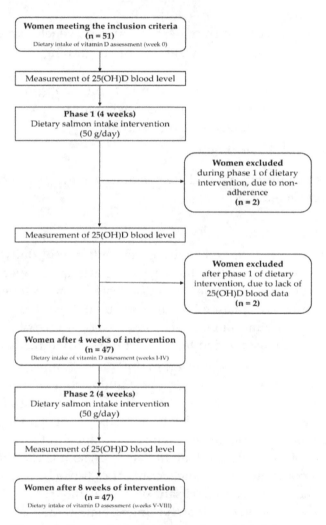

Figure 1. Study design and number of participants.

2.4. Measurements

Anthropometric measurements included body mass and height. Body mass was assessed using a calibrated weighing scale with an accuracy of ± 0.1 kg and body height was assessed using a stadiometer with an accuracy of ± 0.5 cm. The measurements were conducted by a professional nutritionist, according to the recommended procedure [35]. Afterward, the BMI was calculated, based on the Quetelet equation and interpreted to verify the exclusion criteria [30].

Vitamin D status was assessed based on the total serum 25(OH)D level at baseline (t0), after 4 weeks of dietary intervention (t4), and after 8 weeks of intervention (t8). For the analysis, venous blood samples were drawn by a qualified nurse in a certified medical analysis laboratory in Warsaw,

Poland; participants did not have to be in a fasting state before blood collection. Total serum 25(OH)D level tests were performed on BS Mindray BS-200 Chemistry Analyzer using Diazyme EZ Vitamin D assay and the dual vial liquid stable (latex enhanced immunoturbidimetric) method, which enabled determination of total 25(OH)D in the range of 19.0–369.5.8 nmol/L. For this method, a comparison of the EZ Vitamin D assay (y) using samples measured with LC–MS (liquid chromatography–mass spectrometry)/MS in the validation gave the following correlation: y = 1.0297x−0.813 for R^2 = 0.9622 and a comparison of the EZ Vitamin D assay using samples measured with a commercially available 25(OH)D immunoassay in the validation gave the following correlation: y = 1.1537x−1.2321 for R^2 = 0.9716. The assay precision for the method is defined by percent coefficient of variation (%CV) lower than 5% at 75 nmol/L. The method is certified by the Centers for Disease Control and Prevention within the Vitamin D Standardization–Certification Program (CDC VDSCP) and meets the performance target set by the Vitamin D External Quality Assessment Scheme (DEQAS) advisory panel.

Each sample was assessed by the same person, in the same conditions, with the same equipment, and using exactly the same methodology, for each sample within 1 hour from drowing the blood samples. The obtained results of total serum 25(OH)D level were compared to the following reference values: <50 nmol/L—inadequate, 50–250 nmol/L—adequate, >250 nmol/L—potentially toxic [11,25,36,37].

In order to provide the necessary safety precautions, during the whole experiment, participants had their vitamin D intake and blood pressure controlled. As the intervention comprised additional intake of vitamin D, the total vitamin D intake was controlled throughout the experiment, using the Vitamin D Estimation Only—Food Frequency Questionnaire (VIDEO-FFQ), which was previously validated in a group of Polish women aged 20–30 years [38]. Afterward, the obtained vitamin D intake was compared with the upper intake level (UL) of 100 µg [14]. At the same time, because of high salt content in smoked salmon (1.5 g/50 g), the blood pressure of participants was controlled throughout the experiment (once a week), using Omron Healthcare BP 710N blood pressure monitor, according to the recommended procedure [39]. The observed systolic and diastolic blood pressure values were compared with the standard reference values of 140 mmHg and 90 mmHg, respectively, and the increase of blood pressure above the recommended values observed during two following weeks was decided to be interpreted as a reason to suspend participation in the experiment. Neither for vitamin D intake, nor for blood pressure were the excessive values stated, so it was interpreted as obtaining the required safety of dietary intervention.

2.5. Statistical Analysis

The distribution of the obtained values was verified using Shapiro–Wilk test. The groups were compared using the t-Student test (for parametric distributions), the Mann–Whitney U test, and the Kruskal–Wallis Analysis of Variance (ANOVA) with multiple comparisons (for non-parametric distributions), as well as the chi^2 test. The correlations were verified using Pearson correlation coefficient (for parametric distributions) and Spearman rank correlation coefficient (for non-parametric distributions).

The accepted level of significance was $p \leq 0.05$. The following software was used: Statistica 8.0 (Statsoft Inc., Tulsa, OK, USA), Statgraphics Plus for Windows 4.0 (Statgraphics Technologies Inc., The Plains, VA, USA).

3. Results

The characteristics of the studied group and vitamin D status throughout the intervention are presented in Table 1. While comparing the serum 25(OH)D level, it was stated that the results differed throughout the study ($p = 0.0001$; Kruskal-Wallis ANOVA), but for multiple comparisons, it was stated that the results after 8 weeks of intervention did not differ from the results at baseline ($p = 0.7964$). However, during phase I of the study (weeks I-IV), a significant decrease of serum 25(OH)D level was observed ($p < 0.0001$ for the comparison of t0 and t4 results), whereas during phase II (weeks V-VIII), a significant increase was noted ($p = 0.0005$ for the comparison of t4 and t8 results).

Table 1. Characteristics of the studied group accompanied by vitamin D status throughout the study.

	Variables	Mean ± SD	Median (Min–Max)
Characteristics	Age (years)	22.9 ± 1.6	23.0 * (20.0–28.0)
	BMI (kg/m^2)	21.43 ± 2.49	21.27 (16.92–27.96)
Serum level	25(OH)D for t0 (nmol/L)	58.4 ± 20.2	57.1 (19.7–93.1)
	25(OH)D for t4 (nmol/L)	41.2 ± 14.1	39.9 * (15.1–90.3)
	25(OH)D for t8 (nmol/L)	52.5 ± 11.4	54.1 (25.6–72.9)
Change of serum level	25(OH)D change from t0 to t4 (nmol/L)	−17.2 ± 13.9	−16.7 (−44.5–11.4)
	25(OH)D change from t4 to t8 (nmol/L)	11.3 ± 12.2	8,8 * (−18.5–38.8)
	25(OH)D change from t0 to t8 (nmol/L)	−6.0 ± 16.1	−8.1 (−36.8–24.2)

* non-parametric distribution (verified using Shapiro–Wilk test; $p \leq 0.05$); 25(OH)D—25-hydroxyvitamin D; t0—baseline; t4—after 4 weeks of intervention; t8—after 8 weeks of intervention.

The dietary vitamin D intake in the studied group throughout the study is presented in Table 2. It was observed that, at baseline, the dietary intake of vitamin D was significantly lower than those assessed after four weeks of intervention ($p < 0.0001$), as well as after eight weeks of intervention ($p < 0.0001$). At the same time, the dietary intakes of vitamin D assessed after four weeks of intervention and those assessed after eight weeks of intervention did not differ ($p = 0.5758$).

Table 2. Dietary vitamin D intake in the studied group throughout the study.

	Variables	Mean ± SD	Median (Min–Max)	p
Intake for t0 (μg)	Total	3.02 ± 1.36	2.79 * (1.04–7.74)	-
	25(OH)D < 50 nmol/L	3.02 ± 1.63	2.66 * (1.04–7.74)	0.7510
	25(OH)D ≥ 50 nmol/L	3.02 ± 1.20	2.90 * (1.09–5.99)	
Intake for t4 (μg)	Total	10.08 ± 1.34	9.82 * (7.84–13.91)	-
	25(OH)D < 50 nmol/L	9.88 ± 1.26	9.73 * (7.84–13.15)	0.0515
	25(OH)D ≥ 50 nmol/L	10.73 ± 1.46	10.65 * (8.62–13.91)	
Intake for t8 (μg)	Total	9.85 ± 1.96	9.71 * (1.81–13.82)	-
	25(OH)D < 50 nmol/L	9.44 ± 2.43	9.59 * (1.81–13.82)	0.3411
	25(OH)D ≥ 50 nmol/L	10.11 ± 1.60	9.89 * (6.77–12.78)	

* non-parametric distribution (verified using Shapiro–Wilk test; $p \leq 0.05$); t0—baseline; t4—after 4 weeks of intervention; t8—after 8 weeks of intervention.

The assessment of the adequacy of vitamin D status throughout the study is presented in Table 3. While comparing the share of participants characterized by adequate and inadequate serum 25(OH)D level, it was stated that it did not differ between baseline assessment and assessment after eight weeks of intervention ($p = 1.0000$, chi^2). At the same time, the share of respondents characterized by inadequate serum 25(OH)D levels increased during phase I of the study ($p = 0.0002$ for the comparison of t0 and t4 shares, chi^2), but decreased during phase II ($p = 0.0002$ for the comparison of t4 and t8 shares, chi^2).

Table 3. Assessment of adequacy of vitamin D status throughout the study while compared with the reference values.

	Serum 25-hydroxyvitamin D level		
	Inadequate (<50 nmol/L) *	Adequate (50–250 nmol/L) *	Potentially Toxic (> 250 nmol/L) *
Before intervention	18 (38%)	29 (62%)	0 (0%)
After 4 weeks of intervention	36 (77%)	11 (23%)	0 (0%)
After 8 weeks of intervention	18 (38%)	29 (62%)	0 (0%)

* reference values of 50 nmol/L and 250 nmol/L [11,25,36,37].

The analysis of correlations between age or BMI and vitamin D status throughout the study are presented in Table 4. It was stated that neither age nor BMI influenced 25(OH)D level during the study, as well as its changes, apart from the baseline 25(OH)D level which was significantly correlated with BMI ($p = 0.0419$; R = -0.2980).

Table 4. Analysis of correlations between age, or BMI, and vitamin D status throughout the study.

Variables		Age		BMI	
		p	R	p	R
Serum level	25(OH)D for t0 (nmol/L)	0.0880	−0.2512	0.0419 *	−0.2980
	25(OH)D for t4 (nmol/L)	0.5753 *	−0.0838	0.0663 *	−0.2701
	25(OH)D for t8 (nmol/L)	0.2300	−0.1786	0.4038 *	−0.1247
Change of serum level	25(OH)D change from t0 to t4 (nmol/L)	0.4630	0.1096	0.3604 *	0.1364
	25(OH)D change from t4 to t8 (nmol/L)	0.4387 *	0.1157	0.4776 *	0.1062
	25(OH)D change from t0 to t8 (nmol/L)	0.2010	−0.1897	0.1362 *	0.2206

* Spearman rank correlation coefficient for non-parametric distributions (for parametric distribution Pearson correlation coefficient applied); 25(OH)D—25-hydroxyvitamin D; t0—baseline; t4—after 4 weeks of intervention; t8—after 8 weeks of intervention.

The analysis of correlations between vitamin D status before the intervention and its later levels and changes throughout the study is presented in Table 5. It was stated that the 25(OH)D level before the intervention was associated with its levels during the intervention ($p < 0.0001$). It was also associated with 25(OH)D changes from the baseline values ($p < 0.0001$), but not with the following changes (from t4 to t8).

Table 5. Analysis of correlations between vitamin D status before intervention and its later levels and changes throughout the study.

Variables		p	R
Serum level	25(OH)D for t4 (nmol/L)	< 0.0001 *	0.7474
	25(OH)D for t8 (nmol/L)	< 0.0001	0.6089
Change of serum level	25(OH)D change from t0 to t4 (nmol/L)	< 0.0001	−0.7154
	25(OH)D change from t4 to t8 (nmol/L)	0.1257 *	−0.2265
	25(OH)D change from t0 to t8 (nmol/L)	< 0.0001	−0.8274

* Spearman rank correlation coefficient for non-parametric distributions (for parametric distribution Pearson correlation coefficient applied); 25(OH)D—25-hydroxyvitamin D; t0—baseline; t4—after 4 weeks of intervention; t8—after 8 weeks of intervention.

The comparison of the vitamin D status throughout the study between subgroups characterized by various baseline statuses is presented in Table 6.

For the serum 25(OH)D level, all differences were statistically significant ($p \leq 0.05$) and higher levels were observed for participants characterized by adequate baseline status. However, for the serum 25(OH)D level, for participants characterized by inadequate baseline status, lower decreases from t0 to t4 and higher increases from t4 to t8 were noted, when compared to participants with adequate baseline status.

Table 6. Comparison of vitamin D status throughout the study between subgroups characterized by various baseline statuses.

Variables		Inadequate 25(OH)D (<50 nmol/L)		Adequate 25(OH)D (≥50 nmol/L)		p**
		Mean ± SD	Median (Min–Max)	Mean ± SD	Median (Min–Max)	
Serum level	25(OH)D for t0 (nmol/L)	37.5 ± 10.1	36.4 (19.7–48.9)	71.4 ± 12.4	70.4 (53.2–93.1)	<0.0001
	25(OH)D for t4 (nmol/L)	31.2 ± 8.8	29.2 (15.1–51.7)	47.4 ± 13.3	45.4 (28.5–90.3) *	<0.0001
	25(OH)D for t8 (nmol/L)	47.1 ± 12.7	46.8 (25.6–72.9) *	55.8 ± 9.2	56.8 (39.8–72.1)	0.0198
Change of serum level	25(OH)D change from t0 to t4 (nmol/L)	−5.0 ± 13.0	−8.5 (−21.2–32.9)	−17.4 ± 24.0	−16.4 (−53.7–41.3)	0.0501
	25(OH)D change from t4 to t8 (nmol/L)	15.9 ± 12.6	15.1 (−0.3–37.0)	8.4 ± 11.2	7.1 (−18.5–38.8)	0.0381
	25(OH)D change from t0 to t8 (nmol/L)	10.9 ± 16.0	8.2 (−10.3–52.0)	−9.0 ± 23.9	−6.7 (−52.0–53.0) *	0.0005

* Distribution differs from normal; ** Student's t-test (if normal distribution) or Mann–Whitney U test (if distribution differs from normal); 25(OH)D—25-hydroxyvitamin D; t0—baseline; t4—after 4 weeks of intervention; t8—after 8 weeks of intervention.

4. Discussion

In the presented study, it was stated that the baseline 25(OH)D level had a profound impact on the changes observed after the intervention, so it may be concluded that the efficacy of the applied dietary intervention of daily salmon intake depends on the baseline 25(OH)D level. In the total studied group, the median change after eight weeks of intervention was −8.1 nmol/L. Taking into consideration only the participants with adequate 25(OH)D level (≥ 50 nmol/L), it was also a drop (median = −6.7 nmol/L), whereas in participants with inadequate 25(OH)D level (< 50 nmol/L) after eight weeks of intervention, a median rise of 8.2 nmol/L was observed.

While comparing the obtained results with studies by other authors, it may be supposed that it is a general association—in groups with lower baseline 25(OH)D levels, a higher effect of intervention may be observed, so it may confirm that the baseline 25(OH)D blood level plays an important role. In a similar intervention study, lasting eight weeks, conducted in wintertime in Iceland, Spain, and Ireland, in which its participants (mean baseline 25(OH)D blood level 61.9 nmol/L) consumed around 450 g of salmon per week (compared to 350 g per week in our study), the mean rise of 25(OH)D was 8.4 nmol/L [40]. In another study conducted in Finland, on participants with a much higher mean baseline 25(OH)D level of 124.0 nmol/L, no significant increase in 25(OH)D was noted when consuming 300–600 g of fatty fish (including salmon) per week after eight weeks of intervention [41]. However, most studies of this type are small intervention studies with a limited number of participants, therefore comparing the obtained results with the conclusions from a meta-analysis summarizing the most essential results may be helpful. According to a meta-analysis of Lehmann et al. [29], the mean change in 25(OH)D for interventions lasting 4–8 weeks was 3.8 nmol/L; whereas in intervention groups with an inadequate mean 25(OH)D baseline (< 50 nmol/L), the change was 6.1 nmol/L; and in groups with adequate (≥ 50 nmol/L) 25(OH)D blood levels at baseline, the change was only 3.9 nmol/L. Taking into account the results of various studies, presented above, it may be confirmed that the most important factor for the efficacy may be the baseline 25(OH)D level.

However, in the presented study in the total study group, a decrease of 25(OH)D level after four weeks of intervention (phase I) was stated and later on (phase II), an increase (compared to the level after four weeks) was observed. The decline is very surprising, as salmon is known to be a good source of vitamin D [22,26], and in numerous studies, an increase in 25(OH)D was shown, also only after a four-week-long intervention of salmon intake in the winter [29]. Nevertheless, a recent intervention study conducted in Norway indicated that even a weekly consumption of 750 g (compared to 350 g in the presented study) of salmon for eight weeks was not sufficient to prevent the 25(OH)D decrease [42]. In that study, 25(OH)D decreased both in the salmon intervention and the control group, however, the decline was significantly lower in the salmon group compared to the control group.

What should be underlined is that the intervention period in the referred study [42] was from August/September to October/November, whereas in the presented study it was later—from October to December. Therefore, the decline seen in the presented study from week 0 to week 4 may correspond to the one observed in the Norwegian study (after eight weeks of intervention). In the presented study, the amount of salmon consumed was more than two times less than in the Norwegian one, and still in the second part of the intervention (from week four to week eight) an increase in 25(OH)D was observed. Unfortunately, to our best knowledge, the reason why such high amounts of dietary vitamin D from fish do not contribute to an immediate increase of 25(OH)D is not yet described or well known. There are only some hypotheses that could be listed.

The decrease may be attributed to a lack of adequate sunshine exposure to cause skin vitamin D synthesis during the time of intervention. From other studies, it is known that in countries such as Poland, skin synthesis of vitamin D is observed only from April to October (if the sunshine exposure is adequate) [20]. According to a recent Polish study [43], the highest levels of 25(OH)D in Poland are observed in August and the lowest in January, which could be explained by insolation and 25(OH)D synthesis during summer and the mobilization of vitamin D stored in the body (during the summer months) in the winter.

This hypothesis corresponds to the initial decrease of 25(OH)D levels in the presented study, which was performed from October to December. Therefore, it may be assumed that in phase I (until week four), not dietary vitamin D, but rather previously stored vitamin D had been used by the organism causing the constant decrease of 25(OH)D blood level, but in phase II (from week five to week eight), in which not enough vitamin D was stored in the body anymore, the daily intake of 50 g of fish and dietary vitamin D provided must have been intensively stored and metabolized to improve its status. Other studies also emphasize that it is the vitamin D storage from sun exposure and its release from fat tissues in winter that is the major factor contributing to 25(OH)D levels throughout the year [44], and dietary vitamin D is associated with 25(OH)D mainly in the winter and spring [45], in which perhaps not that much vitamin D is left in fat tissue and, therefore, the body must depend on vitamin D from food sources. This could also be the reason for the observed impact of baseline vitamin D status (adequate or inadequate) on 25(OH)D changes throughout the study, since in participants with inadequate vitamin D status, a lower decrease in phase I and a higher increase in phase II were seen, compared to participants with adequate vitamin D status at baseline.

Another possible explanation could be the difference in dietary vitamin D intake throughout the study. The statistical comparison of participants with inadequate (<50 nmol/L) and adequate ($\geq$50 nmol/L) 25(OH)D levels at baseline showed that at week four (t4) participants with adequate ($\geq$50 nmol/L) 25(OH)D levels had a close to significant ($p = 0.0515$) higher vitamin D intake (median = 10.65 µg) in the first part of the intervention (t0 to t4) compared to the ones with inadequate (<50 nmol/L) 25(OH)D levels (median intake = 9.73 µg). What should be noted is that the dietary vitamin D intake of participants with adequate ($\geq$50 nmol/L) 25(OH)D levels covered the recommended 10 µg [11,25,26], which was also assumed in the study. Therefore, the possible reason for having an adequate 25(OH)D level at week four could be the higher (covering the recommended 10 µg) intake of vitamin D in the preceding time (from t0 to t4). At week eight, the vitamin D intakes did not differ significantly between groups ($p = 0.3411$), therefore the influence of vitamin D intake on different outcomes at that time remains unclear.

There are also other possible explanations, such as levels of other nutrients, that are related to the metabolism of vitamin D [46]. It may have been influenced by a decreased calcium level prior to the study [47]. This level was not assessed in the presented study, but if this level is lowered, it may influence not only the 25(OH)D level, but also the 1,25(OH)2D level, and the activity of 25(OH)D 1-α-hydroxylase [47].

Therefore, it may be hypothesized that apart from baseline 25(OH)D levels, it is the starting time of intervention and the dietary intake of vitamin D and other nutrients that play a big role in the outcomes of such short-term fish intake interventions. Other hypothetical possible explanations to such intervention outcomes could be that the digestive system needs to adjust to utilize vitamin D from the smoked salmon. Another could be that the metabolism and storage of vitamin D are faster in the case of high dietary supply (intervention) while the levels of vitamin D stored from the summer exposure are also high, which might lead to lower 25(OH)D levels because the whole consumed amount of vitamin D is converted to other metabolites in a dynamic way. Last but not least, the vitamin D metabolism might have other various pathways depending on the baseline level, exposure, dietary intake, and applied intervention, which we are not yet aware of.

Based on the meta-analysis of Lehmann et al. [29], it may be indicated that the length of the intervention also plays a key role—short-term studies (4-8 weeks) revealed a mean difference of 3.8 nmol/L and long-term studies (around six months) revealed 8.3 nmol/L [29]. If the interventions last longer, there is more time for a change of 25(OH)D. Therefore, it may be supposed that prolonging the conducted intervention would have resulted in a higher increase of 25(OH)D in the studied group. It may be hypothesized especially for participants with an adequate 25(OH)D level for whom after eight weeks of intervention a lower 25(OH)D serum level (median 56.8 nmol/L) than at baseline (median 70.4 nmol/L) was still observed, but higher than after four weeks of intervention (median 45.4 nmol/L), so a progressive increase may be supposed.

In spite of a number of arguments for choosing salmon as a vitamin D source in diet, namely its availability, as well as the content of nutrients and other components, the price of this product is quite high [48], contributing to low consumption in Poland (in 2017: 0.63 kg/person/year) [48]. There are other much cheaper (and therefore consumed more often) fish species available in Poland with significant vitamin D and combined EPA and DHA content, as well as low mercury and dioxin levels, that should be recommended, such as herring and rainbow trout. Both those species have very high (>15 mg/g) combined EPA and DHA content, low ($\leq 0.1\mu g/g$) mercury levels, as well as quite low (0.5–4 pg TEQ/g) dioxin content, so according to the FAO and WHO [32] similarly as in the case of farmed Atlantic salmon, consuming them has more benefits than risks. Their prices in Poland are lower than salmon and they are as follows: rainbow trout—23.62 PLN/kg (approx. 5.42 €/kg) and herring—16.03 PLN/kg (approx. 3.68 €/kg), compared to 57.90 PLN/kg for salmon (approx. 13.28 €/kg) [48].

Moreover, herring and rainbow trout contain significant amounts of vitamin D: 19.00 µg and 13.50 µg/100 g, respectively [22]. There were some single intervention studies in which participants consumed 750 g/week of herring for six weeks [49] or 400–600 g/week of fatty fish including herring and rainbow trout for eight weeks [42] which did not reveal any significant effect of the applied dietary intervention on total 25(OH)D. What should be pointed out is that the study groups consisted of only 32 [49] and 11 participants [41]. Therefore, similar intervention studies with herring or rainbow trout should be conducted in larger study groups in order to explore the influence of consumption of those species on 25(OH)D blood levels. If they also have a positive influence on vitamin D status, it could have a greater impact on Polish people to recommend them to eat more of those fish species, as they can afford and already consume them in higher amounts than salmon (herring 2.56, trout 0.50 kg/person/year in 2017) [48].

According to the American Heart Association [50], as well as the Polish National Food and Nutrition Institute [51], the recommended fish (especially fatty fish) intake is at least two times a week. Fatty fish include the fish species mentioned above such as salmon, herring, and rainbow trout, which are good sources of vitamin D, and should not be reduced from the diet, as they contain less mercury and dioxins than other species. From other studies, it is known that fish intake is the most influential food source contributor to vitamin D intake [52,53], and there may be a strong correlation between fatty fish intake and 25(OH)D levels [54]. A recent cross-sectional Norwegian study revealed that for a median vitamin D intake of 10.3 µg, the mean 25(OH)D level was 64.0 nmol/L, while the prevalence of deficiency (defined as 25(OH)D < 50 nmol/L) was only 24.7%; in such a situation, a low vitamin D deficiency prevalence was observed, even in winter [55]. This suggests that with an adequate vitamin D intake (at least 10 µg/day) it is possible to achieve an adequate 25(OH)D level, maybe even in seasons with limited sunshine exposure such as the autumn and winter seasons in Poland. It corresponds with the obtained results, as even as short a period as eight weeks of dietary intervention of adequate vitamin D intake influenced the 25(OH)D levels and was revealed to be a promising option.

The novelty of the study is the fact that it was the first fish intake intervention study to asses 25(OH)D levels in a Polish population. Moreover, it is one of the first fish intake intervention studies in which a decrease in 25(OH)D was observed despite high salmon intake. The hypothetical explanations for that are listed above. Nevertheless, this indicates that this matter needs further study.

Although the presented study was the first fish intervention study conducted in Poland to access the efficacy of such intervention on 25(OH)D serum levels, limitations of the study should be indicated. The most important issue is associated with the fact that in the presented study there was no control group with no dietary intervention, which makes it more challenging to draw conclusions. However, this study is not the first one to analyze the influence of fish intake intervention on vitamin D status (measured as total 25(OH)D). Similar studies were conducted in other countries (but not in Poland), such as one in Iceland, Spain, and Ireland [39], with similar dietary intervention and similar results. Thus, it can be hypothesized that in the present study, the reason for the recovery of 25(OH)D was also due to fish intake and the presented study is of great importance, despite the lack of a control group.

Taking it into account, it must be stated that this matter needs further study, as there are not many studies conducted on that topic, and, what has to be underlined, there were no such studies in Poland, so far. Finally, the studies conducted in other countries are frequently carried out on very specific participants such as prisoners [56], sex offenders [57], overweight individuals following a low-calorie diet [39], and 8-9-year-old children [58], therefore, there is a great need for similar intervention studies (conducted using various, but defined fish species) on larger, homogenous study groups covering various ages for both male and female participants.

5. Conclusions

Although in the presented intervention study an initial decline of 25(OH)D blood level was observed (probably connected to the starting time of the study), afterward the daily salmon intake contributed to its increase. Based on the observed results, it may be stated that the baseline 25(OH)D status is an important determinant of the intervention effectiveness during the autumn season, as it is more effective in participants with inadequate vitamin D status. It may be concluded that an increase in farmed Atlantic salmon intake may be a good way to improve vitamin D status, especially in vitamin D-deficient individuals.

Author Contributions: D.G. made conception; Z.U. and D.G. performed the research; Z.U. and D.G. analyzed the data; and Z.U. and D.G. wrote the paper. All authors have read and agreed to the published version of the manuscript.

References

1.	Lips, P.; van Schoor, N.M. The effect of vitamin D on bone and osteoporosis. *Best Pract. Res. Clin. Endocrinol. Metab.* **2011**, *25*, 585–591. [PubMed]
2.	Gao, Q.; Kou, T.; Zhuang, B.; Ren, Y.; Dong, X.; Wang, Q. The Association between Vitamin D Deficiency and Sleep Disorders: A Systematic Review and Meta-Analysis. *Nutrients* **2018**, *10*, 1395.
3.	Wang, J.; Lv, S.; Chen, G.; Gao, C.; He, J.; Zhong, H.; Xu, Y. Meta-analysis of the association between vitamin D and autoimmune thyroid disease. *Nutrients* **2015**, *7*, 2485–2498. [PubMed]
4.	Estébanez, N.; Gómez-Acebo, I.; Palazuelos, C.; Llorca, J.; Dierssen-Sotos, T. Vitamin D exposure and Risk of Breast Cancer: A meta-analysis. *Sci. Rep.* **2018**, *8*, 9039.
5.	Keum, N.; Lee, D.H.; Greenwood, D.C.; Manson, J.E.; Giovannucci, E. Vitamin D supplementation and total cancer incidence and mortality: A meta-analysis of randomized controlled trials. *Ann. Oncol.* **2019**, *30*, 733–743.
6.	Wang, M.; Liu, M.; Wang, C.; Xiao, Y.; An, T.; Zou, M.; Cheng, G. Association between vitamin D status and asthma control: A meta-analysis of randomized trials. *Respir. Med.* **2019**, *150*, 85–94.
7.	Christakos, S.; Dhawan, P.; Verstuyf, A.; Verlinden, L.; Carmeliet, G. Vitamin D: Metabolism, Molecular Mechanism of Action, and Pleiotropic Effects. *Physiol. Rev.* **2016**, *96*, 365–408.
8.	Caprio, M.; Infante, M.; Calanchini, M.; Mammi, C.; Fabbri, A. Vitamin D: Not just the bone. Evidence for beneficial pleiotropic extraskeletal effects. *Eat. Weight Disord. Stud. Anorex. Bulim. Obes.* **2017**, *22*, 27–41. [CrossRef]
9.	Durazo-Arvizu, R.A.; Tian, L.; Brooks, S.P.J.; Sarafin, K.; Cashman, K.D.; Kiely, M.; Merkel, J.; Myers, G.L.; Coates, P.M.; Sempos, C.T. The Vitamin D Standardization Program (VDSP) Manual for Retrospective Laboratory Standardization of Serum 25-Hydroxyvitamin D Data. *J. AOAC Int.* **2017**, *100*, 1234–1243. [CrossRef]
10.	Cashman, K.D.; Dowling, K.G.; Škrabáková, Z.; Gonzalez-Gross, M.; Valtueña, J.; De Henauw, S.; Moreno, L.; Damsgaard, C.T.; Michaelsen, K.F.; Mølgaard, C.; et al. Vitamin D deficiency in Europe: Pandemic? *Am. J. Clin. Nutr.* **2016**, *103*, 1033–1044. [CrossRef]
11.	National Academies Press (US). Institute of Medicine (US) Committee to Review Dietary Reference Intakes for Vitamin D and Calcium. In *Dietary Reference Intakes for Calcium and Vitamin D*; Ross, A.C., Taylor, C.L., Yaktine, A.L., Del Valle, H.B., Eds.; The National Academies Collection: Reports funded by National Institutes of Health; National Academies Press (US): Washington, DC, USA, 2011.

12. Aspray, T.J.; Bowring, C.; Fraser, W.; Gittoes, N.; Javaid, M.K.; Macdonald, H.; Patl, S.; Selby, P.; Tanna, N.; Francis, R.M. National Osteoporosis Society Vitamin D Guideline Summary. *Age Ageing* **2014**, *43*, 592–595. [PubMed]

13. Holick, M.F.; Binkley, N.C.; Bischoff-Ferrari, H.A.; Gordon, C.M.; Hanley, D.A.; Heaney, R.P.; Murad, M.H.; Weaver, C.M. Evaluation, Treatment, and Prevention of Vitamin D Deficiency: An Endocrine Society Clinical Practice Guideline. *J. Clin. Endocrinol. Metab.* **2011**, *96*, 1911–1930.

14. EFSA Panel on Dietetic Products. N. and A. (NDA) Dietary reference values for vitamin D. *EFSA J.* **2016**, *14*, e04547. [CrossRef]

15. Rusińska, A.; Płudowski, P.; Walczak, M.; Borszewska-Kornacka, M.K.; Bossowski, A.; Chlebna-Sokół, D.; Czech-Kowalska, J.; Dobrzańska, A.; Franek, E.; Helwich, E.; et al. Vitamin D Supplementation Guidelines for General Population and Groups at Risk of Vitamin D Deficiency in Poland-Recommendations of the Polish Society of Pediatric Endocrinology and Diabetes and the Expert Panel with Participation of National Specialist Consultants and Representatives of Scientific Societies-2018 Update. *Front. Endocrinol.* **2018**, *9*, 246.

16. Gordon, C.M.; DePeter, K.C.; Feldman, H.A.; Grace, E.; Emans, S.J. Prevalence of Vitamin D Deficiency among Healthy Adolescents. *Arch. Pediatr. Adolesc. Med.* **2004**, *158*, 531–537.

17. Andersen, R.; Mølgaard, C.; Skovgaard, L.T.; Brot, C.; Cashman, K.D.; Chabros, E.; Charzewska, J.; Flynn, A.; Jakobsen, J.; Kärkkäinen, M.; et al. Teenage girls and elderly women living in northern Europe have low winter vitamin D status. *Eur. J. Clin. Nutr.* **2005**, *59*, 533–541.

18. ENNS: Étude Nationale Nutrition Santé /Enquêtes et Etudes /Nutrition et Santé /Maladies Chroniques et Traumatismes /Dossiers Thématiques /Accueil. Available online: http://invs.santepubliquefrance.fr/Dossiers-thematiques/Maladies-chroniques-et-traumatismes/Nutrition-et-sante/Enquetes-et-etudes/ENNS-etude-nationale-nutrition-sante (accessed on 25 June 2019).

19. Calvo, M.S.; Whiting, S.J.; Barton, C.N. Vitamin D Intake: A Global Perspective of Current Status. *J. Nutr.* **2005**, *135*, 310–316.

20. Webb, A.R.; Kline, L.; Holick, M.F. Influence of season and latitude on the cutaneous synthesis of vitamin D3: Exposure to winter sunlight in Boston and Edmonton will not promote vitamin D3 synthesis in human skin. *J. Clin. Endocrinol. Metab.* **1988**, *67*, 373–378.

21. Schmid, A.; Walther, B. Natural Vitamin D Content in Animal Products. *Adv. Nutr.* **2013**, *4*, 453–462. [PubMed]

22. Kunachowicz, H.; Przygoda, B.; Nadolna, I.; Iwanow, K. *Tabele Składu i Wartości Odżywczej Żywności*; Wydawnictwo Lekarskie PZWL: Warsaw, Poland, 2017.

23. National Food Institute. *European Nutrition and Health Report 2009*; Elmadfa, I., Ed.; Karger: Basel, Switzerland; New York, NY, USA, 2009.

24. Flynn, A.; Hirvonen, T.; Mensink, G.B. Intake of selected nutrients from foods, from fortification and from supplements in various European countries. *Food Nutr. Res.* **2009**, *12*, 2038.

25. Nordiska Ministerrådet. *Nordic Nutrition Recommendations 2012. Part 3: Vitamins A, D, E, K, Thiamin, Riboflavin, Niacin, Vitamin B6, Folate, Vitamin B12, Biotin, Pantothenic acid and Vitamin C*; Nordisk Ministerråd: Copenhagen, Denmark, 2014.

26. SACN Vitamin D and Health Report. Available online: https://www.gov.uk/government/publications/sacn-vitamin-d-and-health-report (accessed on 20 May 2019).

27. Bailey, R.L.; Dodd, K.W.; Goldman, J.A.; Gahche, J.J.; Dwyer, J.T.; Moshfegh, A.J.; Sempos, C.T.; Picciano, M.F. Estimation of Total Usual Calcium and Vitamin D Intakes in the United States. *J. Nutr.* **2010**, *140*, 817–822. [CrossRef] [PubMed]

28. Benjamin, R.M. Bone health: Preventing osteoporosis. *J. Am. Diet. Assoc.* **2010**, *110*, 498. [CrossRef] [PubMed]

29. Lehmann, U.; Gjessing, H.R.; Hirche, F.; Mueller-Belecke, A.; Gudbrandsen, O.A.; Ueland, P.M.; Mellgren, G.; Lauritzen, L.; Lindqvist, H.; Hansen, A.L.; et al. Efficacy of fish intake on vitamin D status: A meta-analysis of randomized controlled trials. *Am. J. Clin. Nutr.* **2015**, *102*, 837–847. [CrossRef] [PubMed]

30. Body Mass Index—BMI. Available online: http://www.euro.who.int/en/health-topics/disease-prevention/nutrition/a-healthy-lifestyle/body-mass-index-bmi (accessed on 24 July 2019).

31. United States Department of Agriculture, Agricultural Research Service. Available online: https://fdc.nal.usda.gov/fdc-app.html#/food-details/337771/nutrients (accessed on 24 March 2020).

32. Fish and Seafood Market in Poland. 2018. Available online: https://gain.fas.usda.gov/Recent%20GAIN%

20Publications/2017%20Fish%20and%20Seafood%20Market%20in%20Poland_Warsaw_Poland_2-21-2018. pdf (accessed on 19 August 2019).

33. FAO/WHO. *Report of the Joint FAO/WHO Expert Consultation on the Risks and Benefits of Fish Consumption: Rome, 25–29 January 2010*; FAO: Rome, Italy, 2011.

34. Jakobsen, J.; Smith, C.; Bysted, A.; Cashman, K.D. Vitamin D in Wild and Farmed Atlantic Salmon (Salmo Salar)—What Do We Know? *Nutrients* **2019**, *11*, 982. [CrossRef] [PubMed]

35. International Society for Advancement of Kinanthropometry. *International Standards for Anthropometric Assessment*; International Society for the Advancement of Kinanthropometry: Potchefstroom, South Africa, 2001.

36. Reference Values DACH-Referenzwerte. Available online: http://www.sge-ssn.ch/grundlagen/lebensmittel-und-naehrstoffe/naehrstoffempfehlungen/dachreferenzwerte/ (accessed on 2 August 2019).

37. Jarosz, M.; Rychlik, E.; Stoś, K.; Wierzejska, R.; Wojtasik, A.; Charzewska, J.; Mojska, H.; Szponar, L.; Sajór, I.; Kłosiewicz-Latoszek, L.; et al. *Normy Żywienia dla Populacji Polski*; Instytut Żywności i Żywienia: Warszawa, Poland, 2017.

38. Głąbska, D.; Guzek, D.; Sidor, P.; Włodarek, D. Vitamin D Dietary Intake Questionnaire Validation Conducted among Young Polish Women. *Nutrients* **2016**, *8*, 36. [CrossRef]

39. Ogedegbe, G.; Pickering, T. Principles and techniques of blood pressure measurement. *Cardiol. Clin.* **2010**, *28*, 571–586. [CrossRef]

40. Lucey, A.J.; Paschos, G.K.; Cashman, K.D.; Martínéz, J.A.; Thorsdottir, I.; Kiely, M. Influence of moderate energy restriction and seafood consumption on bone turnover in overweight young adults. *Am. J. Clin. Nutr.* **2008**, *87*, 1045–1052. [CrossRef]

41. Erkkilä, A.T.; Schwab, U.S.; de Mello, V.D.F.; Lappalainen, T.; Mussalo, H.; Lehto, S.; Kemi, V.; Lamberg-Allardt, C.; Uusitupa, M.I.J. Effects of fatty and lean fish intake on blood pressure in subjects with coronary heart disease using multiple medications. *Eur. J. Nutr.* **2008**, *47*, 319–328. [CrossRef]

42. Bratlie, M.; Hagen, I.V.; Helland, A.; Midttun, Ø.; Ulvik, A.; Rosenlund, G.; Sveier, H.; Mellgren, G.; Ueland, P.M.; Gudbrandsen, O.A. Five salmon dinners per week was not sufficient to prevent the reduction in serum vitamin D in autumn at 60° north latitude: A randomised trial. *Br. J. Nutr.* **2019**, *123*, 1–21. [CrossRef]

43. Smyczyńska, J.; Smyczyńska, U.; Stawerska, R.; Domagalska-Nalewajek, H.; Lewiński, A.; Hilczer, M. Seasonality of vitamin D concentrations and the incidence of vitamin D deficiency in children and adolescents from central Poland. *Pediatr. Endocrinol. Diabetes Metab.* **2019**, *25*, 54–59. [CrossRef]

44. Diffey, B.L. Modelling vitamin D status due to oral intake and sun exposure in an adult British population. *Br. J. Nutr.* **2013**, *110*, 569–577. [CrossRef] [PubMed]

45. Macdonald, H.M.; Mavroeidi, A.; Barr, R.J.; Black, A.J.; Fraser, W.D.; Reid, D.M. Vitamin D status in postmenopausal women living at higher latitudes in the UK in relation to bone health, overweight, sunlight exposure and dietary vitamin D. *Bone* **2008**, *42*, 996–1003. [CrossRef] [PubMed]

46. Dale, M.M.; Haylett, D.G. *Rang and Dale's Pharmacology Flashcards*; Churchill Livingstone Elsevier: London, UK, 2014.

47. Kumar, R. The metabolism and mechanism of action of 1,25-dihydroxyvitamin D3. *Kidney Int.* **1986**, *30*, 793–803. [CrossRef]

48. Rynek Ryb [Fish Market]. Available online: https://www.ierigz.waw.pl/publikacje/analizy-rynkowe/rynek-ryb/20873,3,3,0,nr-26-2017-rynek-ryb.html (accessed on 16 August 2019).

49. Scheers, N.; Lindqvist, H.; Langkilde, A.M.; Undeland, I.; Sandberg, A.-S. Vitamin B12 as a potential compliance marker for fish intake. *Eur. J. Nutr.* **2014**, *53*, 1327–1333. [CrossRef] [PubMed]

50. Fish and Omega-3 Fatty Acids. Available online: https://www.heart.org/en/healthy-living/healthy-eating/eat-smart/fats/fish-and-omega-3-fatty-acids (accessed on 19 August 2019).

51. Piramida Zdrowego Żywienia i Aktywności Fizycznej [Food and Physical Activity Pyramid]. Available online: http://www.izz.waw.pl/attachments/article/7/Piramida%20Zdrowego%20%C5%BBywienia%20i%20Aktywno%C5%9Bci%20Fizycznej%20Broszura.pdf (accessed on 19 August 2019).

52. Becker, W.; Pearson, M.; Metod Och Resultatanalys. Livsmedelsverket, Uppsala, Sverige, Riksmaten 1997–1998. Available online: https://www.livsmedelsverket.se/globalassets/matvanor-halsa-miljo/kostrad-matvanor/matvaneundersokningar/riksmaten-1997-1998-resultat-och-metodrapport.pdf (accessed on 19 August 2019).

53. Pietinen, P.; Paturi, M.; Reinivuo, H.; Tapanainen, H.; Valsta, L.M. FINDIET 2007 Survey: Energy and nutrient

intakes. *Public Health Nutr.* **2010**, *13*, 920–924. [CrossRef] [PubMed]

54. Van der Meer, I.M.; Boeke, A.J.P.; Lips, P.; Grootjans-Geerts, I.; Wuister, J.D.; Devillé, W.L.J.M.; Wielders, J.P.M.; Bouter, L.M.; Middelkoop, B.J.C. Fatty fish and supplements are the greatest modifiable contributors to the serum 25-hydroxyvitamin D concentration in a multiethnic population. *Clin. Endocrinol. Oxf.* **2008**, *68*, 466–472. [CrossRef]

55. Petrenya, N.; Lamberg-Allardt, C.; Melhus, M.; Broderstad, A.R.; Brustad, M. Vitamin D status in a multi-ethnic population of northern Norway: The SAMINOR 2 Clinical Survey. *Public Health Nutr.* **2019**, 1–15. [CrossRef]

56. Hansen, A.L.; Dahl, L.; Bakke, L.; Frøyland, L.; Thayer, J.F. Fish consumption and heart rate variability: Preliminary results. *J. Psychophysiol.* **2010**, *24*, 41–47. [CrossRef]

57. Hansen, A.L.; Dahl, L.; Olson, G.; Thornton, D.; Graff, I.E.; Frøyland, L.; Thayer, J.F.; Pallesen, S. Fish Consumption, Sleep, Daily Functioning, and Heart Rate Variability. *J. Clin. Sleep Med.* **2014**, *10*, 567–575. [CrossRef]

58. Vuholm, S.; Teisen, M.N.; Buch, N.G.; Stark, K.D.; Jakobsen, J.; Mølgaard, C.; Lauritzen, L.; Damsgaard, C.T. Is high oily fish intake achievable and how does it affect nutrient status in 8-9-year-old children? The FiSK Junior trial. *Eur. J. Nutr.* **2019**, *59*, 1205–1218. [CrossRef]

5

Paucity of Nutrition Guidelines and Nutrient Quality of Meals Served to Kenyan Boarding High School Students

Kevin Serrem [1], Anna Dunay [1], Charlotte Serrem [2], Bridget Atubukha [3], Judit Oláh [4,5,*] and Csaba Bálint Illés [1]

[1] Institute of Business Economics, Leadership and Management, SzentIstván University, 2100 Gödöllő, Hungary; kevin.serrem@phd.uni-szie.hu (K.S.); dunay.anna@gtk.szie.hu (A.D.); illes.b.csaba@gtk.szie.hu (C.B.I.)

[2] Department of Consumer Sciences, School of Agriculture and Biotechnology, University of Eldoret, Eldoret 1125-30100, Kenya; charlottejes@gmail.com

[3] Faculty of Bioscience Engineering, Katholieke Universitiet Leuven, 3001 Leuven, Belgium; Bridget.atubukha@student.kuleuven.be

[4] Institute of Applied Informatics and Logistics, Faculty of Economics and Business, University of Debrecen, 4032 Debrecen, Hungary

[5] TRADE Research Entity, Faculty of Economic and Management Sciences, North-West University, Vanderbijlpark 1900, South Africa

* Correspondence: olah.judit@econ.unideb.hu

Abstract: Adequate nutrition is vital for the optimal growth, development, and general well-being of adolescents. A lack of nutritional guidelines for school meals poses a major challenge in the provision of nutritious meals to students in Kenyan boarding high schools. The aim of the study was to investigate the nutrient quality and portion sizes of meals served to students and the adequacy of the meals in meeting students' health requirements. A cross-sectional study was carried out among 50 catering or kitchen managers of 50 high schools in Kenya. Data were obtained through researcher-assisted questionnaires. It was established that menus were simplistic in nature, lacked variety, and were repetitive. With regard to nutrients, menus offered to students were excessively highin dietary fiber, containing three or five times more than the recommended daily intake. In most cases, students were underfed on nutrients such as carbohydrates, vitamin A, folic acid, potassium, calcium, proteins, and vitamins B1–12, resulting in low energy provision. It is concluded that a majority ofthe Kenyan high schools studied do not provide nutritionally adequate meals. The government of Kenya should have nutrition guidelines to ensure that schools provide diets with high foodand nutrient quality to students.

Keywords: adolescents; high schools; nutrition guidelines; meals; Kenya

1. Introduction

The burden of malnutrition, which includesunder- and over-nutrition, is an emerging crisis in developing countries. Adequate nutrition is vital for optimal growth, development, and general well-being, particularly of children and adolescents [1]. Availability of adequate nutrition, either at home or through the education system, contributes to thereduction of malnutrition, especially among children who attend school [2]. There is evidence that educational institutions in developing countries are grappling with malnutrition that could have far-reaching effects on the health of school-goers, ultimately compromising an entire generation's health. For example, in Africa chronic and acute under-nutrition and micronutrient deficiency of iron, iodine, zinc, and vitamin A persist among children and adolescents, according to Gegios et al. [3]. Additionally, overweight and obesity have increasingly become epidemics in most countries [4]. All forms of malnutrition negatively impact the ability of children to stay in school and learn throughout the year, and affect health by creating deficiency diseases such as protein energy malnutrition, as well as predisposing children to chronic lifestyle diseasesin adulthood [5,6].

Schools provide a perfect opportunity for the prevention of malnutrition, as they provide the best access to a large number of people, including family and community members, school staff, and young people. In most developed countries with well-established school feeding programmes, such as Britain, France, the USA, and Italy, school meals and school feeding have been used as an effective mechanism to address child nutrition, education enrolment, school retention, and hygiene issues [7]. Additionally, they provide income-generation, employment, and economic integration benefits to the communities in which they are implemented [8]. This demonstrates that provision of food in the right portion sizes translates into improved nutrition, nutrition education, and adoption of health measures for the sustained provision of adequate quality, quantity, and composition of the meals and snacks provided [9]. Hence, there is a need for country-specific guidelines and menu designs thataddress the nutrition priorities of the target populations, and the objectives of the feeding programs [9].

In Kenya, students in boarding high schools are a vulnerable group as they depend on meals provided by the school as the main source of all their nutrient needs. Studies carried out in Kenya show that there are nolegislated or advised nutrition guidelines for use in school feeding programmes [10].Therefore, school feeding programmes in the country are guided by other factors, not necessarily nutrition guidelines. For example, onecurrent school feeding programme geographically targets regions with the highest poverty rates, the lowest education achievement rates, or the highest numbers ofchildren residing in highly marginalized areas, for provision of free food [10]. This is unfortunate because Kenya faces numerousnutrition deficienciesdue to inadequate protein, vitamin A, and iron intake [11] among children and adolescents. Eventually these mayinfluence children's cognitive development, lower school performance, limit adult productivity, reduce immunity, and eventually contribute to a high burden of morbidity and mortality [10].

The lack of an adequate school feeding policy and nutrition guidelines indicatesthe lack of adequate knowledge and precise benchmarking with regard to food rations, nutrient content, and feeding patterns when administering food to children and adolescents who attend school. Additionally it implies compromise in the provision of quality nutritious meals to school-going adolescents. Hence, there is a heavy reliance on the World Health Organization's (WHO's) recommendeddaily allowance (RDA) as itis an internationally accepted standard [11]. Therefore, the study investigated the nutrient quality, portion size, and suitability to meet health requirements of meals offered to students in Kenyan boarding high schools.

2. Materials and Methods

2.1. Study Area and Target Population

A cross-sectional study was conducted in May through to July 2019, to assess the portion size and composition of meals offered to Kenyan boarding high school students aged 15 to 18 years, on a daily and weekly basis. The target population was catering managers or head cooks who were in charge of kitchens in high schools. The study was conducted in eight counties, namely Nakuru, UasinGishu, Nandi, Kakamega, Nairobi, Kisumu, Laikipia, and ElgeyoMarakwet, where 50 selected boarding high schools categorized by the Kenyan Ministry of Education as national, extra county, county, and private, were included.

2.2. Sampling Design

Purposive judgmental sampling was used to select eight out of the 47 counties, and 50 out of over 3000 high schools in Kenya were selected toparticipate in the study. This was done based on preliminary knowledge of the various counties and the number of schools they had in each of the Ministry of Education school categories. One catering manager/head cook from each of the 50 schools was interviewed.

2.3. Food Frequency Questionnaire

Food frequency questionnaires were used to collect dietary information from the school catering managers or cooks about the school menus and meals that were provided to students in the high schools. This information was based on the types of foods and meals offered to the students on both a daily and a weekly basis, the frequency of food provision, and the portion size of meals offered to students, all based on the school menus. This was necessary to establish the types of meals offered in the high schools, and the meal frequency and portion sizes served to students to provide nutrients in comparison to the WHO's 2006 guidelines for RDA.

2.4. Ethical Approval

Approval for conducting the study was granted by the National Commission for Science, Technology and Innovation (NACOSTI) in Kenya, under permit number NACOSTI/P/1981086/28440. Permission was also granted by the education sections of all the participating counties and the head teachers of each of the 50 schools. Informed consent was also obtained from the respondents, whose participation was voluntary after assurances of anonymity and confidentiality.

2.5. Data Analysis

NutriSurvey Software for Windows (2007) was used to generate dietary intake data. Many of the foods in the school menus were native Kenyan local foods; therefore, additional nutrient composition was obtained from the Kenya Food Composition Tables 2018, compiled by the FAO/Government of Kenya [12], which has a collection of the foods consumed. The nutrient profile of each food was then added from the Food Composition Tables into the nutrient database of NutriSurvey Software for Windows. Additionally, Minitab 18 Statistical Software for analysis was used to separate the means ANOVA, as the data being analyzed had more than two groups from which comparison was to be made. The level of significance was $p < 0.05$.

3. Results

3.1. Mean Quantity of Each Food Consumed Daily among School Categories

The results in Table 1 show the quantities of each type of food provided daily to individual students in the different school categories. Githeri (a mixture of maize and beans) was providedin the highest (377 g) quantities in county schools, while private schools providedthe least. The highest (166 g) provision of ugali (stiff porridge) was by national schools, while private schools servedthe least. Compared to all other school types the highest (278 g) amount of rice was servedby the private schools. Legume servicewas highest (90 g) in county schools and lowest in private schools, and vegetables were provided most (121 g) in county schools and least in private schools. Bread and potatoes were servedacross all school types. The least providedfoods were spreads, tea, coffee, and milk. Animal-source foods such as eggs, sausage, and beef mainly appeared in the menus of private schools compared to other school types.

Table 1. The mean amounts in grams of the foods provided daily to students aged 15–18 years in the selected high schools.

Categories	Main Foods (g)	Extra County Schools	National Schools	Private Schools	County Schools
Starchy Staples	Rice	158.57 ± 51.9	173.25 ± 39.3	277.61 ± 121.9	225 ± 45.5
	Potatoes	7.95 ± 18.4	44.91 ± 41.8	35.31 ± 29.6	19 ± 26.2
	Chapatti	0.00	22.50 ± 45.0	15.61 ± 29.6	0
	[1] Githeri	335.79 ± 172.2	375.43 ± 197.6	123.75 ± 51.0	377 ± 48.5
	[2] Ugali	146.56 ± 46.0	166.61 ± 66.4	78.36 ± 14.1	141 ± 2.02
	Bread	61.07 ± 38.6	100.00 ± 0.0	100.00 ± 0.0	100 ± 0.0
	Scones	27.57 ± 50.6	0.00	17.86 ± 35.7	0
Beverages	White tea	109.00 ± 48.5	150.00 ± 0.0	132.86 ± 64.2	201 ± 30.3
	Black tea	58.57 ± 77.58	0.00	0.00 ± 0.0	0
	Cocoa	2.43 ± 4.9	5.71 ± 6.6	3.04 ± 6.1	0
	[3] Uji	111.43 ± 105.0	0.00	87.50 ± 71.5	36 ± 50.5
	Coffee	0.00	3.04 ± 6.0	4.29 ± 4.9	0
Dairy	Milk	0.00	26.79 ± 53.57	0.00 ± 0.0	0
Fruits and Vegetables	Fruit	0.70 ± 0.78	0.50 ± 0.6	1.75 ± 0.9	0
	Cabbage	40.50 ± 29.8	42.29 ± 48.9	35.96 ± 15.3	121 ± 71.7
	Kales	47.15 ± 25.6	52.00 ± 66.17	8.34 ± 16.69	0
Legumes	Green grams	0.00	0.00	15.00 ± 17.4	16 ± 22.7
	Beans	71.74 ± 54.8	78.21 ± 13.1	29.57 ± 34.7	64 ± 90.9
Meat	Beef	12.17 ± 10.6	13.79 ± 13.1	16.07 ± 14.48	8.0 ± 1.0
	Sausage	0.00	0.00	2.86 ± 3.3	0
Eggs	Egg	0.50 ± 1.5	6.25 ± 4.8	30.00 ± 40.0	0
Spreads	Spreads	0.00	0.00	15.00 ± 0.0	0
	Total	1191	1261	1030	1308

Figures are means ± standard deviation; [1] githeri is a meal prepared from a stewed mixture of dry maize and beans; [2] ugali is stiff porridge prepared from maize meal; [3] uji is a drinking porridge prepared from maize or millet meal.

3.2. Food Groups Provided by the Different Categories of High Schools

Foods appearing in the school menu were classified into eightfood groups (Table 2). Results showed that the most providedgroup was the starchy staples which contributed 64 to 68% of the total diet. The highest servers of the staples were county schools (898 g) and the lowest were private schools. Dairy products (milk) were only served in national schools. Private schools were the leading providers of fruits, while county schools' menus did not feature fruits. Legumes were the main source of proteins across all the school types, with a high of 6% of the diet, except for the private schools, which included more animal proteins in their menus than all the other school types.

Table 2. The mean amount in grams of the food groups served daily to students aged 15–18 years old in different school types and their percentage contributions to the total dietary intake (g/student/day).

Food Group	Extra County Schools	National Schools	Private Schools	County Schools
Starchy Staples	849 ± 47 (68.69)	883 ± 61 (67.66)	736 ± 35 (64.50)	898 ± 26 (68.65)
Beverages	170 ± 25 (13.75)	159 ± 1 (12.18)	52 ± 21 (4.56)	201 ± 30 (15.37)
Dairy	0 (0)	27 ± 54 (2.07)	0 (0)	0 (0)
Fruits	80 ± 92 (6.47)	60 ± 69 (4.60)	200 ± 105 (17.53)	0(0)
Vegetables	88 ± 16 (7.12)	78 ± 9 (5.98)	44 ± 12 (3.86)	121 ± 21 (9.25)
Legumes	36 ± 29 (2.91)	78 ± 13 (5.98)	45 ± 36 (3.94)	80 ± 68 (6.12)
Meat	12 ± 11 (0.97)	14 ± 13 (1.07)	19 ± 17 (1.67)	8 ± 1 (0.61)
Eggs	1 ± 2 (0.08)	6 ± 5 (0.46)	30 ± 40 (2.63)	0 (0)
Spreads	0 (0)	0 (0)	15 ± 10 (1.31)	0 (0)
Total	1236 (100)	1305 (100)	1141 (100)	1308 (100)

Values are means ± standard deviations. Values in parenthesis are percentage contributions to total dietary intake.

3.3. Mean Amount of Nutrients Provided by the Diet

Table 3 shows an estimate of the mean amount of nutrients providedby the diet for each student in the four school categories. Results indicate that protein intake is highest in national schools (62 g) and lowest in private schools. Retinol and Vitamin A intake were significantly higher in private schools compared with other school types. Retinol intake was lowest (26 μg) in extra county schools while vitamin A provision in extra county and county schools ranged from 114 to 40 μg. The highest (8.15 μg) intake of vitamin B_1 was in county schools, which also exceeded the required daily intake, compared to the lowest intake in extra county schools, which hadan almost eight-times lower intake. Vitamin B_{12} provision is highest in private schools (1.55 μg) and lowest in county schools (0.7 μg). Its provision in extra county schools and private schools was significantly different.

Table 3. The mean amount of nutrients provided daily by the total diet of the high-school students in the sampled schools.

Nutrients	Units	School Type			
		Extra County Schools	National Schools	Private Schools	County Schools
Energy	K Cal	$1453.1^a \pm 267.3$	$1729.0^a \pm 309.0$	$1494.0^a \pm 294$	$1576.0^a \pm 201$
Minerals	g	$12.46^{ab} \pm 2.8$	$13.95^a \pm 3.5$	$9.73^b \pm 0.9$	$13.40^{ab} \pm 2.12$
Proteins	g	$52.98^a \pm 9.4$	$61.80^a \pm 12.9$	$44.80^a \pm 7.2$	$56.35^a \pm 11.9$
Fat	g	$27.38^a \pm 5.9$	$37.27^a \pm 12.8$	$38.80^a \pm 16.7$	$28.75^a \pm 2.9$
Carbohydrates	g	$227.8^a \pm 40.9$	$264.8^a \pm 35.3$	$228.8^a \pm 33.7$	$249.0^a \pm 26.3$
Dietary fiber	g	$45.83^a \pm 11.9$	$48.63^a \pm 13.1$	$26.13^{ab} \pm 3.6$	$45.80^{ab} \pm 10.0$
Retinol	μg	$26.08^b \pm 10.6$	$66.10^b \pm 31.9$	$191.6^a \pm 129.0$	$35.40^b \pm 4.5$
Vitamin A	μg	$114.6^b \pm 43.8$	$169.8^{ab} \pm 148.5$	$253.2^a \pm 160.4$	$39.5^b \pm 4.74$
β-Carotene	mg	$882^a \pm 488$	$975^a \pm 1205$	$167^a \pm 310$	$33.75^a \pm 4.88$
Vitamin B_1	mg	$0.80^c \pm 0.2$	$0.97^{bc} \pm 0.3$	$7.55^a \pm 8.0$	$8.15^{ab} \pm 10.2$
Vitamin B_2	mg	$0.56^a \pm 0.0$	$0.75^a \pm 0.2$	$0.6^a \pm 0.2$	$0.6^a \pm 0.1$
Vitamin B_{12}	μg	$0.74^b \pm 0.5$	$1.12^{ab} \pm 0.8$	$1.55^a \pm 0.9$	$0.7^{ab} \pm 0.1$
Vitamin B_6	mg	$0.21^c \pm 0.1$	$0.43^{ab} \pm 0.2$	$0.48^a \pm 0.2$	$0.15^{bc} \pm 0.0$
Folic acid	μg	$32.89^b \pm 14.2s$	$38.13^b \pm 15.6$	$68.50^a \pm 34.9$	$19.95^b \pm 3.9$
Vitamin C	mg	$67.74^a \pm 26.8$	$44.5^a \pm 37.6$	$63.0^a \pm 17.9$	$31.60^a \pm 6.6$
Sodium	mg	$1631.0^a \pm 471$	$1851.0^a \pm 430$	$1592.9^a \pm 179.3$	$1978.0^a \pm 195$
Potassium	mg	$1542.0^a \pm 610$	$1863.0^a \pm 537$	$1279^a \pm 345$	$1657^a \pm 345$
Calcium	mg	$455.4^a \pm 91.8$	$474.0^a \pm 233$	$254.9^b \pm 29.6$	$292.8^{ab} \pm 56.4$
Magnesium	mg	$295.3^a \pm 67.5$	$330.2^a \pm 80.4$	$210.9^b \pm 25.2$	$290.1^{ab} \pm 43.5$
Phosphorus	mg	$1336.0^a \pm 351$	$1551.0^a \pm 399$	$900.6^b \pm 116.7$	$1456.0^{ab} \pm 254$
Iron	mg	$19.52^a \pm 8.0$	$19.55^a \pm 7.4$	$12.00^a \pm 4.0$	$14.25^a \pm 1.9$
Zinc	mg	$8.32^{ab} \pm 1.4$	$9.55^a \pm 1.9$	$6.78^b \pm 1.0$	$8.80^{ab} \pm 1.7$

Values are means ± standard deviation; values with the same letter (a to c) superscript on the same row are not significantly different at $p < 0.05$ as assessed by Fisher's least significant difference.

Intake of vitamin B_6 and folic acid were significantly higher in private schools compared to all other school categories, whilenational schools had the highest calcium and magnesium intakes. Only the private schools' phosphorous intake was very low—900.6 mg—when compared to other schools. There was a significant difference in zinc provision between national and private schools at 9.55 and 6.8 mg, respectively (Table 3).

3.4. Percentage Fulfillment of Different Nutrients

The percentage fulfillment of the recommended daily allowance of nutrients for students in the four categories of schools is shown in Table 4. Results show that school meals did not meet 100% of the energy needs of the students. Carbohydrates, proteins, and fats provided 57.6, 11.8, and 30.5% of energy requirements, respectively. The meals met more than 75% of the nutritional requirements for proteins, fats, and carbohydrates. The diets fell short in vitamins A, C, and B_{12}. In terms of minerals, calcium was the nutrient for which the requirements were least fulfilled (3–37%). Themineral requirements that were adequately fulfilled were phosphorous, iron, and zinc.

Table 4. Percentage fulfilment of the different nutrients according to school type.

Nutrients	Units	Recommended Value/Day [1]	Percentage (%) Fulfilment			
			Extra County Schools	National Schools	Private Schools	County Schools
Energy	kCal	2036.3	71.35	84.91	73.37	77.40
Proteins	g	60.1	88.15	102.83	74.54	93.76
Fat	g	69.1	39.62	53.94	56.15	41.61
Carbohydrates	g	290.7	78.36	91.09	78.71	85.66
Dietary fiber	g	30	152.77	162.10	87.10	152.67
Vitamin A	µg	1000	11.46	16.98	25.32	3.95
Vitamin B_1	mg	1.15	69.57	84.35	656.52	708.70
Vitamin B_2	mg	1.35	41.48	55.56	44.44	44.44
Vitamin B_{12}	µg	3	24.67	37.33	51.67	23.33
Vitamin B_6	mg	1.4	15.00	30.71	34.29	10.71
Folic acid	µg	400	8.22	9.53	9.53	4.99
Vitamin C	mg	100	67.74	44.50	63.00	31.60
Sodium	mg	2000	81.55	92.55	79.65	98.90
Potassium	mg	3500	44.06	53.23	36.54	47.34
Calcium	mg	1200	37.95	3.95	21.24	24.40
Magnesium	mg	375	78.75	88.05	56.24	77.36
Phosphorus	mg	1250	106.88	124.08	72.00	116.48
Iron	mg	13.5	144.59	144.81	88.89	105.56
Zinc	mg	8.5	97.88	112.35	79.76	103.53

[1] WHO, [11].

4. Discussion

4.1. Diversity of Foods Provided

The low diversity of food items in the schools categories other than private schools may be a result of limited government funding in public schools, while the more diverse foods in private school menus might be influenced by the purchasing power of the parents, who pay higher school fees. A study by Zuilkowski et al. [13] on primary schools in Nairobi showed that low-cost private schools (LCPS) spent an average of Ksh 1962 per annum on school meals, while public schools spent almost half of this. This can be extrapolated to secondary schools, which are guided by the same document, i.e., the National School Meals and Nutrition Strategy 2017–2022 [14]. Furthermore, studies show that parents with students in private schools are more likely to pay more money to ensure that their children are well-fed [13]. This may be the reason for the increased diversity of school meals in private schools. Additionally, private schools have been shown to have better autonomy to make decisions regarding the foods that should appear in school menus compared to public schools [9].Private schools alsoserved the lowest amount of starchy staples such as ugali and githeri. Their main source of carbohydrates is rice. This could be explained by the high amount of energy required to cook maize and beans, making it expensive, hence their preference to cook rice.

4.2. Food Groups Provided

The starchy staples, which contributed more than 60% of the entire diet, have been found to be the major foods fed to school children in developing countries [15] because they are easilyavailable and are a cheap source of energy compared to other food groups. Thompson and Amoroso [16] indicate a shift from a varied diet rich in micronutrients to one derived from predominantly high-carbohydrate starchy staples. Animal-source foods arethe least provided foods in school menus because they are expensive, in the context of school feeding in developing countries. A study by Cornelsen et al. [17] in Kenya found that price was the most frequently reported barrier to the provision of animal-source Foods (ASF), therefore affordability is the main hindrance to their service. Additionally, the study revealed that milk was providedonly by the national schools because they reared dairy animals in their school farms and therefore had a constant supply of milk. Azzarri et al. [18] suggest that livestock ownership confers on households more opportunities to increase the consumption of animal-source food as it translates into cheaper or more reliable access to animal-source food supplies. A study by Kabunga et al. [19] showed that the adoption of improved dairy cow breeds at farm level led to increased consumption of milk in Ugandan households and translated to improved dietary quality among young children. Meat and eggs, the least-provided foods, contributed an estimated 2% to the total dietary intake among the high-school students. The major source of proteins were legumes, which are known to have lower protein quality compared to animal sources [16]. From the foregoing, school feeding policies should advocate for boarding schools to rear animals for milk, eggs, and meat for the cheaper supply of high-quality protein.

The provision of fruitsin private schools and not at all in county schools may be explained by the purchasing ability of the parents of children attending private schools, compared to those attending county schools. Fruits are known to be an expensive source of energy and therefore this could be the reason for the high intake in private schools. Review studies on the determinants of fruit and vegetables consumption among adolescents consistently show that high income levels largely influence fruit intake, often because of affordability [20–22].

The high provision of vegetables in county schools could be attributed to the parents who sometimes supply vegetables from their farms as a form of school fees payment, since most of these types of schools are commonly found in rural areas. A review on the determinants of vegetable consumption among adolescents by Rasmussen et al. [20] showed that availability of vegetables is one of the determinants that greatly influence vegetable intake, which may be true in this study. Furthermore, low vegetable provision in private schools could be attributed to the children'slack of engagement in school gardening activities owing to limited land sizes in private schools. Despite epidemiological evidence by He et al. [23] for the health benefits of a diet rich in fruits and vegetables, such as reduced risk of chronic diseases later in life, large numbers of adolescents do not meet the WHO requirement of a daily intake of at least 400 g (roughly equivalent to five servings per day) of fruits and vegetables [24]. A study by Peltzer and Pengpid [25] in seven African countries, including Kenya, found that most adolescents (77.5%) did not consume the recommended daily servings of vegetables. Similar results to this have been reported by Doku et al. [26], who found that 56% and 48% of adolescents in Ghana rarely consumed fruits and vegetables. According to South African dietary guidelines, it is recommended that adolescents consume at least one fruit and/or vegetable in a meal [24]. This recommendation was not met in this study. This could be due to the limited variety of fruits (banana and orange) and vegetables (kale and cabbage) that appear on most of the school menus. This study is in agreement with other studies that fruit and vegetable consumption is consistently low in many low middle income countries [27].This implies that nutrition guidelines for school feeding in Kenya should recommend the minimum fruit and vegetable portions that should be provided to students.

4.3. The Mean Amount of Nutrients Provided Daily from the Total Diet

The higher consumption of proteins by the national compared to private schools may be explained by the fact that the major sources of proteins in the diets are cereals and legumes. The low values reported in private schools could be due to their low service of these grains. Private schools provided more rice and potatoes, which are poor sources of protein. However, they provided more beef, eggs, and sausages than the other schools— implying that though their protein intake is low, it is of higher quality, since it comes from animal sources.

The national schools' diet had high mineral content, probably because most of the meals served were wholemeal, and therefore the presence of bran in the foods leads to a high mineral content in the diet. In contrast, most of the foods servedin private schools such as rice, chapatti, and potatoes have low mineral content, and are processed with husks removed. For example, the processing of maize by dry milling and fractionation results in the removal of bran, which is the major constituent of the pericarp, and which contains B-vitamins and minerals [28]. Cornbran contains more iron, zinc, and phosphorous than corn starch [29]. This shows that the micronutrient content of cereals is greatly reduced after processing, and therefore serviceof processed foods provides a small proportion of the daily requirements for most vitamins and minerals. Nutrition guidelines should emphasize the use of whole grain for food and food products in adequate portions in school meals to contribute to mineral provision.

The service of higher amounts of fruits, and also spreads on bread that are fortified with vitamin A, to students in private schools may have resulted in their intake of higher amounts of retinol and vitamin A than the other types of school. Margarine has been found to be one of the most suitable vehicles for vitamin A [30]. A study involving Filipino pre-school children who consumed 27 g of vitamin A-fortified margarine per day for a period of six months showed that there was a reduced prevalence of low serum retinol from 26% to 10%, hence consumption of the margarine significantly improved the vitamin A status of the children [31]. Foods high in vitamin A should be deliberately included among those in school menus based on Kenyan nutrition guidelines.

Furthermore, the significantly higher vitaminsB_{12}, B_6, and folic acid in private school menus compared to the other school types may be attributed to the higher service of animal-source foods such as eggs and beef. In contrast, the lowcalcium, phosphorous, and magnesium provision in private schools compared to other school types may be explained bythe low serviceof cereals and legumes, which are rich sources of magnesium and phosphorous [32]. The low intake of milk and milk products could further explain the low calcium levels in the students' diets. In national schools the provision of zinc was higher compared to private schools. This could be explained by the fact that they were served with more beans compared to other schools. They were also provided with milk, which is a rich source of zinc.

4.4. Fulfillment of the Different Nutrients

The study's findings showed that school meals did not meet 100% of the energy needs of the students. This findings contradict those of Buluku [7] who found that energy was adequately met in the diets of girls in selected boarding schools in Nairobi, Kenya. However, this finding concurs with that of Nhlapo et al. [33] who found that the meals served under a South African school feeding scheme did not meet the energy needs of school children aged 11–18 years. Carbohydrates, proteins, and fats provided 57.6%, 11.8%, and 30.5% of energy, respectively. According to the School Food Trust [34], a minimum of 50% of the energy provided should be sourced from carbohydrates, less than 35% should be met from fats and the rest from proteins. The results of this study indicate that the meals provided met the recommendations regarding the contributions of the three macronutrients to the energy needs of the high school students. This finding is similar to that of Charrondiere et al. [35] on numerous food items from different countries in which the total carbohydrate content supplied 50–80% of energy, and 7–11% of energy was from protein. Nutrition guidelines should recommend portion sizes for energy-providing foods that enable students in boarding high schools to meet all their energy requirements.

The ability of national schools to meet the required protein intake while the other school types fell short may be attributed to the provision of more food groups rich in proteins, and that this was the only school type that served milk. Protein is important in adolescent nutrition as it provides structure for the body and major components of the bones, blood, muscle, cell membranes, enzymes, and immune factors [32].

The high intake of dietary fiber in the public schools may be explained by the provision of starchy unrefined grains and pulses in meals such as githeri (a stewed mixture of maize and beans), wholemeal ugali, and vegetables such as kale, which are rich in fiber. This is an advantage, as high-fiber diets have a low glycemic index, therefore keeping students full for a longer time, which reduces the rate of hunger. For instance, a study by Ye et al. [36] showed that consumption of dietary fiber decreased hunger and increased satiety hormones in humans when ingested with a meal. This implies that that the students can concentrate in class for longer without feeling hungry. However, high-fiber diets tend to be rich in phytates, and therefore could bind key mineral elements such as calcium and also reduce absorption of minerals such as iron and zinc [37]. Private schools reported a low intake of fiber compared to the other schools because of the low provision of githeri, ugali, and kale, and the provision of potatoes, which are low in fiber as stated earlier.

The percentage fulfillment of Vitamin A was high in private schools and lowin county schools due to the high intake of fruits, and Vitamin A-fortified spreads on bread. The low value reported in county schools is a result of the non-provision of fruits and spreads. This finding is similar to that of Buluku [7] on Kenyan boarding school meals that did not meet the nutrient requirements for adolescent girls. Vitamin A is a key micronutrient in adolescent nutrition. Vitamins play important roles in the body such as aiding vision, immunity (vitamin A), calcium absorption (vitamin D), and anti-oxidative protection in cell membranes (vitamin E) [32].

Calcium intake was highest in the national schools probably because of milk provision to students in national schools, compared to private schools, which did not providemilk. The ability of most school types to fully meet their phosphorous, zinc, and iron requirements in most school types could be attributed to the consumption of cereals and pulses, which are rich sources of these minerals [32]. The intake of iron and zinc appears to be overestimated because the diet was mainly composed of cereals and pulses. These minerals were mostly found in githeri, ugali, millet porridge, maize porridge, beans, and green grams, which were frequentlyprovided.

However, cereals and legumes contain phytates which chelate with minerals and metals such as calcium, magnesium, zinc, and iron, forming insoluble salts that are not easily absorbed by humans [38]. Phytates particularly form complexes which can severely impair the availability of zinc and iron [39]. Furthermore, most schools took beverages in the form of tea, coffee, or cocoa, which may also contain tannins that could further bind the minerals [40]. Deficiencies of iron and zinc are a public health problem in developing countries, particularly among adolescents and women of reproductive age. A systematic review to evaluate iron and zinc intakes in adolescents from Ethiopia, Kenya, Nigeria and South Africa concluded that diet-related anaemia and zinc deficiencies are problems of public health significance [41]. Zinc is important in adolescence because of its critical function in growth and sexual maturation [42]. However, adolescents are at high risk of deficiency, often related to the consumption of plant-based diets, which have low zinc and iron bioavailability reported by Gibson et al. [43], which are similar to the diets providedin this study. From the forgoing it is noted that the studyonly focused on whatschools provided to their students and did not consider how much the adolescents consumed.

5. Conclusions

Kenyan high schools fail to provide students with adequate nutritious foods. In comparison to the FAO [9], which acts as the worldwide benchmark for the development of school dietary guidelines, the study notes that none of the high schools in Kenya adequately meet the nutritional requirements of meals served to school-goers. Meal menus lack variation in food options, and are very repetitive and simplistic in nature. It observed that the foods most providedin schools are highfiber foods, githeri (a mixture of maize and beans), and starchy foods such as ugali (stiff porridge) and rice, while the least-consumed food are fruits, proteins—especially during breakfasts—and vegetables. National schools consume the highest quantities of dietary fiber and starch in comparison to other high schools whilst county schools consume the highest quantities of vegetables. Private schools consume the highest quantities of breakfast proteins, while county schools consume the highest quantities of legumes. The majority of Kenyan high schools fail to attain nutrient requirements in meals offered to students, while in some cases schools surpass the recommended amount, such as for dietary fiber, where schools offer three times more than the recommended amounts. The Kenyan government should work with various stakeholders to ensure the development of an adequate school feeding policy to develop and implement nutrition guidelines in the county. Additionally, an area of further research would be how much food the adolescents in Kenyan high schools consumed.

Author Contributions: Conceptualization by C.B.I. and K.S.; methodology designed by C.B.I., K.S. and A.D.; Data compilation by K.S. and C.S.; Data analysis by K.S. and B.A.; validation by C.B.I., A.D. and J.O.; original draft prepared by K.S.; review and editing by K.S., C.B.I., A.D. and J.O. All authors have read and agreed to the published version of the manuscript.

References

1. Afoakwa, E.O. *Enhancing the Quality of School Feeding Programmes in Ghana—Developments and Challenges*; University of Ghana: Accra, Ghana, 2005; Available online: http://www.academia.edu/download/52162306/fulltext_stamped.pdf (accessed on 20 February 2020).

2. Essuman, E.; Walker, L.R.; Maziasz, P.J.; Pint, B.A. Oxidation behaviour of cast Ni–Cr alloys in steam at 800 °C. *Mater. Sci. Technol.* **2013**, *29*, 822–827. [CrossRef]

3. Gegios, A.; Amthor, R.; Maziya-Dixon, B.; Egesi, C.; Mallowa, S.; Nungo, R.; Gichuki, S.; Mbanaso, A.; Manary, M.J. Children consuming cassava as a staple food are at risk for inadequate zinc, iron, and vitamin A intake. *Plant Foods Hum. Nutr.* **2010**, *65*, 64–70. [CrossRef]

4. Muthuri, S.K.; Wachira, L.-J.M.; Leblanc, A.G.; Francis, C.E.; Sampson, M.; Onywera, V.O.; Tremblay, M.S. Temporal trends and correlates of physical activity, sedentary behaviour, and physical fitness among school-aged children in Sub-Saharan Africa: A systematic review. *Int. J. Environ. Res. Public Health* **2014**, *11*, 3327–3359. [CrossRef]

5. Crookston, B.T.; Schott, W.; Cueto, S.; Dearden, K.A.; Engle, P.; Georgiadis, A.; Lundeen, E.A.; Penny, M.E.; Stein, A.D.; Behrman, J.R. Postinfancy growth, schooling, and cognitive achievement: Young Lives. *Am. J. Clin. Nutr.* **2013**, *98*, 1555–1563. [CrossRef]

6. Nyaradi, A.; Li, J.; Hickling, S.; Foster, J.; Oddy, W.H. The role of nutrition in children's neurocognitive development, from pregnancy through childhood. *Front. Hum. Neurosci.* **2013**, *7*, 97. [CrossRef]

7. Buluku, E. *An Assessment of the Adequacy of School Meals in Meeting the Nutritional Requirements of Girls in Boarding Secondary Schools in Nairobi*; Kenyatta University: Nairobi, Kenya, 2012; Available online: http://ir-library.ku.ac.ke/handle/123456789/2516 (accessed on 20 February 2020).

8. Moriset, B. Building new places of the creative city: The rise of coworking spaces. *Territ. Mov. J. Geogr. Plan.* **2017**. [CrossRef]

9. FAO. *Nutrition Guidelines and Standards for School Meals: A Report from 33 Low and Middle-Income Countries*; Food and Agriculture Organization of the United Nations: Rome, Italy, 2019; pp. 10–11. Available online: http://www.fao.org/3/CA2773EN/ca2773en.pdf (accessed on 20 February 2020).

10. Kisa, S.; Zeyneloğlu, S.; Yilmaz, D.; Güner, T. Quality of sexual life and its effect on marital adjustment of Turkish women in pregnancy. *J. Sex Marital. Ther.* **2014**, *40*, 309–322. [CrossRef]

11. WHO. *Adolescent Nutrition: A Review of the Situation in Selected South-East Asian Countries*; WHO Regional Office for South-East Asia: New Delhi, India, 2006; Available online: https://apps.who.int/iris/handle/10665/204764 (accessed on 20 February 2020).

12. FAO/Government of Kenya. *Food Composition Tables. Government of Kenya, Nairobi*; The Food and Agriculture Organization of the United Nations, The Ministry of Health, Republic of Kenya and The Ministry of Agriculture and Irrigation, Republic of Kenya: Kenya, Africa, 2018; pp. 12–56. Available online: http://www.fao.org/3/I8897EN/i8897en.pdf (accessed on 20 February 2020).

13. Zuilkowski, S.S.; Piper, B.; Ong'ele, S.; Kiminza, O. Parents, quality, and school choice: Why parents in Nairobi choose low-cost private schools over public schools in Kenya's free primary education era. *Oxf. Rev. Educ.* **2018**, *44*, 258–274. [CrossRef]

14. Ministry of Education; Ministry of Health; Ministry of Agriculture, LAF. *National School Meals and Nutrition Strategy 20172–022*; Ministry of Education; Ministry of Health Ministry of Agriculture, Livestock and Fisheries: Nairobi, Republic of Kenya, 2018; pp. 1–64. Available online: https://planipolis.iiep.unesco.org/sites/planipolis/files/ressources/kenya_school_meals_nutrition_strategy_20172--022.pdf (accessed on 20 February 2020).

15. Kennedy, G.L.; Pedro, M.R.; Seghieri, C.; Nantel, G.; Brouwer, I. Dietary diversity score is a useful indicator of micronutrient intake in non-breast-feeding Filipino children. *J. Nutr.* **2007**, *137*, 472–477. [CrossRef]

16. Thompson, B.; Amoroso, L. *Combating Micronutrient Deficiencies: Food-Based Approaches*; CABI: Wallingford, UK, 2011. [CrossRef]

17. Cornelsen, L.; Alarcon, P.; Häsler, B.; Amendah, D.D.; Ferguson, E.; Fèvre, E.M.; Grace, D.; Dominguez-Salas, P.; Rushton, J. Cross-sectional study of drivers of animal-source food consumption in low-income urban areas of Nairobi, Kenya. *BMC Nutr.* **2016**, *2*, 70. [CrossRef]

18. Azzarri, C.; Cross, E.; Haile, B.; Zezza, A. Does livestock ownership affect animal source foods consumption and child nutritional status? Evidence from rural Uganda. *J. Dev. Stud.* **2014**, *51*, 1034–1059. [CrossRef]

19. Kabunga, N.S.; Ghosh, S.; Webb, P. Does ownership of improved dairy cow breeds improve child nutrition? A pathway analysis for Uganda. *PLoS ONE* **2017**, *12*, e0187816. [CrossRef] [PubMed]

20. Rasmussen, M.; Krølner, R.; Klepp, K.-I.; Lytle, L.; Brug, J.; Bere, E.; Due, P. Determinants of fruit and vegetable consumption among children and adolescents: A review of the literature. Part I: Quantitative studies. *Int. J. Behav. Nutr. Phys. Act.* **2006**, *3*, 22. [CrossRef] [PubMed]

21. Kiss, A.; Popp, J.; Oláh, J.; Lakner, Z. The reform of school catering in hungary: Anatomy of a health-education attempt. *Nutrients* **2019**, *11*, 716. [CrossRef] [PubMed]

22. Kiss, A.; Pfeiffer, L.; Popp, J.; Oláh, J.; Lakner, Z. A blind man leads a blind man? Personalised nutrition-related attitudes, knowledge and behaviours of fitness trainers in Hungary. *Nutrients* **2020**, *12*, 663. [CrossRef]

23. He, F.; Nowson, C.; Lucas, M.; MacGregor, G. Increased consumption of fruit and vegetables is related to a reduced risk of coronary heart disease: Meta-analysis of cohort studies. *J. Hum. Hypertens.* **2007**, *21*, 717–728. [CrossRef]

24. Kimmons, J.; Gillespie, C.; Seymour, J.; Serdula, M.; Blanck, H.M. Fruit and vegetable intake among adolescents and adults in the United States: Percentage meeting individualized recommendations. *Medscape J. Med.* **2009**, *11*, 26.

25. Peltzer, K.; Pengpid, S. Fruits and vegetables consumption and associated factors among in-school adolescents in seven African countries. *Int. J. Public Health* **2010**, *55*, 669–678. [CrossRef]

26. Doku, D.; Koivusilta, L.; Raisamo, S.; Rimpelä, A. Socio-economic differences in adolescents' breakfast eating, fruit and vegetable consumption and physical activity in Ghana. *Public Health Nutr.* **2013**, *16*, 864–872. [CrossRef]

27. Darfour-Oduro, S.A.; Buchner, D.M.; Andrade, J.E.; Grigsby-Toussaint, D.S. A comparative study of fruit and vegetable consumption and physical activity among adolescents in 49 low-and-middle-income countries. *Sci. Rep.* **2018**, *8*, 1–12. [CrossRef]

28. Gwirtz, J.A.; Garcia-Casal, M.N. Processing maize flour and corn meal food products. *Ann. N. Y. Acad. Sci.* **2014**, *1312*, 66. [CrossRef] [PubMed]

29. USDA. *Food Composition Databases*; USA Department of Agriculture Agricultural Research Service: Washington, DC, USA, 2018. Available online: https://ndb.nal.usda.gov/ (accessed on 20 February 2020).

30. Dary, O.; Mora, J.O. Food fortification to reduce vitamin A deficiency: International Vitamin A Consultative Group recommendations. *J. Nutr.* **2002**, *132*. [CrossRef] [PubMed]

31. Solon, F.; Solon, M.; Mehansho, H.; West, J.K.; Sarol, J.; Perfecto, C.; Nano, T.; Sanchez, L.; Isleta, M.; Wasantwisut, E. Evaluation of the effect of vitamin A-fortified margarine on the vitamin A status of preschool Filipino children. *Eur. J. Clin. Nutr.* **1996**, *50*, 720–723. [PubMed]

32. Morris, E.; Whitney, E.N.; Cataldo, C.B. Understanding normal and clinical nutrition. *Am. J. Nurs.* **1984**, *84*, 146. [CrossRef]

33. Nhlapo, N.; Lues, R.J.; Kativu, E.; Groenewald, W.H. Assessing the quality of food served under a South African school feeding scheme: A nutritional analysis. *South Afr. J. Sci.* **2015**, *111*, 1–9. [CrossRef]

34. School Food Trust. *A Guide to Introducing the Government's Food-Based and Nutrient-Based Standards for School Lunches*; School Food Trust: London, UK, 2013; Available online: https://sheu.org.uk/content/guide-introducing-governments-food-based-and-nutrient-based-standards-school-lunches-0 (accessed on 20 February 2020).

35. Charrondiere, U.; Chevassus-Agnes, S.; Marroni, S.; Burlingame, B. Impact of different macronutrient definitions and energy conversion factors on energy supply estimations. *J. Food Compos. Anal.* **2004**, *17*, 339–360. [CrossRef]

36. Ye, Z.; Arumugam, V.; Haugabrooks, E.; Williamson, P.; Hendrich, S. Soluble dietary fiber (Fibersol-2) decreased hunger and increased satiety hormones in humans when ingested with a meal. *Nutr. Res.* **2015**, *35*, 393–400. [CrossRef]

37. Prosky, L. *Dietary Fiber | Effects of Fiber on Absorption*; Elsevier: Oxford, UK, 2003; pp. 1838–1844.

38. Al Hasan, S.M.; Hassan, M.; Saha, S.; Islam, M.; Billah, M.; Islam, S. Dietary phytate intake inhibits the

bioavailability of iron and calcium in the diets of pregnant women in rural Bangladesh: A cross-sectional study. *BMC Nutr.* **2016**, *2*, 24. [CrossRef]

39. Murphy, K.J.; Marques-Lopes, I.; Sánchez-Tainta, A. Cereals and legumes. In *The Prevention of Cardiovascular Disease Through the Mediterranean Diet*; Sánchez-Villegas, A., Sánchez-Tainta, A., Eds.; Academic Press: Cambridge, UK, 2018; Chapter 7; pp. 11–132.

40. Delimont, N.M.; Haub, M.D.; Lindshield, B.L. The impact of tannin consumption on iron bioavailability and status: A narrative review. *Curr. Dev. Nutr.* **2017**, *1*, 1–12. [CrossRef]

41. Harika, R.; Faber, M.; Samuel, F.; Mulugeta, A.; Kimiywe, J.; Eilander, A. Are low intakes and deficiencies in iron, vitamin A, zinc, and iodine of public health concern in Ethiopian, Kenyan, Nigerian, and South African children and adolescents? *Food Nutr. Bull.* **2017**, *38*, 405–427. [CrossRef]

42. Kawade, R. Zinc status and its association with the health of adolescents: A review of studies in India. *Glob. Health Action* **2012**, *5*. [CrossRef] [PubMed]

43. Gibson, R.S.; Raboy, V.; King, J.C. Implications of phytate in plant-based foods for iron and zinc bioavailability, setting dietary requirements, and formulating programs and policies. *Nutr. Rev.* **2018**, *76*, 793–804. [CrossRef] [PubMed]

6

Quantification of Household Food Waste in Hungary: A Replication Study using the FUSIONS Methodology

Gyula Kasza [1],*, Annamária Dorkó [1], Atilla Kunszabó [1] and Dávid Szakos [2]

[1] Risk Management Directorate, National Food Chain Safety Office, 1024 Budapest, Hungary; dorkoa@nebih.gov.hu (A.D.); kunszaboa@nebih.gov.hu (A.K.)

[2] Department of Veterinary Forensics and Economics, University of Veterinary Medicine Budapest, 1078 Budapest, Hungary; szakos.david@univet.hu

* Correspondence: kaszagy@nebih.gov.hu

Abstract: Household food waste accounts for the most significant part of total food waste in economically developed countries. In recent times, this issue has gained recognition in the international research community and policy making. In light of the Sustainable Development Goals of FAO, mandatory reporting on food waste has been integrated into European legislation, as a basis of preventive programs. The paper presents the results of research that aimed to quantify the food waste generated by Hungarian households. Research methodology was based on the EU compliant FUSIONS recommendations. In total, 165 households provided reliable data with detailed waste logs. Households were supported by kitchen scales, measuring glasses, and a manual. Based on the extrapolation of the week-long measurement, the average food waste was estimated to be 65.49 kg per capita annually, of which the avoidable part represented 48.81%. Within the avoidable part, meals, bakery products, fresh fruits and vegetables, and dairy products are accountable for 88% of the mass. This study was a replication of the first Hungarian household food waste measurement conducted in 2016 with the same methodology. Between the two periods, a 4% decrease was observed. The findings, for instance the dominant share of meals in food waste, should be put in focus during preventive campaigns. National level food waste measurement studies using the FUSIONS methodology should be fostered by policy makers to establish the foundations of effective governmental interventions and allow for the international benchmarking of preventive actions.

Keywords: household food waste; food waste measurement; food waste composition; sustainable consumption; consumer research; consumer behavior

1. Introduction

In recent times, food wastage has become a frequently investigated issue. In countries with developed economic status, the largest quantities of food waste are generated at the consumer level [1]. The ratio of household food waste in the EU is estimated to be about 53% of the amount produced within the entire food chain, which equals 92 kg per capita annually [2]. Reported numbers are mainly originating from calculations based on general waste databases of the member countries. In the last ten years, research activity has been accelerated in this field, delivering a variety of—sometimes contradictory—results. The greatest research activity in the field is focusing on Northern and Western Europe (prominently Denmark, Finland, Germany, Italy, Norway, UK) [3], but there are several examples outside the continent as well (USA, China, Canada) [4].

The majority of the variation in reported empirical data can be explained by the wide range of methods used by the studies. There are quite a few direct research methods that can be applied in research practice to acquire primary data on the extent of household food waste [5,6]. Several reports used questionnaires, where the focus is on self-reporting [7–10]. Physical measurements, such as diaries or composition analysis, are also widely used methods [11–17]. Nevertheless, we must take into consideration that these methods entail a great deal of uncertainty [18]. For instance, separately collected organic stream could contain other biological waste besides food (flowers, green waste from the garden or the street), while other elements could be missing (food put in general domestic waste, poured into the sink, or fed to animals) [19]. A significant gap can be detected between the amounts resulted from physical measurements and self-report surveys [17,20,21]. Self-reported numbers tend to be significantly lower than physically measured values. Physical measurements, however, require a serious commitment from family members and a great amount of trust invested in the researchers. Moreover, gathering an appropriate sample size in physical measurement surveys is a very challenging endeavor. Despite the known hindrances of the discussed research methods, it must be noted that acquiring reliable and detailed information on household food waste is of great importance [22]. A recent study argues that physical measurements for data collection instead of self-reporting are preferred [23], because of their higher reliability.

In the case of Hungary, the only study based on the physical measurement of household food waste was conducted in 2016 [24]. The measurement involved 100 households during a one-week period. Besides solid food items, liquid waste was also measured, which was later considered to be an essential element of food waste accounting [6]. Based on the results of this study, 68.04 kg of food were wasted in average by a Hungarian person annually, and out of it 33.14 kg would have been avoidable. Similar results have been found by a Greek study, involving 101 urban households [15]. After the 14-day measurement period, the assessment of the total per capita food waste resulted in 76.1 kg, of which 25.9 kg considered as avoidable. The proportion of the avoidable part was proven to be relevant in Finland too, 23 kg per person annually [12]. The differentiation between avoidable and unavoidable food waste is recommended and applied in the vast majority of measurements [11,12,16,17,23,25,26].

The composition of household food waste varies from country to country. However, it is found that generally the most perishable food items are thrown away the most frequently [27]. With respect to the avoidable category, these food items include fresh fruits and vegetables, bakery products, and dairy products. According to the previous Hungarian study, the main types were meals, bakery products, fresh vegetables, dairy products, and fresh fruits [24]. In Serbia, though in reverse order, bakery products and ready-to-eat food items were on the top of the list as well [28]. Bread and bakery products were observed to have the highest ratio in Norway as well, based on a waste composition analysis [14]. Contrastively, fresh vegetables and drinks have been found to be the most prevalent types in the UK [11]. Vegetables were observed to be the most commonly wasted food items in Denmark, Greece, and Israel [13,15,16].

In order to reach the Sustainable Development Goal 12.3, halving the amount of food waste by 2030 [29], it would be essential to determine the magnitude of the problem that we face. According to a recent study, 15-16% of the total environmental impact of the food supply chain is derived from food waste [30]. It can also be stated that food wastage is in constant increase (especially in developed countries), and the actual quantity of food waste seems to be twice the amount of the results from previous global estimates [31]. Physical measurements, especially in households—a sector that accounts for the most significant part of food waste in the food chain—play a central role in refining national statistics [24]. However, this poses a particular challenge, since the lifestyle, consumption trends, and purchasing habits are in continuous change [32].

Recently, the importance of food waste measurement has become a major concern also at the European Union legislative level. Until 2020, the member states have to integrate a food waste reduction strategy into the national waste reduction plan. Furthermore, food waste prevention campaigns have to be established at the national level [33]. The European Commission has recently issued a decision establishing a common methodology and determining the minimum quality requirements for food waste data collection [34,35]. The national food waste reduction strategies should be based on the actual numbers, which have to be updated every four years. Measures aiming at food waste reduction may pose an elevated health risk to consumers (for instance eating expired food or offering it to charity, feeding potentially infectious food to livestock or companion animals). Therefore, communication campaigns, governmental institutions, and business organizations should handle food safety questions as a priority during food waste reduction efforts [36].

The aim of this research was to estimate the amount of food waste that an average Hungarian household generates based on a one-week period. This investigation was a replication of a household food waste measurement conducted in 2016 with the same methodology, within the boundaries of the Wasteless (Maradék nélkül) food waste prevention campaign, organized by the National Food Chain Safety Office in Hungary.

2. Materials and Methods

2.1. Theoretical Background

Household food waste measurement is an important step in the global efforts to reduce food losses [37]. However, a sound theoretical background, experience in practical organization, and comparable research data are scarce in this field. From the 1990s, a methodological evolution can be observed in this field, which allowed us to apply a standardized research methodology in 2016 and in our actual study (Figure 1).

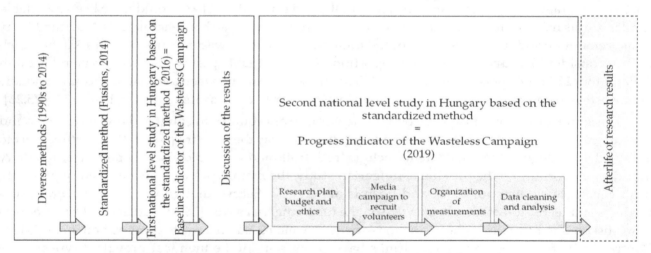

Figure 1. Theoretical framework and organization of the study.

2.1.1. Diverse Methods (1990s to 2014)

Despite the growing number of studies carried out to investigate the quantity of household food waste, the diversity of methodological approaches that different researchers applied limited the comparability of most of the findings, dating back to the 1990s [18], which might have resulted in different interpretations of the research data. The following methods were used most frequently [38]:

• Questionnaire-based survey
• Food waste diary
• Waste composition analysis
• Interview
• Mass and energy balance
• Statistics from authorities or waste management companies

Besides the basic methodological approach itself, a further significant bias resulted from small differences in the considerations applied by the research teams. For instance, the recorded material streams and their classification were also divergent. Typical differences found in the literature:

- Inedible part of foodstuffs are recorded / not recorded [3]
- Differences in the definition of the avoidable (in some studies: edible) food waste category [16]
- Potentially avoidable food waste is classified / not classified as an individual category [26]
- Liquids are measured / not measured [12,15,17]
- Food waste going to a valorization stream (e.g., composting, feed) are not recorded / recorded [5, 12,17]
- Food made for human consumption but ultimately fed to animals is classified / not classified as food waste [39]
- The part of food eaten that exceeds physiological nutritional needs of a person is recorded / not recorded as food waste [40]
- Food packaging is recorded / not recorded [41]
- Different measurement tools and units that could not be reliably converted to mass units [42,43]
- Different observation units (e.g., per person, per household) [13–15,17,20,24]

2.1.2. Standardized Method (FUSIONS, 2014)

Considering these barriers, the FUSIONS project, funded by the European Commission recognized the importance of establishing a standard for measurement methodology. Between 2012 and 2016, FUSIONS consulted researchers, industrial stakeholders, and governments about the harmonization of food waste monitoring. One of the most significant deliverables of FUSIONS was a well-balanced, widely accepted standardized methodology (that we refer to as FUSIONS methodology) to measure food waste [38], which provides the opportunity to compare country-level data. The most important considerations for household food waste measurement studies, according to the FUSIONS methodology, are the following:

- Food waste should be recorded in a diary by household members
- A description of the food or drink waste should be given in the diary
- The amount of waste should be measured and recorded
- Throughout the fieldwork, it is suggested that the researcher maintains regular contact with each household to resolve any issues, encourage participation, and ensure the accurate completion of the diary
- The sample needs to include at least a few hundred households

2.1.3. First National-Level Study in Hungary Based on the FUSIONS Methodology

In 2015 the National Food Chain Safety Office of Hungary (Nébih) started a public awareness campaign on household food waste prevention, called "Wasteless", which received support in 2016 from the LIFE Framework program of the EU. Influential European food waste prevention projects were: the EU-FP7 funded FUSIONS and Horizon 2020 funded "Refresh", the "Love Food Hate Waste" campaign by WRAP, and the "Every Crumb Counts" campaign organized by different entities from the industry, academics, and NGOs. Wasteless aimed to decrease avoidable food waste in Hungarian households. In order to measure the efficiency of the communication efforts, a baseline was set based on the FUSIONS recommendations. This has been the first empirical study carried out with this methodology in the Central and Eastern European region [24]. Focus group interviews and a preliminary, representative quantitative consumer survey covering 1000 respondents were conducted to receive research insights before the first household study.

2.1.4. Discussion of the Experiences of the First Study

The results of the survey were published in national and international academic journals and discussed at national and international platforms (including the EU Platform on Food Losses and Food Waste, operated by the Directorate-General for Health and Food Safety of the EU Commission) and scientific conferences [24,37,44–47]. Public dissemination activities through the Wasteless communication campaign were also conducted, presenting the results to consumers. Until 2018 the campaign managed to achieve a reach of almost 50 million via different media platforms [37], and the research results are widely cited by journalists even today.

2.1.5. Second National-Level Study in Hungary Based on the FUSIONS Methodology

A second study for household food waste measurement using the FUSIONS methodology took place in 2019. The research was organized to deliver comparable data to monitor changes in food waste quantities and composition. According to the analysis of the literature, this study has most probably been the first one to repeat a previous measurement based on the FUSIONS methodology. Although the research was not conducted as a validation action, its experiences are still valuable to indicate the replicability of the methodology.

The research was advertised in a media campaign in October 2019. The households were selected on the basis of voluntary registration. The advertisement included radio interviews (6); television interviews (3); press releases (7); online press releases (45). In total, 200 households applied to participate in the research. The sampling was conducted in November and December of 2019. The sampling period avoided all national and religious holidays.

During the measurement period, households received a kitchen scale (accuracy in grams) and a measuring glass from the research team. The measuring glass served for liquid wastes. Participants were also provided with a manual which described the scope of the research and explained clearly the difference between avoidable and unavoidable food waste. The following definitions were used for avoidable, unavoidable, and potentially avoidable food waste [11]:

- Avoidable—food and drink thrown away that was, at some point prior to disposal, edible (e.g., slice of bread, apples, meat)
- Unavoidable—waste arising from food or drink preparation that is not, and has not been, edible under normal circumstances (e.g., meat bones, egg shells, pineapple skin, tea bags)
- Potentially avoidable: food and drink that some people eat and others do not (e.g., bread crusts), or that can be eaten when a food is prepared in one way but not in another

For the whole duration of the investigation, e-mail and telephone support was provided to the participants. Attention was paid to reduce bias by explaining participants that the data collection was conducted anonymously, which impeded the identification of the household. The data was administered into a unified waste log. Participating households could use an online platform or a printed sheet for this purpose. The weight and exact type of each unit of food waste had to be recorded for one week (seven days). This time frame is assumed to be long enough to reduce influences caused by the participants' compulsion to conform, originating from the fact of observation. Data on solid and liquid waste were documented in mass (grams). Weighing had to be performed prior to disposal.

A supporting unit was set up to help participants of the household measurement study taking care of:

- Procurement of the tools for the measurement (scales and measuring glasses)
- Compilation of the questionnaire and log-book structure, maintaining the online platform for online data registration, and printing the hard copy version of the diary
- Receiving volunteer applications and answering the questions of the prospective households
- Distributing measurement tools, guides, and diaries to the participants
- Operating a call center for the participants to resolve any issues that arose during the measurement
- Online messages to keep contact with participants and maintain motivation

A total of 200 households applied for participation in the survey during the media campaign, but only 165 households provided a reliable and complete data set for the analysis. The most common reasons for excluding households were interrupted communication, logs not returned by the participants, an inappropriate or non-consecutive data recording period, missing data in the logs, or inadequate details on food types that made the classification impossible.

Although there have been several studies on the topic with similar or even smaller sample sizes ([15] n = 101; [16] n = 192; [19] n = 61; [41] n = 61; [48] n = 13), a post-hoc calculation was performed to determine the power of the test, for which data from the previous Hungarian study [24] served as basis. The power of the test proved to be 87.5%, which meets the criterion of being above 80% [49–51].

Based on the household data, a detailed classification was made by the research team (e.g., meals, bakery products, fresh vegetables, etc.). Each item had to be recorded by the participants as accurately as possible (e.g., not 'bread' but 'bread crust', not 'chicken' but 'chicken skin', etc.). Avoidable and unavoidable food waste units were recorded separately by the households. Subsequently to the measurement, the research team conducted a post-hoc validation of the categorization performed by the consumers.

After data cleaning, the analysis was conducted for each household. Data on food waste per person were calculated in each food waste category. The results were extrapolated to one year by multiplying the results of the one week by 52.

2.1.6. Afterlife of Research Results

Planning the afterlife of the research data has been a part of the theoretical framework. This inspection is considered to be an element in a time series study, with the objective of providing information to policy makers on a regular basis and also helping to optimize public awareness-raising activities. Since the reporting on the food waste situation became a compulsory activity of all EU member states [35], the study will also serve to deliver national level data to the EU Commission with respect to Hungarian households. Experiences with the FUSIONS methodology are still scarcely available, and therefore sharing the findings of this research may help other research teams to start their own activities in this field.

2.2. Sample Description

The socio-demographic characteristics were collected by an initial questionnaire at the households (Table 1). This characterization covered the geographical location, the size and level of income of the household, and the age, sex, and qualification of the respondent. The income level of the household had to be estimated subjectively by the inhabitants, compared to the average income level in Hungary. A total of 200 households applied, and 165 households provided appropriate data sets. These 165 households represented 452 consumers in total.

Table 1. Socio-demographic composition of the sample.

Sex of respondents generally responsible for food purchasing	Sample
Female	83.03%
Male	16.97%
Regions (NUTS1)	
Central Hungary	42.42%
Transdanubia	32.73%
Great Plain and North	24.85%
Geographical location	
Capital city (Budapest)	27.88%
Other city	55.76%
Village	16.36%

Table 1. *Cont.*

Household size	
1	13.94%
2	33.94%
3	26.06%
4	17.58%
≥5	8.48%
Average household size	2.7 people
Age of the person generally responsible for food purchasing	
Under 30 years	11.52%
Between 30 and 39 years	20.61%
Between 40 and 59 years	50.91%
Above 60 years	16.97%
Qualification of the person generally responsible for food purchasing	
Elementary	6.0%
High school graduation	38.2%
Higher education	61.2%
Income level of the household	
Low	1.21%
Below average	14.55%
Average	65.45%
Above average	16.97%
Very high	1.82%

Number of households: 165; Total participants: 452.

3. Results

During the measurement period, the 165 participant households generated 532.76 kg of food waste (Table 2). After extrapolation of the one week data per capita, the total per capita food waste was estimated to be 65.49 kg annually (Table 3).

Table 2. Quantity of food waste categories.

Food Waste Categories	Amount of Waste of 165 Households during the One-Week Period (kg)		
	Solid	Liquid	Total = Solid + Liquid
Unavoidable food waste	239.33	7.19	246.52
Potentially avoidable food waste	17.40	3.27	20.67
Avoidable food waste	218.27	47.29	265.56
Total food waste	475.00	57.75	532.75

Table 3. Estimation of annual household food waste generation per capita.

Food Waste Categories	Extrapolated Data for One Year per Capita (kg)		
	Solid	Liquid	Total = Solid + Liquid
Unavoidable food waste	29.91 (51.08%)	0.90 (12.99%)	30.81 (47.04%)
Potentially avoidable food waste	2.30 (3.93%)	0.42 (6.06%)	2.72 (4.16%)
Avoidable food waste	26.35 (45.00%)	5.61 (80.95%)	31.97 (48.81%)
Total food waste	58.56 (100.00%)	6.93 (100.00%)	65.49 (100.00%)

The proportion of the avoidable fraction (which means the real wastage) was the highest, with 48.81% (31.97 kg annually) of the total food waste. The unavoidable food waste category accounted for 47.04% (30.81 kg annually). The most prevalent food types within this category were coffee grounds, inedible fruit and vegetable parts (inedible peels such as banana peels, citrus fruit peels, onion peels, and inedible stalk, woody parts, seeds, shell of nuts, etc.), bones, eggshell, and teabags. The potentially avoidable part, such as edible fruit and vegetable peels (e.g., apple, pear, cucumber, tomato, mushroom, zucchini), chicken skin, bread crust, greasy pieces of meat, pickling liquid, juices of canned food, and oil of canned fish represented the smallest proportion, with 4.16% (2.72 kg annually). Liquid food, such as soft drinks, coffee, tea, and soup appeared in the logs in significantly lower amounts compared to solid foodstuffs (Table 3).

Eighteen different food categories were defined based on the recorded elements within avoidable food waste. Results suggest that consumers discard perishable foodstuffs more frequently and in larger quantities than durable products. Table 4 shows the mass and proportion of the established food waste categories.

Table 4. Quantity and proportion of food categories within avoidable food waste based on the one-week measurement in 165 households.

Avoidable Food Waste	Total Weight	Proportion (%)
Meals (home-made and ready-to-eat)	118.80	44.74
Bakery products	46.79	17.62
Fresh vegetables	26.55	10.00
Dairy products	21.12	7.95
Fresh fruits	20.03	7.54
Mineral water, soft drinks, coffee, tea	6.93	2.61
Canned foods, pickles	5.37	2.02
Processed animal products	4.77	1.80
Grain products (flour, semolina, oat)	4.25	1.60
Raw meat	4.06	1.53
Sauces, toppings (ketchup, mustard, salad dressings) mayonnaise)	2.87	1.08
Marmalades, jams	1.37	0.52
Confectionery, snacks	1.20	0.45
Yeast, muesli, corn flakes, raisins, puffed rice, baking mixtures	0.77	0.29
Eggs	0.49	0.19
Frozen meats, vegetables	0.10	0.04
Fats (butter, margarine, lard, etc.)	0.08	0.03
Packed spices (rosemary, marjoram, parsley, etc.)	0.01	0.01
Total	265.56	100.00

Meals (including homemade and ready-to-eat) account for the highest proportion (44.74%) of the total avoidable food waste. Bakery products (typically breads and bread rolls) are the second most common food items in this category (17.62%). Fresh vegetables and fresh fruits have also significant proportions, 10.00% and 7.54 %, respectively. Dairy products ranked in the fourth place of the list (7.95%). These five product categories altogether represent almost 88% of the total avoidable food waste.

Based on the actual data collected from the households, the quantity of annual avoidable wastage in the particular food categories has been estimated with extrapolation (Figure 2).

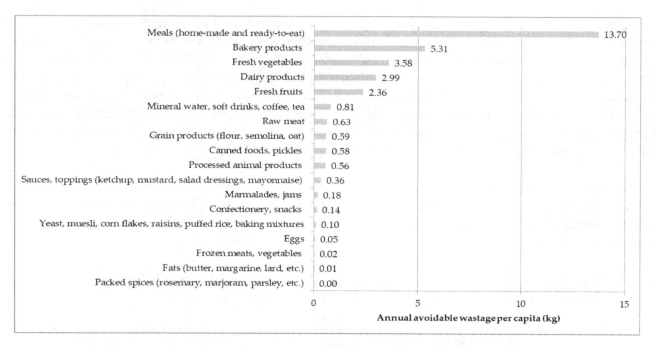

Figure 2. Annual avoidable wastage in different food categories per capita.

4. Discussion

Due to the role of consumers in food wastage in economically developed countries, gathering actual data on food waste generation from households is a major issue and a prerequisite for successful prevention campaigns [22]. Since the reliability of physical measurements is higher [23], our research placed the focus on determining the exact quantities of food waste that Hungarian households generate. The study illustrates several similarities and also some contradictions when compared to the international literature.

The actual study is a replication of the first Hungarian measurement-based research conducted in 2016 [24], which we regarded as a baseline for comparison. A decrease of 4% in the annual food waste per capita estimation was observed between the two periods (65.49 kg compared to 68.04 kg from 2016). The constitution of food waste in the households was found to be very similar, indicating a dominant rate of avoidable food waste (an actual share of 48.82% compared to 48.70% in 2016), followed by the unavoidable part (47.05% compared to 47.13% in 2016). Potentially avoidable food waste remained under 5% in both periods. (Table 5). Although the second household food waste measurement study was not organized and not suitable to be a validation of the replicability of the FUSIONS methodology [38], the results suggest that it might deliver reliable data for decision makers.

Table 5. Proportion of food waste categories (actual data compared to the findings of the 2016 study [24]).

Food Waste Categories	2019	2016
Unavoidable food waste	47.05%	47.13%
Potentially avoidable food waste	4.15%	4.16%
Avoidable food waste	48.82%	48.70%
Total food waste	**100.00%**	**100.00%**

As already discussed, a number of methodological differences can be observed in the international literature that limits the comparability of the studies [16,26]. The proportion of avoidable food waste observed by the two Hungarian studies is higher than in Greece, where 25.9 kg from the total 76.1 kg household food waste (n = 101) was reported based on the diary method [15]. In Finland, a significantly lower amount of avoidable food waste was also found (23 kg in total) [12]. Similarly, a lower proportion

(35%) of avoidable waste was noted in Sweden, as a result of a waste composition analysis with 486 households [52], in which the number of individual consumers remained unknown. The differences might partially be explained by the fact that the Hungarian studies have also recorded liquids besides solid food waste, and their contribution to the avoidable part was found to be significant. However, an Italian study—applying the diary method with 388 families—demonstrated that a person generates 27.5 kg of avoidable food waste annually in Italy, including liquids [17]. They also pointed out that, based on the studies that they had reviewed, the average mass of edible food waste was between 27.5 and 33 kg. The diary method was also employed in the UK, in a one-week measurement survey involving 13 households [48]. The study revealed that the total measured food waste (including liquids) was 0.199 kg per person per day. This would be equal to 72.63 kg annually, which is higher, but still similar to our results. In Israel 573 g/day per capita of food waste was measured involving 192 households, based on a new method of physical measurement [16]. Scaling up this outcome to one year, we would get 209 kg, which is much higher compared to the above-mentioned studies. A further hindrance to comparing international food waste research data are the units that researchers use to present their results. While some studies—including the present paper—provide the amount of generated food waste "per person" [12,15–17,48], others presented their results "per household" [13,14,17,20,41]. The one-week long household food waste measurement study in Israel, applying waste sorting analysis, resulted in 3.012 kg avoidable food waste per household per week (without liquids) [20]. In Denmark, the outcomes of a waste composition analysis (involving 1474 households) showed that a household generates 183 kg of food waste, of which the avoidable food waste was 103 kg per household per year [13].

Parfitt et al. made the general observation that perishable food items are the most frequently discarded ones [27], which has been confirmed by this paper as well. Results indicate that meals, bakery products, fresh vegetables, dairy products, and fresh fruits are the most frequently discarded food types within the avoidable part. This also correlates with the outcomes of the first Hungarian study [24], which indicated the same order. However, the share of meals appeared to be higher (44.73%) than in the previous study (40.08%), which suggests that Hungarian consumers' cooking habits and leftover storage practices should be addressed in food waste prevention awareness campaigns. Similarly to our study, bakery products and ready-to-eat food items were found to be the most prevalent food waste types in Serbia [28]. In contrast, fresh vegetables were on the top of the list in the UK, Greece, Denmark, and Israel [11,13,15,16,52]. The discovered differences in the composition of household food waste may be explained by the eating habits of different nations. It has been assumed that the season when the survey period is conducted also affects the composition of discarded food types [24].

Concerning the limitations of the general interpretation of the data, it has to be mentioned that the recruitment of households was challenging, as well as achieving participant engagement and constant activity, even for a period as short as seven days. An additional potential source of error is the fact that, since the research team was not capable of offering a financial incentive, enthusiastic participants with a more conscious behavior regarding food handling were more likely to get involved in the study [12]. However, we assume that the other family members compensated this behavior to some extent. Furthermore, in the measurement period, participants might have changed their general wasting behavior and become more conscious, knowing that they were being observed, which could have resulted in lower amounts of weighed total food waste [15]. It could also have happened that, due to the compulsion to conform, originating from the fact of observation, some participants did not record all of the discarded food items. On the other hand, the duration of the survey could be long enough to alleviate this phenomenon to some extent. An additional limitation of the study was that income level is difficult to assess, compared to other socio-demographic parameters [53]. Asking the respondents' income directly would have been indiscreet, and therefore a subjective ordinal scale was applied in the questionnaire. All of the mentioned limitations could result in an underestimation of the actual food waste amount generated in the household. However, by the application of the standardized methodology developed by FUSIONS, research data from different countries could be compared, and the possibility to observe tendencies in the different countries was also provided.

The importance of a common methodology is especially highlighted by the recent changes in EU legislation. According to the latter, all member states should conduct food waste measurements in the food chain covering agriculture, food processing, retail, catering, and households from 2020. Considering the major contribution of households to the total food waste production, this subject is expected to be a focus area in most of the countries during the next years. The EU Platform on Food Losses and Food Waste made the recommendation that in implementing national strategies to prevent food waste, member states should make full use of the latest findings of behavioral science research [37]. The experiences of the first academic studies in this field have certainly provided important input for the officers responsible for national level food waste measurements and preventive campaigns. Even more importantly, time series studies will help decision makers and communication experts to measure the performance of public awareness campaigns within a few years.

As a conclusion, the results of this research point out that a very significant part of food waste could be avoided. It is too early to reliably estimate the results of public campaigns before having longer time series data, but the first replication study using the FUSIONS methodology provided promising results. The 4% drop in avoidable food waste in a three-year period is notable, especially in a period of economic expansion. However, the fact that the list is still dominated by meals (incorporating the highest level of energy and natural resources amongst all food types) suggests that future public awareness campaigns should aim at this subject more efficiently.

Based on the results, some policy recommendations could be also formulated. Fostering national-level food waste measurement research with standardized methodologies is essential to establish the fundamentals of effective governmental interventions and to allow for the international benchmarking of preventive actions. It is also clear that food waste is a problem, which will persist for a long period in developed countries. A very intense campaign, which was observed in Hungary between the dates of the two studies, could contribute to a 4% drop in food waste, with all its novelty to society. While awareness raising should be continued, a major effort has to be placed on the integration of food waste prevention principles into children education. To be sure, we will never reach a zero level of avoidable food waste, but changing the habits of the new generation of food consumers can bring profound and long-lasting changes.

Author Contributions: G.K. and D.S. conceived the study; A.D. contributed to the data curation and formal analysis; G.K. contributed to funding acquisition; A.D. and A.K. contributed to the investigation, G.K. and D.S. contributed to the methodology and the supervision; G.K., A.D., and A.K. wrote the original draft; D.S. contributed to the review and editing of the manuscript. All authors have read and agreed to the published version of the manuscript.

References

1.　Food and Agricultural Organization. Global Food Losses and Food Waste. 2011. Available online: http://www.fao.org/3/a-i2697e.pdf (accessed on 14 February 2020).

2.　FUSIONS. Estimates for European Food Waste Level. 2016. Available online: https://www.eufusions.org/phocadownload/Publications/Estimates%20of%20European%20ood%20waste%20levels.pdf (accessed on 14 February 2020).

3.　Schneider, F. Review of food waste prevention on an international level. *Waste Res. Manag.* **2013**, *166*, 187–203. [CrossRef]

4.　Do Carmo Stangherlin, I.; de Barcellos, M.D. Drivers and barriers to food waste reduction. *Br. Food J.* **2018**, *120*, 2364–2387. [CrossRef]

5.　Quested, T.E.; Parry, A.D.; Easteal, S.; Swannell, R. Food and drink waste from households in the UK. *Nutr. Bull.* **2011**, *36*, 460–467. [CrossRef]

6. Corrado, S.; Caldeira, C.; Eriksson, M.; Hanssen, O.J.; Hauser, H.E.; van Holsteijn, F.; Liu, G.; Östergren, K.; Parry, A.; Secondi, L.; et al. Food waste accounting methodologies: Challenges, opportunities, and further advancements. *Glob. Food Secur.* **2019**, *20*, 93–100. [CrossRef]

7. Jorissen, J.; Priefer, C.; Brautigam, K.R. Food waste generation at household level: Results of a survey among employees of two European research centers in Italy and Germany. *Sustainability* **2015**, *7*, 2695–2715. [CrossRef]

8. Lorenz, B.A.S.; Hartmann, M.; Langen, N. What makes people leave their food? The interaction of personal and situational factors leading to plate leftovers in canteens. *Appetite* **2017**, *116*, 45–56. [CrossRef]

9. Ponis, S.T.; Papanikolaou, P.A.; Katimertzoglou, P.; Ntalla, A.C.; Xenos, K.I. Household food waste in Greece: A questionnaire survey. *J. Clean. Prod.* **2017**, *149*, 1268–1277. [CrossRef]

10. Stefan, V.; van Herpen, E.; Tudoran, A.A.; Lähteenmäki, L. Avoiding food waste by Romanian consumers: The importance of planning and shopping routines. *Food Qual. Prefer.* **2013**, *28*, 375–381. [CrossRef]

11. Quested, T.; Johnson, H. *Household Food and Drink Waste in the UK*; Final Report. WRAP: Banbury, UK, November 2009. Available online: www.wrap.org.uk/sites/files/wrap/Household_food_and_drink_waste_in_the_UK_-_report.pdf (accessed on 14 February 2020).

12. Koivupuro, H.K.; Hartikainen, H.; Silvennoinen, K.; Katajajuuri, J.M.; Heikintalo, N.; Reinikainen, A.; Jalkanen, L. Influence of socio-demographical, behavioural and attitudinal factors on the amount of avoidable food waste generated in Finnish households. *Int. J. Consum. Stud.* **2012**, *36*, 183–191. [CrossRef]

13. Edjabou, M.E.; Petersen, C.; Scheutz, C.; Astrup, T.F. Food waste from Danish households: Generation and composition. *Waste Manag.* **2016**, *52*, 256–268. [CrossRef]

14. Hanssen, O.J.; Syversen, F.; Stø, E. Edible food waste from Norwegian households—Detailed food waste composition analysis among households in two different regions in Norway. *Resour. Conserv. Recycl.* **2016**, *109*, 146–154. [CrossRef]

15. Abeliotis, K.; Lasaridi, K.; Boikou, K.; Chroni, C. Food waste volume and composition in households in Greece. *Glob. Nest J.* **2019**, *21*, 399–404. [CrossRef]

16. Elimelech, E.; Ofira, A.; Eyal, E. What gets measured gets managed: A new method of measuring household food waste. *Waste Manag.* **2018**, *76*, 68–81. [CrossRef]

17. Giordano, C.; Alboni, F.; Falasconi, L. Quantities, determinants, and awareness of households' food waste in Italy: A comparison between diary and questionnaires quantities. *Sustainability* **2019**, *11*, 3381. [CrossRef]

18. Bräutigam, K.R.; Jörissen, J.; Priefer, C. The extent of food waste generation across EU-27: Different calculation methods and the reliability of their results. *Waste Manag. Res.* **2014**, *32*, 683–694. [CrossRef]

19. Parizeau, K.; von Massow, M.; Martin, R. Household-level dynamics of food waste production and related beliefs, attitudes, and behaviours in Guelph. Ontario. *Waste Manag.* **2015**, *35*, 207–217. [CrossRef]

20. Elimelech, E.; Ert, E.; Ayalon, O. Exploring the Drivers behind Self-Reported and Measured Food Wastage. *Sustainability* **2019**, *11*, 5677. [CrossRef]

21. Van Herpen, E.; van der Lans, I.A.; Holthuysen, N.; Nijenhuis-de Vries, M.; Quested, T.E. Comparing wasted apples and oranges: An assessment of methods to measure household food waste. *Waste Manag.* **2018**, *88*, 71–84. [CrossRef]

22. Corrado, S.; Sala, S. Food waste accounting along global and European food supply chains: State of the art and outlook. *Waste Manag.* **2018**, *79*, 120–131. [CrossRef]

23. Schanes, K.; Dobernig, K.; Gözet, B. Food waste matters-A systematic review of household food waste practices and their policy implications. *J. Clean. Prod.* **2018**, *182*, 978–991. [CrossRef]

24. Szabó-Bódi, B.; Kasza, G.; Szakos, D. Assessment of household food waste in Hungary. *Br. Food J.* **2018**, *120*, 625–638. [CrossRef]

25. Schneider, F.; Obersteiner, G. Food waste in residual waste of households—Regional and socio-economic differences. In Proceedings of the Eleventh International Waste Management and Landfill Symposium, Sardinia, Italy, 1–5 October 2007; pp. 469–470.

26. Lebersorger, S.; Schneider, F. Discussion on the methodology for determining food waste in household waste composition studies. *Waste Manag.* **2011**, *31*, 1924–1933. [CrossRef]

27. Parfitt, J.; Barthel, M.; MacNaughton, S. Food waste within food supply chains: Quantification and potential for change to 2050. *Philos. Trans. R. Soc. B Biol. Sci.* **2010**, *365*, 3065–3081. [CrossRef]

28. Djekic, I.; Miloradovic, Z.; Djekic, S.; Tomasevic, I. Household food waste in Serbia–Attitudes, quantities and global warming potential. *J. Clean. Prod.* **2019**, *229*, 44–52. [CrossRef]

29.   United Nations. *Transforming Our World: The 2030 Agenda for Sustainable Development*; United Nations: New York, NY, USA, 2015.

30.   Scherhaufer, S.; Moates, G.; Hartikainen, H.; Waldron, K.; Obersteiner, G. Environmental impacts of food waste in Europe. *Waste Manag.* **2018**, *77*, 98. [CrossRef]

31.   van den Bos Verma, M.; de Vreede, L.; Achterbosch, T.; Rutten, M.M. Consumers discard a lot more food than widely believed: Estimates of global food waste using an energy gap approach and affluence elasticity of food waste. *PLoS ONE* **2020**, *15*, e0228369. [CrossRef]

32.   Oláh, J.; Zéman, Z.; Balogh, I.; Popp, J. Future challenges and areas of development for supply chain management. *LogForum* **2018**, *14*, 127–138. [CrossRef]

33.   European Union. Directive (EU) 2018/851 of the European Parliament and of the Council of 30 May 2018 Amending Directive 2008/98/EC on Waste. Available online: https://eur-lex.europa.eu/legal-content/EN/TXT /PDF/?uri=CELEX:32018L0851&from=EN (accessed on 14 February 2020).

34.   European Commission. Commission Delegated Decision (EU) 2019/1597 of 3 May 2019 Supplementing Directive 2008/98/EC of the European Parliament and of the Council as Regards a Common Methodology and Minimum Quality Requirements for the Uniform Measurement of Levels of Food Waste. Available online: https://eur-lex.europa.eu/legal-content/EN/TXT/PDF/?uri=CELEX:32019D1597&from=en (accessed on 14 February 2020).

35.   European Commission. Commission Implementing Decision (EU) 2019/2000 of 28 November 2019 Laying Down a Format for Reporting of Data on Food Waste and for Submission of the Quality Check Report in Accordance with Directive 2008/98/EC of the European Parliament and of the Council. Available online: https://eur-lex.europa.eu/legal-content/EN/TXT/PDF/?uri=CELEX:32019D2000&from=EN (accessed on 14 February 2020).

36.   Kasza, G.; Szabó-Bódi, B.; Lakner, Z.; Izsó, T. Balancing the desire to decrease food waste with requirements of food safety. *Trends Food Sci. Technol.* **2019**, *84*, 74–76. [CrossRef]

37.   Caldeira, C.; De Laurentiis, V.; Sala, S. *Assessment of Food Waste Prevention Actions: Development of an Evaluation Framework to Assess the Performance of Food Waste Prevention Actions*; Publications Office of the European Union: Luxembourg, 2019; ISBN 978-92-76-12388-0. [CrossRef]

38.   FUSIONS. Report on Review of (Food) Waste Reporting Methodology and Practice. 2014. Available online: http://www.eu-fusions.org/index.php/download?download=7%3Areport-on-review-of-food-wast e-reporting-methodology-and-practice (accessed on 14 February 2020).

39.   Food Loss + Waste Protocol. Available online: https://flwprotocol.org/ (accessed on 3 April 2020).

40.   Blair, D.; Sobal, J. Luxus consumption: Wasting food resources through overeating. *Agric. Hum. Values* **2006**, *23*, 63–74. [CrossRef]

41.   Williams, H.; Wilkström, F.; Ottebring, T.; Löfgren, M.; Gustaffson, A. Reasons for household food waste with special attention to packaging. *J. Clean. Prod.* **2012**, *24*, 141–148. [CrossRef]

42.   Van Dooren, C.; Janmaat, O.; Snoek, J.; Schnirijnen, M. Measuring food waste in Dutch households: A synthesis of three studies. *Waste Manag.* **2019**, *94*, 153–164. [CrossRef] [PubMed]

43.   Van Geffen, L.E.J.; van Herpen, E.; van Trijp, H. Quantified Consumer Insights on Food Waste: Pan-European Research for Quantified Consumer Food Waste Understanding. 2017. Available online: https://eu-refresh.org/sites/default/files/REFRESH%202017%20Quantified%20consumer%20insight s%20on%20food%20waste%20D1.4_0.pdf (accessed on 2 April 2020).

44.   Kunszabó, A.; Szakos, D.; Kasza, G. Food waste—A general overview and possible solutions. *Hung. Agric. Res.* **2019**, *28*, 14–19.

45.   Szakos, D.; Szabó-Bódi, B.; Kasza, G. Consumer awareness campaign to reduce household food waste based on PLS-SEM behavior modeling. In Proceedings of the 7th International Conference on Sustainable Solid Waste Management, Heraklion Crete Island, Greece, 26–29 June 2019.

46.   Doma, E.; Szakos, D.; Kasza, G.; Szabó-Bódi, B.; Bognár, L. Food waste measurement and prevention in Hungarian households. In Proceedings of the 17th Annual STS Conference, Graz, Austria, 7–8 May 2018.

47.   Kunszabó, A.; Szabó-Bódi, B.; Szakos, D.; Doma, E.; Kasza, G. Education campaign based on household food waste measurement study. In Proceedings of the Reduce Food Waste Conference on Food Waste Prevention and Management, Vienna, Austria, 25–26 April 2019.

48.   Langley, J.; Yoxall, A.; Heppel, G.; Rodriguez, E.M.; Bradbury, S.; Lewis, R.; Luxmoore, J.; Hodzic, A.; Rowson, J. Food for Thought? A UK pilot study testing a methodology for compositional domestic food waste analysis. *Waste Manag. Res.* **2010**, *28*, 220–227. [CrossRef] [PubMed]

49.   Kane SP. Post. ClinCalc. Available online: https://clincalc.com/stats/Power.aspx (accessed on 31 March 2020).

50.   Rosner, B. *Fundamentals of Biostatistics*, 7th ed.; Brooks/Cole: Boston, MA, USA, 2011; ISBN 978-0-538-73349-6.

51.   Levine, M.; Ensom, M.H. Post hoc power analysis: An idea whose time has passed? *Pharmacotherapy* **2001**, *21*, 405–409. [CrossRef]

52.   Schott, A.B.S.; Andersson, T. Food waste minimization from life-cycle perspective. *J. Environ. Manag.* **2015**, *147*, 219–226. [CrossRef]

53.   Patten, M.L. *Questionnaire Research: A Practical Guide*; Routledge: Abingdon, UK, 2017; ISBN 1936523310.

7

Losses in the Grain Supply Chain: Causes and Solutions

Ákos Mesterházy [1], Judit Oláh [2,3,*] and József Popp [3,4]

[1] Cereal Research Non-Profit Ltd., 6701 Szeged, Hungary; akos.mesterhazy@gabonakutato.hu
[2] Faculty of Economics and Business, University of Debrecen, 4032 Debrecen, Hungary
[3] Faculty of Economic and Management Sciences, TRADE Research Entity, North-West University, Vanderbijlpark 1900, South Africa; Popp.Jozsef@gtk.szie.hu
[4] Faculty of Economics and Social Sciences, Szent István University, 2100 Gödölő, Hungary
* Correspondence: olah.judit@econ.unideb.hu

Abstract: Global grain production needs a significant increase in output in the coming decades in order to cover the food and feed consumption needs of mankind. As sustainability is the key factor in production, the authors investigate global grain production, the losses along the value chain, and future solutions. Global wheat, maize, rice, and soybean production peaked at 2.102 million tons (mt) of harvested grain in 2018. Pre-harvest losses due to diseases, animal pests, weeds, and abiotic stresses and harvest destroy yearly amount to about 35% of the total possible biological product of 3.153 mt, with 1051.5 mt being lost before harvest. The losses during harvest and storage through toxin contamination are responsible for 690 mt, with a total of 1.741 mt or 83% of the total newly stored grain. Limited cooperation can be experienced between scientific research, plant breeding, plant protection, agronomy, and society, and in addition, their interdependence is badly understood. Plant breeding can help to reduce a significant part of field loss up to 300 mt (diseases, toxins, water and heat stress) and up to 220 mt during storage (toxin contamination). The direct and indirect impact of pest management on production lead to huge grain losses. The main task is to reduce grain losses during production and storage and consumption. Better harvest and storage conditions could prevent losses of 420 mt. The education of farmers by adopting the vocational school system is a key issue in the prevention of grain loss. In addition, extension services should be created to demonstrate farmers crop management in practice. A 50% reduction of grain loss and waste along the value chain seems to be achievable for the feeding 3–4 billion more people in a sustainable way without raising genetic yields of crop cultivars.

Keywords: preharvest losses; postharvest losses; prevention of losses; plant breeding solution; sustainability

1. Introduction

Grain production is the basis of global food security and is indispensable for feeding the world. In 1798, Malthus argued that the global population increases more rapidly than global food supply until war, disease, or famine reduces the number of people [1]. The failure thus far of Malthus's prediction has not prevented others from promoting similar scenarios in more recent decades. For example, Paddock [2] forecasted a worldwide famine by 1975 and stated that within the short-term, it would be impossible to feed the population. Ehrlich and Ehrlich [3] predicted worldwide famine in the 1970s and 1980s due to overpopulation and urged action to control population growth. Just like Malthus [1] and Paddock [2], Ehrlich and Ehrlich [3] failed to appreciate the creativity of humanity. Romer [4] highlighted that a sustainable economy may be introduced in the future. Therefore, the green revolution introduced agricultural technologies that resulted in a doubling of grain production globally.

More than 50 years have passed since the predictions of Paddock and Ehrlich, but mass starvation has not become widespread, although the number of undernourished and malnourished people has risen.

Nevertheless, many of Ehrlich and Ehrlich's [3] arguments related to resource scarcity appear to be close to reality because there are physical limits on natural resources. This has led Diamond [5] and other experts to shift the blame from "population" to "consumption." Fifty years ago, a growing population was highlighted as the main challenge facing humanity. Today, what really matters is people's consumption and production, which creates a resource problem. The principal challenge today is not population but total world consumption—namely, the product of local population times the local per capita consumption rate and it is not even consumption itself that is the major issue—only irresponsible and excessive waste, for which efficiency is the solution. The rational use of natural resources is the only way to avoid the global collapse of food, energy, and environmental security. Now we must take another step forward. We hypothesize that the problem is deeper and needs an even wider consideration to find more effective solutions or ways that will lead us towards a significant improvement.

1962 was the year of peak population growth, with a growth rate of 2.1%. Subsequently, population growth slowed to 1.1% by 2015. Global population will continue to grow until 2100; however, the rate of growth is expected to fall gradually to 0.1% annually [6]. The world's population reached 7.8 billion in 2020 and is projected to increase to 11.2 billion by 2100 (Figure 1). Can it realistically cover its grain needs?

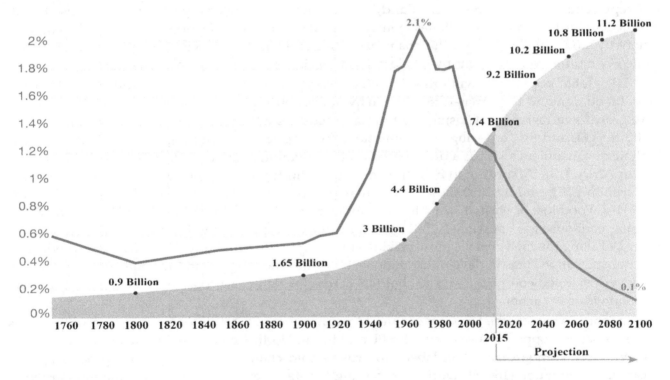

Figure 1. World population and growth rate, 1750–2015, and projections until 2100. Source: Roser and Ortiz-Ospina [7].

The goal of this study is to summarize losses along the grain value chain and identify more effective solutions. The problems at the pre-harvest and storage stages combined with mycotoxin contamination represent a very strong limitation, leading to huge losses of grains. This paper first presents the literature review on food loss (FL) and food waste (FW), followed by global grain production and losses along the grain chain including field, storage, mycotoxin contamination losses and consumer waste, a discussion of the role of integrated pest management, plant breeding, and agronomy in grain production and the implications for the future supply of grain for food and feed. Overall, this study explains the complexity of the grain value chain in order to make decisions on how best to prevent grain losses across the supply chain.

2. Literature Review

The world's population is predicted to reach 9.8 billion by 2050 and this will require an increase of 70% in food availability [6]. FL may occur due to technical limitations, such as a lack of proper storage facilities, infrastructure, packaging etc. [8]. FW is generated after the food is spoiled or expires due to poor stock management or neglect. FL takes place during the production, post-harvest, and processing stages, while food waste typically occurs at the retail and consumer levels. FL and FW can be evaluated and measured in different ways. There are several definitions of FL and FW taking into consideration pre-harvest and/or post-harvest stages, the use and destination of food, the edible part of food products, or the nutritional value of FL and FW [9]. The inclusion of pre-harvest stages to quantify food loss is essential because the pre-harvest management stage can increase the quantity and quality of tomatoes along the value chain [10]. FL and FW have negative food-security, economic, and environmental impacts. FL and FW represent the natural resources (water, land, energy) used to produce food. Reducing loss and waste throughout the food supply chain is an effective solution to mitigate the GHG emissions of agriculture and improve global food and nutrition security [8].

In 2011, the FAO estimated that annually around one-third of food produced is lost or wasted globally, which amounts to about 1.3 billion tons per year [11]. The 2030 Agenda for Sustainable Development (target 12.3) calls for halving per capita global FW at retail and consumer levels. FW represents a waste of resources (land, water, energy, soil, seeds, pesticides, etc.) used in its production; furthermore, it contributes to increasing GHG emissions. FL originates from decisions and actions by food suppliers in the value chain, excluding retailers, food service providers and consumers, while FW occurs from decisions by retailers, food service providers and consumers [12].

The global population will grow from its current 7.8 billion in 2020 to 9.8 billion in 2050 and global food demand is estimated to increase by at least 50%, but demand for protein rich products may grow even faster [6]. Closing the food gap requires a decreasing rate of growth in demand by cutting FL and FW, reducing GHG emissions from agriculture, shifting the meat-based diets of high meat consumers towards a plant-based diet, innovation and a voluntary reduction of the birth rate in Africa [13]. According to BCG the annual FL and FW has reached 1.6 billion tons, worth ca. $1.2 trillion USD, and by 2030 these figures may go up to 2.1 billion tons, worth about $1.5 trillion USD [14]. Food lost or wasted annually accounts for one-third of global food production and 8% of global greenhouse gas emissions, while over 800 million people worldwide suffer from malnutrition. FL or FW along the food chain is most striking at the beginning and the end, namely in the production and transportation stage in developing regions, while it is more typical in the retail and consumption stage in developed countries. All stakeholders across the value chain can play a crucial role in food loss and waste reduction [15].

In North America (the US, Canada, and Mexico), annual FL and FW amount to 168 million tons. There are several opportunities to address FL and FW in North America including multi-stakeholder collaboration, standardized data labels, improved cold chain management, and processing and packaging innovation [16–18]. During processing, waste is generated by inadequate infrastructure and machinery, contamination, trimming and cutting problems, confusing date labels, food safety issues and cold chain problems [16–18]. About 40% of the annual US food supply is lost and wasted, so action is required across the food supply chain, with collaboration among agencies, businesses, and communities. The United States Government Accountability Office identified three key areas – limited data and lack of awareness about food loss and waste, and limited infrastructure – which should be addressed to cut FL and FW. In 2015, the US announced a goal to reduce national FL and FW by half by 2030 [19].

FL and FW accounted for approximately 20% of food produced in the EU with a value of €143 billion in 2015 [20]. Household expenditure on food indicates how food is valued in different countries. Household income spent on food in the EU is low as a proportion of income, on average 13%. By contrast, in several African countries, almost 50% of income is spent on food. The share of FW in Europe was a few percent in the 1930s but has increased sharply since then to current global levels where one-third of food produced is lost or wasted. The relatively cheap food in the EU gives little economic incentive for consumers to avoid waste. In addition to FW, plastic waste is also a major economic, environmental, and social challenge. The overwhelming majority of plastic packaging is used only once. In the food supply chains, materials, including packaging, should be reduced, reused, and recycled in the framework of the circular economy [21].

According to a market study, up to 10% of food waste generated annually in the EU is linked to date marking; however, the market survey showed a high level of compliance. The authors conclude that FW linked to date marking can be reduced with a clear and legible date mark and consumers can make the distinction between "use by" (indicator of safety) and "best before" (indicator of quality). Nevertheless, significant FW prevention in relation to date labelling can be achieved in the dairy, fresh juices, chilled meat and fish supply chain [22]. Misinterpretation of food date labels is one of the key factors leading to FW. Date labels on food in the USA show a large variety of forms such as "use by," "best before," "sell by," and "enjoy by" dates, which are poorly understood by consumers. This paper makes recommendations on changes to the date labelling system in the USA and addresses actions needed to clarify the issue [23].

A survey conducted in Italy showed three key factors defining the extent of household FW, namely socio-demographic characteristics (household income spent on food), food shopping patterns and consumer behavior. More education and information are needed for the prevention of household FW [24]. Another study identified measures to combat FW along the food value chain in the metropolitan region of Barcelona and stressed the relevance of more research, since stakeholders oppose the introduction of new regulations and policies. Future research on the impact of new regulation including strong FW prevention measures is needed to reach a consensus and willingness among stakeholders of the food supply chain to implement new policies [25].

Conrad et al. [26] analyzed the link between FW, diet quality, nutrient waste and sustainability in the US. Higher quality diets lead to higher FW associated with greater amounts of wasted irrigation water and pesticides but less cropland waste due to the increasing consumption of fruits and vegetables included in higher quality diets, which have lower cropland and higher input needs compared to other crops. The results of the study show that simultaneously improving diet quality and reducing FW is a complex issue. Fanelli [27] found similarities in the structure of the food supply in relation to the quantity of animal-based products. The environmental impact of agriculture depends mainly on the structures of the food supply and agricultural practices applied in the different Member States of the EU. Consequently, large differences can be detected in the food supply of animal-based products and the GHG emissions intensity of the livestock sector. The farming system applied in the Member States should be based on the impact of agricultural practices on the environment to achieve a balance between livestock production and the intensity of GHG emissions. Furthermore, Fanelli [28] investigated the impact of agricultural activities on the environment in the Member States of the EU. She came to the conclusion that several Member States have similar production methods with a high impact on the environment; however, Mediterranean and northern Member States use traditional production methods including livestock grazing. Production methods with a high impact on the environment must be directed towards sustainable intensification.

The prevention of FW has become a global issue. In order to achieve this goal in developing countries a higher budget is needed for education, training and communication, technology implementation, and better infrastructure. Collaboration and dialogue between stakeholders along the food supply chain is crucial. Furthermore, data collection and comparable figures for different countries are needed, as well [29]. Ishangulyyev et al. [30] emphasized that FL and FW are complicated issues involving multi-stakeholders along the food supply chain; therefore, more research, collaboration, and awareness is needed for the prevention of FL and FW. Authors focus on awareness of the impacts of FL and FW leading to changing consumer attitudes and behaviors.

In China, Gao [19] reported an annual loss of 7.9 mt for rice, wheat, and maize, but with advanced storage management, this can be reduced substantially. In India, the annual losses are estimated at 12–16 mt; however, losses are much higher at traditional farms, where 60–70% of the harvested grain is kept in short term storage facilities [31]. According to estimates by the FAO, in India the general FL and FW is around 40%, but for cereals 30% [32]. In the developing countries almost all pre-harvest and and post-harvest operations are conducted manually, therefore post-harvest loss accounts for 15% in the field, 13–20% during processing, and 15–25% during storage [33]. Smallholder farmers generally use conventional grain storage facilities, which are not effective against insects and mold. The replacement of these traditional storage structure with improved storage systems will maintain crop quality, reduce grain losses and food insecurity [34]. Up to 50–60% of cereal grains are lost during storage due to the traditional storage structures. Modern storage structures can reduce these losses up to 98%, thereby increasing food security [35]. In Jordan, the total loss and waste along the wheat supply chain amounts to 34% associated with significant of losses of natural resources. Among postharvest losses, consumer waste ranks first, accounting for 13% [36].

3. Materials and Methods

Several methodological approaches can be used to identify household FW. Multivariate statistical techniques (descriptive statistics, principal component analysis, and hierarchical cluster analysis) were conducted to study the similarities and differences—namely the links between the structure of the food supply and the impact of agricultural practices on the environment between EU Member States [25]. Another study performed an exploratory online survey using a questionnaire adapted for studies on FW [24]. Furthermore, multivariate analyses was carried out for a comparative analysis of the environmental characteristics of the EU Member States with different agricultural systems [26].

The causes and prevention of losses along the grain supply chain is shown, based on the review of relevant literature and combining results from relevant studies and global models. Various combinations of the following terms were used to search in various papers: preharvest losses, postharvest losses, prevention of losses, plant breeding solution, sustainability. The literature on food security is already substantial; however, grain losses across the supply chain have not been addressed in detail. Furthermore, there is a lack of available publications relating to the causes of preharvest and postharvest grain losses. In addition, we also conducted supplementary searches by examining bibliographies of articles for additional references. The references of the paper mainly cover the period 2001 to 2019. References might differ in their focus on potential or realized FL and FW, their use of different baselines for comparisons, and other background conditions.

The FAO, OECD, International Grains Council, EU, and many other institutions publish serial data on grain production, use, trade, and prices. Other international sources issue estimates or data on grain losses before and after harvest. However, these data and information have not been aggregated and combined to make comparisons in order to calculate the benefits and trade-offs of grain production and losses at a global level. Data on food security related to production and losses have not been embedded into a global perspective. This paper attempts to combine all information collected to obtain a clear picture on the issue, which may serve the interests of multi-stakeholders along the grain supply chain. Other losses in the grain supply chain—losses in the conversion of feed into animal products, processing losses and over-consumption—have not been included in the calculation of grain losses and waste, and soybean losses and waste have also been excluded from our calculation.

4. Global Grain Production

Worldwide, wheat, maize, and rice are the most important cereal grains, and soybean is the major oilseed grain. The production volume and projected yield growth of grains are closely related to food security. The consumption of grains is projected to exceed supply, while supply is expected to grow at a higher rate than demand in the midterm (Table 1). Borlaug [37] also reported an increase in the deficit of grain production compared to consumption, primarily in developing countries. According to the forecast of the International International Grains Council [38], yield increase is expected to grow from 0.8% to 1.5% annually from 2013/2018 to 2020/2024.

On the other hand, the increase in consumption is projected to decrease from 2.1% to 1.0% on average in the same period. So, the gap between the supply and demand of grains is shrinking leading to a decreasing accumulated deficit of about 155 million tons (mt) between production and consumption over the period of 2017/2018–2023/2024. The consequence of the downward trend in supply is a fall in the carry-over stock level from 29% to 21% between 2017/2018 and 2023/2024, making supply in critical years more problematic. However, data on supply and demand between years may vary strongly. Grain prices are expected to be higher than in past years in both real and nominal terms due to shrinking stock levels. This means that a robust increase in production cannot be expected; therefore, we should investigate how to feed the global population and livestock. The yearly change for the next years (y/y) is projected at 1.5% for production, 1.0% for consumption and 1.5% for exports accompanied by a slow decrease of carryover stocks. For this reason, much greater attention must be given to losses along the food chain.

Table 1. Forecast for global grain production (wheat, corn, rice and soybean), 2017/2018–2023/2024 (million tons).

Total Grains	17/18	18/19	19/20	20/21	21/22	22/23	23/24	5 y Average 2013/2018	y/y Change, % 2020/2024
Production (M t)	2.102	2.089	2.120	2.149	2.181	2.213	2.246	0.8%	1.5%
Consumption (M t)	2.107	2.137	2.156	2.179	2.201	2.226	2.249	2.1%	1.0%
Exports (Jul/Jun, M t)	367	368	371	377	383	390	396	5.5%	1.5%
Carryover Stocks (M t)	614	566	530	500	480	468	465	3.2%	−3.8
y/y change	−5	−48	−36	−30	−20	−12	−3	-	-
Stocks to use ratio, %	29%	26%	25%	23%	22%	21%	21%		

Source: International Grains Council [38].

On the production side, further substantial increases in yield will be constrained. Obviously, technology, plant breeding, improving agronomy, and new production methods will resolve this phenomenon to a certain extent. Higher yielding cultivars are on the market, but their effect on yield is only moderate as the genetic capacity of these cultivars is just partly exploited. Without making long-term predictions of the global grain supply the problem described above must be approached from a different perspective. We have to focus on causes and solutions of food and feed losses along the grain supply chain.

5. Losses along the Grain Chain

The reduction of grain losses by biotic factors (pests, pathogens and weeds) is a major challenge for food supply [39]. In addition to pre-harvest losses, the losses occurring during transport, pre-processing, storage, processing, packaging, marketing and plate waste are also substantial (Figure 2). Reduction of losses results in a higher revenue than an increase in genetic yield ability [40]. Globally, an average of 35% of potential crop yield is lost to pre-harvest pests [41]. In addition to pre-harvest losses, the losses occurring during transport, pre-processing, storage, processing, packaging, marketing, and plate waste are also substantial [39]. By reducing FL and FW, food security combined with resource efficiency can be enhanced. In 2011 the European Commission set targets to halve the disposal of edible FW by 2020, and in 2012, the European Parliament also issued a resolution to halve FW by 2025 and designated 2014 as the "European Year against Food Waste" [42,43]. The problem is that the EC targeted only edible FW excluding other forms of waste, for example in feeding.

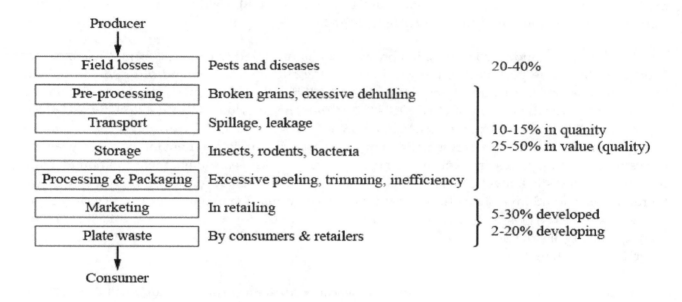

Figure 2. Losses along the food chain. Source: International Water Management Institute [44].

Each stage of the grain value chain is a source of grain losses and waste, each with a different loss ratio. The problem needs a multidisciplinary approach [45,46], but in spite of efforts we are far from the solution. Therefore, our goal was to summarize losses along the grain value chain and identify more effective solutions. The increasing loss and waste of grains reduce food security and also negatively affect sustainable development (natural resources, environment and human health) [47,48]. In addition to the actual loss and waste, resource inputs, e.g. arable land, irrigated water, fertilizer and energy, are also lost and wasted Gustavsson et al. [11] contributing to higher cost for a unit of really consumed product.

In 2018, global grain production accounted for 2.102 million tons [38]. The data we used for losses are based on estimates in the literature. However, the different sources produce comparable numbers. We used the lower estimates for increased reliability. It is well known that the loss expressed in monetary term is very high at lower quality or high toxin contamination even the amount of losses does not change much. For example, toxin contamination can cause 100% income loss at minimal yield reduction. Approximately one-third of potential crop yield is lost to pre-harvest pests, pathogens and weeds. The theoretical yield would account for 3.153 mt per annum and pre-harvest losses amounts to 1.051 million mt. Grain losses at harvest are estimated at 3% or 60 mt annually, with wide regional variations between small and large farms. Total pre-harvest and harvesting losses account for about 1.110 mt per year. In addition, 420 mt of grains are lost during storage, 210 million tons due to field mycotoxin contamination (excluding consumer's waste of 286 mt annually (Table 2). It can be concluded that one third of the possible yield is lost before harvest, another 20% is wasted due to storage and mycotoxin contamination, and only one third of the grain total production potential is really consumed, including consumer waste of around 10%).

Table 2. Losses of grains along the full value chain in 2018 (million tons).

Items	Million Metric Tons	% to Total Capacity	% to Harvest
Total production capacity	3.153	100.00	
Total harvested yield	2.102	66.67	100.00
Losses due to biotic and abiotic factors	1.051	33.33	50.00
Harvest losses (cc 3%)	60	1.90	2.85
Storage losses	421	13.35	20.03
Field mycotoxin contamination	210	6.67	9.99
Consumer's waste	286	9.07	13.61
Total loss	2.028	64.32	96.50
			0.00
Total grain consumed	1125	35.68	53.52

Source: Authors' own calculation based on the production data of International Grains Council [38].

5.1. Field (Pre-Harvest) Losses

Yield losses caused by pests, pathogens, and weeds are major challenges to crop production. Increased use of plant protection increased crop harvests from 42% of the theoretical worldwide yield in 1965 to 70% of the theoretical yield by 1990; however, at least 30% of the theoretical yield was still being lost due to ineffective pest-management methods applied in several regions of the world. Without plant protection, 70% of crop yields could have been lost to pests [41]. Actual losses were estimated at 26–30% for soybean and wheat, and 35%, and 40% for maize and rice, respectively [40]. Cramer [49] estimated crop losses of around 28% due to all pests in North and Central America. Russel (1978) cited over 50% yield losses caused by pests, pathogens and weeds worldwide. Schumann and D'Arcy [50] estimated the loss of yield caused by all pests at some 20% despite the billions of dollars spent on plant protection. According to Ubrizsy [51], in the 1960s mean yield loss for all crops caused by pests, pathogens and weeds stood at 15–20% and 25–30%, respectively. Actual losses were estimated at 36% for wheat and 38% for maize on average during the period 1950–1960 excluding yield loss caused by abiotic factors (temperature, humidity, rain, floods, etc.).

It is well known that plants would not survive a crisis-level water shortage because under a certain level of precipitation plants cannot cope with water stress. In Hungary, yield sensitivity to droughts show that maize and wheat yield reduction may reach up to 50% in drought seasons. The effects of drought play a large role in damage to crops, depending on soil quality. Drought conditions can have a moderate impact on good soil or a profound impact on sand. The yield loss in drought and dry seasons reaches one-third on average. Beyond the 35% loss due to biotic factors, we did not count extra losses for abiotic losses, as in drought years, diseases, insects, and weeds normally cause significantly less damage. The authors aimed to make a conservative estimate that is close to reality. Aflatoxin is an exception in draught and hot years, but in yield reduction it is not important.

5.2. Integrated Crop Management

Integrated crop management (ICM) is an environmentally sensitive and economically viable production system by using the latest available techniques to produce high quality food in an efficient manner [52]. Reductions in pest control costs and in the use of pesticide in ICM programs can be achieved by introducing populations of natural enemies, variety selection, applying alternative pesticides, etc. For farmers, the main benefit of ICM is still a reduction in pesticide use, although most programs still rely heavily on pesticides [39]. The main task of ICM is not to decrease pesticide costs but it contributes to the production of healthy food. The present production level is a consequence of contemporary pesticide use, but the losses that could not be prevented show that a significant development is necessary.

Bajwa and Kogan [53] listed 67 definitions for integrated pest management but did not mention resistance to pests. Ehler [54] focused on integrated pest management but also mentioned integrated pesticide management because it is not the pests themselves, but rather the use of pesticides, that should be better managed in order to reduce the occurrence of pests.

Kumar and Shivay [55] took a step forward by combining pest management with seedling establishment and nutrient management. Vanlauwe et al. [56] highlighted integrated soil fertility management, focusing on the efficiency of fertilizer use. Bottrell [57] reported on integrated pest management, while Barzman et al. [58] summarized the most important principles with the objective of optimizing the use of pesticides but did not mention resistance as a possible influencing agent, only considering pesticide reduction.

Lehoczki-Krsjak et al. [59], Mesterhazy and Bartok [60], Mesterházy [61], Mesterházy [62], Mesterházy et al. [63], Mesterházy et al. [64], and Mesterházy et al. [65] were among the first to indicate the decisive role of the resistance level increasing the effect of fungicides against Fusarium head blight in wheat. Actually, the resistance level regulates the fungicide effect. Lamichhane et al. [66] stressed the role of resistance in sustainable and low-input agricultural systems and the role of breeding cultivars with the resistance traits required for organic production. Nevertheless, in a wider context we need to integrate plant breeding, water management, and storage conditions with an emphasis on the fungi that normally cause less grain loss. It is possible to keep the whole production process under control by the introduction of Intelligent Field Crop Management (ICM).

5.3. Storage Losses

Storage plays a central role in the grain supply chain. Grain storage losses are affected by several factors, including direct and indirect losses. Direct losses are related to the physical loss of grains, and indirect losses occur due to loss in quality and nutrition [35]. Storage losses can be classified into biotic factors (insect, pests, rodents, fungi) and abiotic factors (temperature, humidity, rain). Moisture content and temperature affects storage life. For example, storage molds spread rapidly at higher temperatures and humidity. Damage and losses caused by insects and rodents can refer to physical deterioration (e.g., holes in the grain) and quality (value) loss. Huge direct and indirect storage losses are reported in developing countries. Jayas [67] estimated losses as low as 1–2% in developed countries using metal silos compared to 20–50% in developing countries where grains are generally stored by family farms in traditional storage structures. According to Manandhar et al. [34] in most developing countries up to 80% of cereals are produced by small family farms where grain losses can go up to 15% in the field, 13–20% during processing, and 15–25% during storage, giving total losses of 43–60%. Grain losses in storage account for 10 to 20% of stored products, as a result of damage caused by insects [68]. By calculating an average storage loss and damage of up to 20% of stored grain worldwide, approximately 420 mt of grains are lost during storage annually.

The storage and handling methods should minimize losses. Before storage the grain must be cleaned and contaminants (dust, insects, straw, chaff, weed seeds, etc.) removed. Furthermore, test for toxins is mandatory before storage and grains of low, medium and high contamination level must be stored separately from each other. Critical physiological factors (moisture content and temperature) affect the storability of crops because high moisture content and heat cause fungal and insect problems, therefore humidity temperature and CO_2 control is highly important. Introducing best practices for handling and storage is a cheaper solution compared to the loss of grains during storage.

5.4. Mycotoxin Contamination

Mycotoxins are toxic chemicals unsuitable for animal feed and human consumption. High concentrations of aflatoxin can pose a serious health risk both to humans and livestock. A significant concentration of toxin levels is frequently measured even during harvest, leading to the need for control measures, both pre- and post-harvest. The most important source of the mycotoxin problem is the generally high susceptibility of the grain crops. Large toxin epidemics in the fields are always consequences of an epidemic. WHO estimated that 25% of the world's crops are contaminated by mycotoxins excluding considerable preharvest losses [69]. Up to 25–40% of global cereal grains are contaminated by the mycotoxins produced by fungi Kumar et al. [70]. Dowling [71] reported, based on UNO and FAO data, that 25% of the world grain crop is significantly contaminated by mycotoxins.

McMullen et al. [72] reported that due to nearly yearly epidemics in the USA the acreage of wheat was reduced from 29 million ha (1992) to 21.4 million ha (2010). The same numbers for barley are 2.9 and 1.0 million ha. This is normally not taken into consideration. The high toxin contamination of harvested yield caused additional quality loss and a reduction in prices of 50–90%. The yield and quality losses amounted to billions of dollars. Similar losses were also recorded in epidemic years in Canada, Europe, and China. Maize is a more complicated problem. Szabo et al. [73] estimated losses in maize in 2014 at about 300 million dollars, with nearly no yield reduction. The contaminated grain (deoxynivalenol, zearalenone, fumonisins) decreased the price of harvested grain by about $32 USD/t or 25%, while the rest of the loss was attributed to animal husbandry by lower weight gain, sexual disorders, higher death rates, cost of toxin bindings and antibiotics, etc.

This means that yield loss on its own does not show the significance of the problem. The damage caused by toxins during storage represents about 10% yearly loss at a conservative estimation and removes 210 mt of grain per annum globally. This amount represents stored grains and does not include the infected and light grain part in small grains blown out by the combine at harvest, which cannot be measured, but exists. Due to toxin regulations the human population is well protected in the developed world, but this is not the case in many countries where animal husbandry the situation is similar—contaminated grains are normally used in animal husbandry In small grains cleaning systems, optical selection of infected grains can reduce toxin levels, but the cost is relatively high equivalent to a yield loss of 10–20%. In maize, however, such effective methods are in experimental phase. Most of the toxins are of field origin detectable at harvest, but bad storage conditions can cause significant increases. In order to minimize grain losses, fungi formation must be addressed during storage.

5.5. Consumer Waste

Plate waste of food is as high as 5–20%. In developing countries, FL and FW including plate waste is higher than in developed countries. FL and FW depend on technology and on consumer behavior. 1.3 billion tons of food or 1/3 of all food produced for human consumption is lost or wasted from harvest to consumption annually, without accounting for losses in livestock production worldwide [11,74]. Carrying out evidence-based FL and FW calculations still presents an open challenge. In estimating FW, the most critical research gap is related to the lack of a clear definition of FW and a harmonized FW accounting methodology [75]. In developed countries, consumers throw away 286 mt of cereal products [76]. Just taking into consideration maize, wheat, and rice, at least 200 mt of cereals are wasted by consumers per annum globally (soybean loss and waste is not included).

5.6. Breeding Versus Food Losses

Research has clarified that resistance is the most important toxin regulator [61,62,65,77,78]. However, large resistance differences occur, in wheat Fusarium head blight deoxynivalenol concentration varied between 5 mg/kg and 400 mg/kg at the higher epidemic pressure in 2001–2002 [78]. According to literature sources there is no effective means for solving toxin contamination before harvest Jans et al. [79] stressing the preharvest prevention of disease and toxin by resistance. This is a problem as this also inhibits breeding activity and creates difficulties for stakeholders. The results of the wide international literature do not support this view. Ten- to 20-fold resistance differences also exist in toxin response; therefore, this problem should be exploited.

Resistance also influences further fungicide efficiency and improves the predisposition of plants to previous crops with high pathogen population [80]. Disease and toxin forecasts will be better when resistance levels are considered [81]. Zorn et al. [82] indicated that ploughing was as effective a way to reduce deoxynivalenol (DON) as planting a more resistant variety, and in other diseases the experiences are similar. Breeding for adaptation to different soil and climatic conditions is essential.

Tolerance to acidic soils is also a breeding problem among many others. Minimum tillage and organic production needs plants that are highly resistant against the most important diseases, as in these cases the disease pressure can be significantly higher as effective fungicide are forbidden to use. Resistance to biotic and abiotic factors brings a direct and significant improvement to yield, quality stability and adaptation. Breeding for more efficient fertilizer use in order to improve the photosynthetic activity, adaptation etc. of the plants has also its place. The main problem is that the extensive knowledge available in the scientific community suffers from a bottleneck effect when it should be applied in plant breeding. Most breeding firms are small, with 1–2 breeders for a plant or less, and they lack any laboratory background or support from trained scientists. This is true also for European family companies. The large firms concentrate on high yields but often neglect food safety and other problems, so varieties with high yields often produce severe financial losses.

Unsatisfactory breeding efforts contribute to 210 mt loss due to toxin contaminated grains. Much of the storage microbes are of field origin, so lack of resistance might be partly responsible for storage losses. The devastating effect of storage microbes is characteristic when storage conditions are bad, however, most of the losses could be prevented by advanced storage technology. Storage microbes cause about 50% of storage loss, i.e. 210 million mt of grain per annum; therefore, mycotoxins of field and storage origin are treated separately because they need to be treated using different approaches.

5.7. Agronomy

The keys to the green revolution, whose father was the Nobel laurate Norman Borlaug, were improved seeds, especially the short straw lodging resistant wheat varieties giving higher yields due to a better harvest index and resistance to diseases such as rust. The breeding program was led by Norman Borlaug and supported by inputs (chemicals and fertilizer) and irrigation water [83]. The green revolution spread in all developing and developed regions and helped more than one billion people to survive. The key factor was to find a connection between breeding and agronomy that helped to exploit the greater abilities of the new varieties. This is essentially an update the basic ideas of the Green revolution by Baranski [83] adapted to present needs and balancing the negative effects of climate change.

Breeding for more yield resulted in increasing nitrogen and water dependence. We can mention possible shortage in phosphorus and potassium. Today, several breeding companies are focusing on shorter season crops, less water use, and gene editing, among other things. There is an improving efficiency of utilization of nitrogen, potassium, phosphorus, but insufficient supply of microelements is also a growing problem. The decoupling of nitrogen from yield dependence is another central research area. Nevertheless, breeding for higher yields still has priority. Wang et al. [84] summarized the possibilities for wheat, but similar patterns exist also for other crops. Pest management is also relevant in agronomy to protect grain against pests, pathogens, and weeds. Effective disease control is indispensable in epidemic years. However, the contemporary storage needs also pesticides for insect control and specific fungicides against diseases.

6. Discussion

In order to explain the possible shortage of food globally, Malthus [1] warned about overpopulation. He did not take into consideration the fact that more food required could be produced due to scientific and technological development. The forecasted mass starvation has so far not become a reality. Diamond [5] blamed consumption, especially overconsumption in the developed world. The present global food production is not sustainable and accompanied by huge losses and wastes along the value chain. For this reason, agricultural production needs a reorganization with specific local solutions for particular regions.

Half of total losses occur before the process of harvesting begins representing 1051.5 mt annually. This makes production per unit more expensive. Among the causes of crop losses, the bottleneck effect between basic, applied research, and breeding must be highlighted, which means that most of the available existing knowledge does not reach, for example, those involved in breeding, plant protection, agronomy, etc. Consequently, the effectiveness of the combination and utilization of the knowledge adapted for local and regional breeding, agronomy, etc. is very poor and the results are embarrassing. More knowledge is needed about the interrelations between disease resistance, yield ability, and the efficacy of pesticides and agronomy responses at cultivar level to develop the optimum mix of different procedures for each region and field. At least two third of grain field losses, namely 700 mt, are related to biotic stresses. Plant breeding is supposed to be responsible for about a third of this amount (233 mt). However, it is impossible to breed for resistance against 200 diseases of a crop. Plants are generally treated for the 4–5 most important diseases by farms; therefore, resistance and pesticide treatments should be combined. For the rest of the losses (roughly 500 mt) correct pest management and agronomy measures combined are needed. Suggestions made by Lamichhane et al. [66] for organic culture are also necessary for conventional plant production. Near Szeged (southern Hungary), Mesterházy et al. [64] and Mesterházy et al. [65] reorganized the crop structure and fungicide program and tillage for wheat. Without any additional cost, the wheat yield became much healthier (fewer toxins and leaf diseases), and a 4–5% yield increase could be achieved on about 2000 ha of wheat.

Grain production is expected to expand well below the growth rates of the last decades. Yield increase will not be sufficient to achieve global food security. The potential increase in maize, wheat, rice and soybean yields will be less than 10% in the next five years, resulting at the most in 220 million mt of extra yields. Comparing the projected extra yields of grains for the next five years, with the annual losses and waste of about 980 mt excluding pre-harvest losses, it becomes apparent that reducing food loss and waste can help to enhance food security more efficiently then yield increase can. The slowdown in global annual yield growth is the main challenge facing global food security. Maize, rice, wheat, and soybean together produce about 64% of global agricultural calories (of this maize, rice, and wheat 57%) and decreasing yield gains in these crops will have serious implications for the global grain supply chain [85]. Grain yields are affected by both biophysical and socioeconomic factors worldwide leading to increased yield variability in the future. Climate-change-related heat stress, scarcity of water for irrigation, depletion of soil fertility and salinization, soil erosion, pest and disease build-up and a lack of capital will have a greater impact on yield development than the genetic improvement of grain cultivars. In Europe, yields are affected by climate change in several EU member states.

Moreover, agricultural subsidies are tied to the reduction of environmental burden and agricultural inputs, leading to yield stagnation. For this reason, the reduction of loss and waste is the most important factor in achieving food security with a much larger impact on the grain supply for food and feed than higher yielding crops. Breeding for yield stability and resistance combined with advanced pest management strategies, improved agronomy and storage could reduce grain losses and waste by at least 50% and meet the demand for grain for an additional 3-4 billion people globally.

At least two third of the field losses namely 700 mt are related to biotic stresses. Plant breeding is supposed to be responsible for about a third of this amount (233 mt). However, it is impossible to breed for resistance against 200 diseases of a crop. Plants are generally treated for the 4–5 most important diseases by farms; therefore, resistance and pesticide treatments should be combined. For the rest of the losses (roughly 500 mt), both correct pest management and agronomy measures are needed.

7. Conclusions

Increasing global population and decreasing natural resources associated with growing FL and FW have resulted in an unprecedented challenge. Transforming a wasteful food supply chain into a sustainable food solution needs collaboration between researchers and multi-stakeholders in the food supply system. This conclusion is in line with several studies [16,19,29,30]. We have to understand the key drivers causing FL and FW across the food supply chain and find solutions for reducing these losses and waste. The field of FL and FW lacks appropriate metrics used to calculate benefits and trade-offs. Stakeholders in the food supply chain have incomplete data about how much FL and FW is generated or the total costs of its management to make these comparisons. This outcome is supported by other reviews as well [9,10].

The results show that cumulating harvesting, storage and toxins losses along the grain supply chain may reach up to 690 mt annually, excluding the 1.051 mt of pre-harvest losses. Besides pre-harvest losses, the highest rates of loss are associated with storage and mycotoxin contamination. Breeding, cultivation, plant protection, harvesting, storage, handling, and transportation practices play key roles in the efficiency of the grain supply chain. Grain loss in the future depends on technology and the workforce. In addition, consumer waste is based on consumer behavior and food waste management.

At present, a significant increase in global grain production only by the introduction of new higher yielding varieties is not possible. What are the limitations? Interestingly, it is not the shortage of inputs, such as fertilizers, chemicals, etc., that plays an important role in restricting increases in the yield and quality of cereals. The shortage of water is an acute problem on one side, with research on the other. Cooperation among those involved in breeding to increase yield, to improve resistance to biotic and abiotic stresses and agronomy is poor, as these fields largely work separately from each other and their positive innovative effects are not utilized to the extent that could be possible. Special problems occur in the field of toxigenic fungi. However, in recent decades screening and genetic methods have been used to increase resistance levels. The problems at the harvest and storage stages represent a very strong limitation, leading to losses of several hundred million tons of grains. In many regions of the globe poor infrastructure is also a strong limiting factor, inhibiting the application of modern logistics, installations, machines etc. The lack of special education is a very strong limiting factor causing very high losses before and after harvest. This conclusion is consistent with previous studies [24,29,30].

How do we prevent a significantly higher amount of loss? Plant breeding must consider closely global needs and local activity in order to reach the highest adaptation of the cultivars bred, because cultivars have to resist to different challenges of both biotic and abiotic stresses. We face a special problem in stresses such as drought, heat, toxic fungi, and leaf spots that inherit mostly polygenic traits and so breeding is more complicated than in the case of monogenic traits. We should be aware that each crop has about 200 pathogens, of which 4-5 can be treated in a breeding program, indicating the need for chemical control when a new disease appears in the field. A much higher resistance is needed against toxigenic fungi; therefore, crop production must be much better adapted to local conditions. The finding is in line with previous literature [28].

About 20 years ago, we spoke about integrated crop management to reduce pesticide use. Today, Intelligent Field Crop Management is spreading. It is necessary to evaluate for each field the optimal mix of variety, agronomy, pesticide application, irrigation, and previous crop selection to ensure maximum possible yield by using necessary pesticides. Harvest and storage management should be modernized. The current best available technologies reduce loss during storage by 2–3% without quality reduction. This outcome is in accordance with the studies published by [8,19,34,35].

In order to prevent grain loss farmers should be educated globally; the extension service in the US can be an example to follow. Demonstration farms can show farmers how intelligent field crop management works in practice. The vocational school system should also be adopted to meet these new challenges. The production method used in developed countries should also be applied in the developing world. Developed countries also face new problems, including the ecological crisis, therefore new solutions are needed through global action and scientific innovation. Local agricultural development programs can solve local problems, but international organizations must support and harmonize the global and local network.

Agriculture is a capital and knowledge-intensive sector, so huge agricultural investments must be made in the next few decades to meet the challenges of the growing yield demand for grain, and at the same time to maintain a sustainable environment. Therefore, reducing global losses and waste along the grain supply chain is the most effective way to increase global food and nutrition security. This needs long term thinking and not short run profit at a maximum level. For this reason, national and global regulations, investment and scientific policies are preconditions to provide reasonable and sustainable solutions for the future; however, the largest task is to change the way we think.

Author Contributions: Á.M. and J.P. conceived and designed the experiments. Á.M. and J.P. contributed analysis tools. Á.M., J.O. and J.P. wrote the paper. All authors have read and agreed to the published version of the manuscript.

Acknowledgments: The authors are indebted to the projects MycoRed FP7 (KBBE-2007-2-5-05 (2009–2012), GOP 1.1.1.-11-2012-0159 EU-HU (2012–2013), GINOP-2.2.1-15-2016-00021 (2016–2020) and TUDFO/5157/2019/ITM for financial support. This research was supported by the ÚNKP-19-4-DE-147 New National Excellence Program of the Ministry for Innovation and Technology and by the János Bolyai Research Scholarship of the Hungarian Academy of Sciences.

References

1. Malthus, T.R. *An Essay on the Principle of Population as it Affects the Future Improvement of Society, with Remarks on the Speculations of Mr. Goodwin, M. Condorcet and Other Writers*, 1st ed.; J. Johnson in St Paul's Church-yard: London, UK, 1798.

2. Paddock, W. Famine-975! America's Decision: Who Will Survive? Little, Brown and Company: Boston, MA, USA, 1967.

3. Ehrlich, P.R.; Ehrlich, A.H. The Population Bomb Revisited. *Electron. J. Sustain. Dev.* **2009**, *1*, 63–71.

4. Romer, P.M. Two strategies for economic development: Using ideas and producing ideas. *World Bank Econ. Rev.* **1992**, *6*, 63–91. [CrossRef]

5. Diamond, J. *Collapse: How Societies Choose to Fail or Succeed*; Viking Penguin/Allen Lane: New York, NY, USA; London, UK, 2005.

6. United Nations. *World Population Prospects: The 2017 Revision*; United Nations, Department of Economic and Social Affairs, Population Division: New York, NY, USA, 2017. Available online: https://www.un.org/development/desa/publications/world-population-prospects-the-2017-revision.html (accessed on 15 January 2020).

7. Roser, M.; Ortiz-Ospina, E. World Population Growth, Our World in Data. 2017. Available online: https://ourworldindata.org/world-population-growth (accessed on 15 January 2020).

8. Food and Agriculture Organization of the United Nations. *Global Initiative on Food Loss and Waste Reduction*; Food and Agriculture Organization of the United Nations: Rome, Italy, 2015; pp. 1–8. Available online: http://www.fao.org/3/a-i4068e.pdf (accessed on 20 February 2020).

9. Chaboud, G.; Daviron, B. Food losses and waste: Navigating the inconsistencies. *Glob. Food Secur.* **2017**, *12*, 1–7. [CrossRef]

10. Chaboud, G. Assessing food losses and waste with a methodological framework: Insights from a case study. Resources. *Conserv. Recycl.* **2017**, *125*, 188–197. [CrossRef]

11. Gustavsson, J.; Cederberg, C.; Sonesson, U.; van Otterdijk, R.; Meybeck, A. *Global Food Losses and Food Waste: Extent Causes and Prevention*; Food and Agriculture Organization of the United Nations (FAO): Rome, Italy, 2011; pp. 1–37. Available online: http://www.fao.org/3/mb060e/mb060e.pdf (accessed on 20 February 2020).

12. United Nations Transforming Our World. *The 2030 Agenda for Sustainable Development*; A/70/L.1; United Nations, General Assembly: New York, NY, USA, 2015; pp. 1–35. Available online: https://www.un.org/ga/search/view_doc.asp?symbol=A/RES/70/1&Lang=E (accessed on 20 February 2020).

13. Searchinger, T.; Waite, R.; Hanson, C.; Ranganathan, J.; Dumas, P.; Matthews, E. *Creating a Sustainable Food Future: A Menu of Solutions to Feed Nearly 10 Billion People by 2050*; World Resources Institute: Washington, DC, USA, 2019; pp. 1–556. Available online: https://reliefweb.int/report/world/world-resources-report-creating-sustainable-food-future-menu-solutions-feed-nearly-10 (accessed on 20 February 2020).

14. BCG Food and Agriculture Organization of the United Nations. *Global Food Losses and Food Waste*; BCG FLOW Model. 2015 Findings, in 2015 Dollars; FAOSTAT Database: Boston, MA, USA, 2018. Available online: https://www.consulting.us/news/860/global-food-wastage-could-hit-21-billion-tons-by-2030-in-staggering-crisis (accessed on 20 February 2020).

15.   Hegnsholt, E.; Unnikrishnan, S.; Pollmann-Larsen, M.; Askelsdottir, B.; Gerard, M. *Tackling the 1.6-Billion-Ton Food Loss and Waste Crisis*; The Boston Consulting Group in Collaboration with Food Nation and State of Green: Boston, MA, USA, 2018. Available online: https://www.consulting.us/news/860/global-food-wastage-could-hit-21-billion-tons-by-2030-in-staggering-crisis (accessed on 20 February 2020).

16.   CEC Technical Report. *Quantifying Food Loss and Waste and its Impacts*; Commission for Environmental Cooperation: Montreal, QC, Canada, 2019; pp. 1–129. Available online: http://www3.cec.org/islandora/en/item/11813-technical-report-quantifying-food-loss-and-waste-and-its-impacts (accessed on 20 February 2020).

17.   Popp, J.; Kiss, A.; Oláh, J.; Máté, D.; Bai, A.; Lakner, Z. Network analysis for the improvement of food safety in the international honey trade. *Amfiteatr. Econ.* **2018**, *20*, 84–98. [CrossRef]

18.   Popp, J.; Olah, J.; Fari, M.; Balogh, P.; Lakner, Z. The GM-regulation game-The case of Hungary. *Int. Food Agribus. Manag. Rev.* **2018**, *21*, 945–968. [CrossRef]

19.   Gao, D. *Food Loss and Waste*; Building on Existing Federal Efforts Could Help to Achieve National Reduction Goal; GAO-19-391 United States Government Accountability Office: Washington, DC, USA, 2019; pp. 1–47. Available online: https://www.gao.gov/assets/gao-19-391.pdf (accessed on 20 February 2020).

20.   European Commission. *Frequently Asked Questions: Reducing Food Waste in the EU*; European Commisison: Brussels, Belgium, 2019; pp. 1–5. Available online: https://ec.europa.eu/food/sites/food/files/safety/docs/fs_eu-actions_fwm_qa-fight-food-waste.pdf (accessed on 20 February 2020).

21.   Schweitzer, J.; Gionfra, S.; Pantzar, M.; Mottershead, D.; Watkins, E.; Petsinaris, F.; Ten Brink, P.; Ptak, E.; Lacey, C.; Janssens, C. *Unwrapped: How Throwaway Plastic Is Failing to Solve Europe's Food Waste Problem (and What We Need to Do Instead)*; Institute for European Environmental Policy: Brussels, Belgium, 2018; pp. 1–28. Available online: http://www.foeeurope.org/sites/default/files/materials_and_waste/2018/unwrapped_-_throwaway_plastic_failing_to_solve_europes_food_waste_problem.pdf (accessed on 20 February 2020).

22.   European Commsission. *Market Study on Date Marking and Other Information Provided on Food Labels and Food Waste Prevention, Final Report, Written by ICF in Association with Anthesis, Brook Lyndhurst, and WRAP January 2018*; Directorate-General for Health and Food Safety: Brussels, Belgium, 2018; pp. 1–100. Available online: https://ec.europa.eu/food/sites/food/files/safety/docs/fw_lib_srp_date-marking.pdf (accessed on 14 March 2020).

23.   Leib, E.B.; Gunders, D.; Ferro, J.; Nielsen, A.; Nosek, G.; Qu, J. *The Dating Game: How Confusing Food Date Labels Lead to Food Waste in America*; National Resources Defense Council: New York, NY, USA, 2013; pp. 1–61. Available online: https://cpb-us-e1.wpmucdn.com/blogs.uoregon.edu/dist/a/3266/files/2013/10/dating-game-report-1m6z9a3.pdf (accessed on 20 February 2020).

24.   Fanelli, R.M. Using causal maps to analyse the major root causes of household food waste: Results of a survey among people from Central and Southern Italy. *Sustainability* **2019**, *11*, 1183. [CrossRef]

25.   Diaz-Ruiz, R.; Costa-Font, M.; López-i-Gelats, F.; Gil, J.M. Food waste prevention along the food supply chain: A multi-actor approach to identify effective solutions. *Resour. Conserv. Recycl.* **2019**, *149*, 249–260. [CrossRef]

26.   Conrad, Z.; Niles, M.T.; Neher, D.A.; Roy, E.D.; Tichenor, N.E.; Jahns, L. Relationship between food waste, diet quality, and environmental sustainability. *PLoS ONE* **2018**, *13*, e0195405. [CrossRef]

27.   Fanelli, R.M. The interactions between the structure of the food supply and the impact of livestock production on the environment. A multivariate analysis for understanding the differences and the analogies across European Union countries. *Qual. Acc. Success* **2018**, *19*, 131–139.

28.   Fanelli, R.M. The (un) sustainability of the land use practices and agricultural production in EU countries. *Int. J. Environ. Stud.* **2019**, *76*, 273–294. [CrossRef]

29.   Swedish International Agricultural Network Initiative. *Reducing Food Waste Across Global Food Chains*; Policy Brief: Stockholm, Sweden, 2017; pp. 1–4. Available online: https://www.siani.se/wp-content/uploads/2017/10/policy_brief.pdf (accessed on 20 February 2020).

30.   Ishangulyyev, R.; Kim, S.; Lee, S.H. Understanding Food Loss and Waste—Why Are We Losing and Wasting Food? *Foods* **2019**, *8*, 297. [CrossRef] [PubMed]

31.   Nagpal, M.; Kumar, A. Grain losses in India and government policies. *Qual. Assur. Saf. Crop. Foods* **2012**, *4*, 143. [CrossRef]

32. National Academy of Agricultural Sciences. *National Academy of Agricultural Sciences Saving the Harvest: Reducing the Food Loss and Waste*; Policy Brief No.5; National Academy of Agricultural Sciences: New Delhi, India, 2019; pp. 1–10. Available online: http://naasindia.org/documents/Saving%20the%20Harvest.pdf (accessed on 20 February 2020).

33. Abass, A.B.; Ndunguru, G.; Mamiro, P.; Alenkhe, B.; Mlingi, N.; Bekunda, M. Post-harvest food losses in a maize-based farming system of semi-arid savannah area of Tanzania. *J. Stored Prod. Res.* **2014**, *57*, 49–57. [CrossRef]

34. Manandhar, A.; Milindi, P.; Shah, A. An overview of the post-harvest grain storage practices of smallholder farmers in developing countries. *Agriculture* **2018**, *8*, 57. [CrossRef]

35. Kumar, D.; Kalita, P. Reducing postharvest losses during storage of grain crops to strengthen food security in developing countries. *Foods* **2017**, *6*, 8. [CrossRef]

36. Khader, B.F.Y.; Yigezu, Y.A.; Duwayri, M.A.; Nianed, A.A.; Shideed, K. Where in the value chain are we losing the most food? The case of wheat in Jordan. *Food Secur.* **2019**, *11*, 1009–1027. [CrossRef]

37. Borlaug, N. Increasing and Stabilizing Food Production. In *Plant Breeding II*; Frey, K.J., Ed.; Iowa State University Press: Ames, IA, USA, 1979; pp. 467–492.

38. International Grains Council. *International Grains Council Grain Market Report Five-Year Baseline Projections of Supply and Demand for Wheat, Maize (Corn), Rice and Soyabeans to 2023/24 March 2019*; International Grains Council: London, UK, 2019; pp. 1–4. Available online: http://www.igc.int/en/downloads/gmrsummary/gmrsumme.pdf (accessed on 15 January 2020).

39. Popp, J.; Pető, K.; Nagy, P. Pesticide productivity and food security. A review. *Agron. Sustain. Dev.* **2013**, *33*, 243–255. [CrossRef]

40. Oerke, E.-C.; Dehne, H.-W. Safeguarding production-losses in major crops and the role of crop protection. *Crop Prot.* **2004**, *23*, 275–285. [CrossRef]

41. Oerke, E. Crop losses to pests. *J. Agric. Sci.* **2006**, *144*, 31–43. [CrossRef]

42. European Commission. *European Commission Roadmap to a Resource Efficient Europe COM(2011) 571*; European Commission: Brussels, Belgium, 2011. Available online: https://www.eea.europa.eu/policy-documents/com-2011-571-roadmap-to (accessed on 10 January 2020).

43. European Commission. *European Parliament EP Parliament Calls for Urgent Measures to Halve Food Wastage in the EU—Plenary Sessions*; European Commission: Brussels, Belgium, 2020. Available online: https://www.europarl.europa.eu/news/en/press-room/20120118IPR35648/parliament-calls-for-urgent-measures-to-halve-food-wastage-in-the-eu (accessed on 10 January 2020).

44. International Water Management Institute. *Water for Food, Water for Life: A Comprehensive Assessment of Water Management in Agriculture*; International Water Management Institute: London, UK, 2007; pp. 1–645. Available online: http://www.iwmi.cgiar.org/assessment/Publications/books.htm (accessed on 15 January 2020).

45. Leslie, J.; Logrieco, A. *Mycotoxin Reduction in Grain Chains*; Wiley Blackwell: Hoboken, NJ, USA, 2014.

46. Logrieco, A.; Visconti, A. An Introduction to the MycoRed Project. In *Mycotoxin Reduction in Grain Chains*; Leslie, J., Logrieco, A., Eds.; Wiley Blackwell: Hoboken, NJ, USA, 2014; pp. 1–7.

47. Godfrey, H.; Beddington, J.; Crute, I.; Haddad, L.; Lawrence, D.; Muir, J.; Pretty, J.; Robinson, S.; Thomas, S.; Toulmin, C. The Challenge of Feeding 9 Billion People. *Science* **2010**, *12*, 812–818. [CrossRef] [PubMed]

48. Kummu, M.; De Moel, H.; Porkka, M.; Siebert, S.; Varis, O.; Ward, P.J. Lost food, wasted resources: Global food supply chain losses and their impacts on freshwater, cropland, and fertiliser use. *Sci. Total Environ.* **2012**, *438*, 477–489. [CrossRef] [PubMed]

49. Cramer, H.-H. Plant protection and world crop production. *Bayer Pflanzenschutz-Nachr.* **1967**, *20*, 1–524.

50. Schumann, G.; D'Arcy, C. *Essential Plant Pathology*; APS Press: Sao Paulo, MN, USA, 2006.

51. Ubrizsy, G. *Növényvédelmi Enciklopédia I. Általános Növényvédelem—Szántóföldi Növényvédelem—Encyclopedia of Plant Protection, Volume 1. General Plant Protection—Protection of Field Crops*; Mezőgazdasági Kiadó—Agr Publ House: Budapest, Hungary, 1968.

52. Bradley, B.D.; Christodoulou, M.; Caspari, C.; Di Luc, P. *Integrated Crop Managements Systems in the EU. Amended Final Report for European Commission DG Environment*; Agra CEAS Consulting: Ashford, UK, 2002; pp. 1–141. Available online: https://ec.europa.eu/environment/agriculture/pdf/icm_finalreport.pdf (accessed on 14 March 2020).

53. Bajwa, W.I.; Kogan, M. Compendium of IPM Definitions (CID)—What is IPM and how is it defined in the Worldwide Literature. *IPPC Publ.* **2002**, *998*, 1–14.

54. Ehler, L. Perspective Integrated pest management (IPM): Definition, historical development and implementation, and the other IPM Pest Management. *Science* **2006**, *62*, 787–789.

55. Kumar, D.; Shivay, Y. *Modern Concepts in Agriculture, Integrated Crop Management*; Indian Agricultural Research Institute: New Delhi, India, 2008; pp. 1–30. Available online: https://www.researchgate.net/publication/ 237569012 (accessed on 15 January 2020).

56. Vanlauwe, B.; Bationo, A.; Chianu, J.; Giller, K.E.; Merckx, R.; Mokwunye, U.; Ohiokpehai, O.; Pypers, P.; Tabo, R.; Shepherd, K.D. Integrated soil fertility management: Operational definition and consequences for implementation and dissemination. *Outlook Agric.* **2010**, *39*, 17–24. [CrossRef]

57. Bottrell, D. *Ntegrated Pest Management. Record Number: 19810584265*; United States Government Printing Office: Washington, DC, USA, 1979; pp. 1–120.

58. Barzman, M.; Bàrberi, P.; Birch, A.N.E.; Boonekamp, P.; Dachbrodt-Saaydeh, S.; Graf, B.; Hommel, B.; Jensen, J.E.; Kiss, J.; Kudsk, P. Eight principles of integrated pest management. *Agron. Sustain. Dev.* **2015**, *35*, 1199–1215. [CrossRef]

59. Lehoczki-Krsjak, S.; Varga, M.; Mesterházy, Á. Distribution of prothioconazole and tebuconazole between wheat ears and flag leaves following fungicide spraying with different nozzle types at flowering. *Pest Manag. Sci.* **2015**, *71*, 105–113. [CrossRef]

60. Mesterhazy, A.; Bartok, T. Control of Fusarium head blight of wheat by fungicides and its effect on the toxin contamination of the grains. *Pflanzenschutz-Nachr. Bayer* **1996**, *49*, 181–198.

61. Mesterházy, Á. Breeding Wheat for Fusarium Head Blight Resistance in Europe. In *Fusarium Head Blight of Wheat and Barley*; Bushnell, W., Ed.; APS Press: Sao Paulo, MN, USA, 2003; pp. 211–240.

62. Mesterházy, Á. Control of Fusarium Head Blight of Wheat by Fungicides. In *Fusarium Head Blight of Wheat and Barley*; Leonard, K., Bushnell, W., Eds.; APS Press: Sao Paulo, MN, USA, 2003; pp. 363–380.

63. Mesterházy, Á.; Tóth, B.; Varga, M.; Bartók, T.; Szabó-Hevér, Á.; Farády, L.; Lehoczki-Krsjak, S. Role of fungicides, application of nozzle types, and the resistance level of wheat varieties in the control of Fusarium head blight and deoxynivalenol. *Toxins* **2011**, *3*, 1453–1483. [CrossRef] [PubMed]

64. Mesterházy, Á.; Varga, M.; György, A.; Lehoczki-Krsjak, S.; Tóth, B. The role of adapted and non-adapted resistance sources in breeding resistance of winter wheat to Fusarium head blight and deoxynivalenol contamination. *World Mycotoxin J.* **2018**, *11*, 539–557. [CrossRef]

65. Mesterházy, Á.; Varga, M.; Tóth, B.; Kotai, C.; Bartók, T.; Véha, A.; Ács, K.; Vágvölgyi, C.; Lehoczki-Krsjak, S. Reduction of deoxynivalenol (DON) contamination by improved fungicide use in wheat. Part 1. Dependence on epidemic severity and resistance level in small plot tests with artificial inoculation. *Eur. J. Plant Pathol.* **2018**, *151*, 39–55. [CrossRef]

66. Lamichhane, J.R.; Arseniuk, E.; Boonekamp, P.; Czembor, J.; Decroocq, V.; Enjalbert, J.; Finckh, M.R.; Korbin, M.; Koppel, M.; Kudsk, P. Advocating a need for suitable breeding approaches to boost integrated pest management: A European perspective. *Pest Manag. Sci.* **2018**, *74*, 1219–1227. [CrossRef] [PubMed]

67. Jayas, D.S. Storing grains for food security and sustainability. *Agric. Res.* **2012**, *1*, 21–24. [CrossRef]

68. Philip, T.; Throne, J. Biorational approaches for managing stored-product insect. *Annu. Rev. Entomol.* **2010**, *55*, 375–397. [CrossRef]

69. World Health Organisation. *Aflatoxins, REF. No.: WHO/NHM/FOS/RAM/18*; World Health Organisation, Department of Food Safety and Zoonoses: Geneva, Switzerland, 2018; pp. 1–5. Available online: https: //www.who.int/foodsafety/FSDigest_Aflatoxins_EN.pdf (accessed on 20 February 2020).

70. Kumar, R.; Mishra, A.K.; Dubey, N.; Tripathi, Y. Evaluation of Chenopodium ambrosioides oil as a potential source of antifungal, antiaflatoxigenic and antioxidant activity. *Int. J. Food Microbiol.* **2007**, *115*, 159–164. [CrossRef]

71. Dowling, T. Fumonisin and its toxic effects. *Cereal Foods World* **1997**, *42*, 13–15.

72. McMullen, M.; Jones, R.; Gallenberg, D. Scab of wheat and barley: A re-emerging disease of devastating impact. *Plant Dis.* **1997**, *81*, 1340–1348. [CrossRef]

73. Szabo, B.; Toth, B.; Toth Toldine, E.; Varga, M.; Kovacs, N.; Varga, J.; Kocsube, S.; Palagyi, A.; Bagi, F.; Budakov, D. A New Concept to Secure Food Safety Standards against Fusarium Species and Aspergillus Flavus and Their Toxins in Maize. *Toxins* **2018**, *10*, 372. [CrossRef]

74. Alexander, P.; Brown, C.; Arneth, A.; Finnigan, J.; Moran, D.; Rounsevell, M.D. Losses, inefficiencies and waste in the global food system. *Agric. Syst.* **2017**, *153*, 190–200. [CrossRef]

75. Caldeira, C.; Caldeira, C.; Sara, C.; Serenella, S. *Food Waste Accounting: Methodologies, Challenges and Opportunities. JRC109202; 9279778889*; Publications Office of the European Union: Brussels, Belgium, 2017.

76. Food and Agriculture Organization of the United Nations. *Food Loss and Waste Facts*; Food and Agriculture Organisation of the United Nations: Rome, Italy, 2019. Available online: http://www.fao.org/3/a-i4807e.pdf (accessed on 10 January 2020).

77. Mesterházy, Á. Breeding for Resistance against FHB in Wheat. In *Mycotoxin Reduction in Grain Chains: A Practical Guide*; Logrieco, A.F., Visconti, A., Eds.; Blackwell-Wiley: Hoboken, NJ, USA, 2014; pp. 189–208.

78. Mesterházy, A.; Lehoczki-Krsjak, S.; Varga, M.; Szabó-Hevér, Á.; Tóth, B.; Lemmens, M. Breeding for FHB Resistance via Fusarium Damaged Kernels and Deoxynivalenol Accumulation as Well as Inoculation Methods in Winter Wheat. *Agric. Sci.* **2015**, *6*, 970–1002.

79. Jans, D.; Pedrosa, K.; Schatzmayr, D.; Bertin, G.; Grenier, B. Mycotoxin Reduction in Animal Diets. In *Mycotoxin Reduction in Grain Chains*; Logrieco, A.F., Visconti, A., Eds.; Wiley: Oxford, UK, 2014; pp. 101–110.

80. Edwards, S.G.; Jennings, P. Impact of agronomic factors on Fusarium mycotoxins in harvested wheat. *Food Addit. Contam. Part A* **2018**, *35*, 2443–2454. [CrossRef] [PubMed]

81. Cowger, C.; Weisz, R.; Arellano, C.; Murphy, P. Profitability of integrated management of Fusarium head blight in North Carolina winter wheat. *Phytopathology* **2016**, *106*, 814–823. [CrossRef] [PubMed]

82. Zorn, A.; Musa, T.; Lips, M. Costs of preventive agronomic measures to reduce deoxynivalenol in wheat. *J. Agric. Sci.* **2017**, *155*, 1033–1044. [CrossRef]

83. Baranski, M.R. Wide adaptation of Green Revolution wheat: International roots and the Indian context of a new plant breeding ideal, 1960–1970. *Stud. Hist. Philos. Sci. Part C* **2015**, *50*, 41–50. [CrossRef]

84. Wang, J.; Vanga, S.K.; Saxena, R.; Orsat, V.; Raghavan, V. Effect of climate change on the yield of cereal crops: A review. *Climate* **2018**, *6*, 41. [CrossRef]

85. Tilman, D.; Balzer, C.; Hill, J.; Befort, B.L. Global food demand and the sustainable intensification of agriculture. *Proc. Natl. Acad. Sci. USA* **2011**, *108*, 20260–20264. [CrossRef]

8

The Bioeconomy and Foreign Trade in Food Products—A Sustainable Partnership at the European Level?

Dan Costin Nițescu * and Valentin Murgu

Bucharest University of Economic Studies, Faculty of Finance and Banking, Department of Money and Banking, 010961 Bucharest, Romania; valentin_murgu@hotmail.com
* Correspondence: dan.nitescu@fin.ase.ro

Abstract: This research addresses the problem of the synergistic relationship between the sustainable development of the green economy (bioeconomy) at the European level and the commercial flows with food. Mainly, two components were analyzed and integrated: A qualitative one, on the perspective of the development of the bioeconomy at the European level, and a quantitative one, on the study of the nature of the inter-correlation between the exogenous indicators of foreign food trade (exports and imports) and the relevant endogenous indicators (the labor force, gross added value of agriculture, forestry and fisheries, research and development expenditure, forest area, fossil fuel energy consumption, and renewable energy consumption), for 24 European countries over a 22 year period. Exports and imports of food products are positively influenced by the added value of the agricultural sector and by the share of research and development expenditures, both in the short and long term. Renewable energy consumption influences exports in the short term, but in the long term, the forest area has a significant positive impact. Imports are negatively influenced by renewable energy consumption. The findings of this research can provide support for the future mix of policies.

Keywords: sustainability; bioeconomy; foreign trade in food products; agriculture; labor force; research and development; renewable energy; European Union

1. Introduction

Due to the technological developments, behavioral changes, and the recent orientation towards "green" projects and sectors, currently, the European economy is facing significant changes. In this context, the production, commercialization, and trade of food products at the level of the European countries, interconnected with the renewable energy resources used for food production, together with their transport and distribution routes, create the premises for development of sustainable communities.

Sustainability (sustainable development) is not a new concept. In 1987, the World Commission on Environment and Development (WCED) published the report "Our Common Future" [1], which developed the concept of sustainable development, which involves people's relations with the environment and the responsibilities of present generations to future generations. At the European level, investment programs have been developed to support innovation and research and to provide solutions to the challenges facing national and global food systems, with respect to food consumption and ensuring food security. An overall deterioration of the state of food security at the global level is being witnessing, generated by the emergence of major risk factors, both structural (increasing world population, global warming, degradation of water resources and land with agricultural potential, etc.), and short term (adoption of inadequate policies, erosion of the political–economic role of the states, proliferation of poverty, etc.). To the extent that the current manifestation trends do not change, there are premises for a serious global food crisis, which entails adverse implications for all of the coordinates of global and, implicitly, national and individual security [2]. The gradual transition from linear economy to bioeconomy (the bioeconomy encompasses those parts of the economy that use renewable resources from land and sea, such as crops, forests, fish, animals, and microorganisms, to produce food, materials, and energy [3]) is a strategic goal at the European level. The notion of bioeconomy has grown in importance, both in the research environment, in the public debate, and at the level of political decision-makers, as it is considered as an alternative solution for a different set of problems; strategies/studies were developed as a basis for the construction of a unitary vision on the development, sustainability, and implications of the transition to the bioeconomy. From the European Commission's perspective, the bioeconomy represents an "economy that includes the production of renewable biological resources and their transformation into food, biological, and bioenergy products. This includes the gross added value of agriculture, forestry and fisheries, food, cellulose, and paper production, as well as parts of the chemical, biotechnological, and energy industry" [4]. Other considerations of the bioeconomy are highlighted in certain sectors (e.g., biofuels [5]; biotechnologies [6]; reduced emissions and use of fossil fuels [7]). The studies that were conducted synthesize some views on the bioeconomy, which are included in the following table (Table 1).

Table 1. Key features of the visions regarding the bioeconomy.

	The Vision of Bio-Technologies	The Vision of Bio-Resources	The Vision of Bio-Ecology
Purposes and goals	Economic growth and job creation	Economic growth and sustainability	Sustainability, biodiversity, conservation of ecosystems, avoiding soil degradation
Creation of value	Applications of biotechnology, commercialization of research and technology	Conversion and upgrade of bioresources (process orientation)	Development of integrated production systems and high-quality products with territorial identity
Catalyst and mediators of innovation	Research and Development, patents, Technology Transfers Officers, research and funding councils (focus on science, linear model)	Interdisciplinarity, optimization of land use, inclusion of degraded land in biofuel production, use and availability of bioresources, waste management, engineering, science and market (interactive and network production mode)	Identification of favorable organic agro-ecological practices, ethics, risk, interdisciplinary sustainability, ecological interactions, re-use and re-circulation of waste, land use (circular and self-sustainable production mode)
Space focus	Global clusters/central regions	Rural/ peripheral regions	Rural/ peripheral regions

Source: [8].

The remaining part of this paper is structured in five sections. Section 1 provides a literature review on bioeconomy and foreign trade in food, focusing on EU countries. Section 2 explains the research methodology of the calculation and presents the econometric methodology, specifically the database, variable, and quantitative methods. The third and fourth sections show the discussion and results of the quantitative findings of the study, and the final section provides concluding remarks and policy recommendations.

2. Literature Review

The European Organization for Cooperation and Development (OECD) emphasizes a vision based on bio-technologies (reflected in the vision on bioecology-focused bioeconomy) [9]. From the perspective of the European Commission (2017), a significant variety of research and innovation priorities related to the bioeconomy at the level of the European regions/countries have been identified. Most countries/regions use a mix of thematic areas, from the perspective of both the focus on bioresources and the orientation towards energies obtained from bioresources. At the European level [10], several bioeconomic development initiatives, including with regional orientation, have been identified. The need for each nation to build a competitive advantage (supported by a localized territorial process and allowing it to differentiate from the other nations) has allowed the emphasis on the competitiveness of a nation through its ability to innovate and on the ability to create and assimilate knowledge [11]. Other approaches [12] consider that the focus on the bioeconomy stems from the need to cover the food requirements of a growing population, related to lower yields of agricultural production, or from the need to ensure energy and food security as well as economic prosperity in the face of some new challenges—climate change. The transition to the bioeconomy involves concerted efforts, both on the part of the authorities and on the part of the society, as such a transformation involves substantial changes in the market through the impact of technological development on industrial processes, ultimately affecting the production and consumption patterns. The success of the bioeconomy is dependent on the active involvement of the authorities in the creation of an adequate legislative framework, taking into account that the advanced bioeconomy will become a reality only if the intensification of the research and development efforts will be reflected in the subsequent implementation of the technologies. The bioeconomy can reflect the direct link between innovation and economic growth [10], in the sense that increasing productivity by maximizing the efficiency of the resources used in counterpoint with limiting the impact on the environment can be achieved only through technological research and development. It is worth mentioning that innovation must be accepted by each participant in the economic chain as well as by the society as a whole. Identifying the stimulating factors of the transition to the bioeconomy is a difficult and complex process, given their diversity. The analysis carried out in 2018 by the FAO (Food and Agriculture Organization, a specialized agency of the UN, with the aim to eliminate world hunger, as well as to improve the food, by coordinating the activities of the governments in the field of agriculture, forestry, and the fishing industry [13]) reflects the contribution of the bioeconomy to a country's economic growth. Although the implementation of this concept requires a harmonization with the particularities and priorities of each state, in a general framework, however, certain aspects essential to the development of the bioeconomy can be identified (see Figure 1).

Figure 1. Essential factors for the development of a sustainable bioeconomy. Source: [13].

An empirical study [14] on the EU component states has shown that Finland and Sweden have the lowest levels of environmental pollution due to the rigorous ecological awareness of the population; the focus in these countries is on education and vocational training, with a basis of solid knowledge. In addition, the two states are among the most innovative countries in the EU and are based on rich and diverse natural resources. Denmark, Ireland, the Netherlands, and the UK have similarities in terms of innovation capacity supported by a developed economy, but natural assets are narrower than in the two Nordic countries (much of the countries' areas are used for agriculture, as forests are restricted as a surface), and the quality of the environment is above average. At the opposite end, these states are noted: Bulgaria, Croatia, Cyprus, Czech Republic, Greece, Hungary, Italy, Latvia, Lithuania, Poland, Portugal, Romania, Slovakia, and Spain. Although they have the largest agricultural sectors, innovation activity is relative low, which results in a low employment rate in the technological field. In these countries, the public authorities are less dedicated to education and training, and the population is not so concerned about the environment. According to the studies, the historical, geographical, and cultural factors influence the pro-bioeconomic behaviors adopted by the citizens. The size of the socio-economic context highlighted the most visible differences between countries, leading us to the conclusion that the countries of Central and Eastern Europe are in different stages of development [15].

The economic literature has developed progressively, encompassing the issues of bioeconomy and sustainability, as well as the determining elements that influence the economic growth.

In this section of the research, relevant aspects of some studies and research were analyzed, which include the issue of the sustainability of economic growth, the analysis of its components, analysis of the developments of bioeconomy at the European level, research on the six macroeconomic indicators included in the empirical study, and their correlations. Foreign trade, both export and import, continued to be one of the fundamental factors of economic growth contributing to the growth of national economies. The value of foreign agri-food trade is relevant, considering that, in 2018, the EU maintained its position as a world leader in the global export of agri-food products, with EU exports reaching 138 billion EUR in 2018. The top five destinations for food products exported by the European Union continue to be the United States, China, Switzerland, Japan, and Russia, which account for 40% of EU exports. The EU's common agricultural policy has become increasingly market-oriented, thus contributing to the EU's success in agricultural trade [16]. In 2018, the EU became the second largest importer of agri-food products in the world, the value of its imports amounting to 116 billion EUR, bringing the EU trade balance for this sector to a net positive result of 22 billion EUR. The EU mainly imports three types of products: Products not produced in the EU (or are produced only to a small extent, such as tropical fruits, coffee, and fresh or dried fruits), representing 23.4% of EU imports, products that are intended for animal feed (accounting for 10.8% of EU imports), and products used as ingredients in further processing [16]. Although agri-food trade is shown to benefit from a positive global climate assessment from 2019 [16], substantial future risks remain for trade developments [16]. The biggest threats to trade developments include protectionist political approaches (which are increasingly important for economies), more frequent trade disputes, and possible trade unrest linked to Britain's decision to leave the EU. On the positive side, global demand for food is likely to increase, correlated with population growth, income growth, middle-class expansion, and changes in consumer preferences [17].

Figures 2 and 3 respectively show the evolution of exports and food imports at the level of the 24 countries included in the empirical study, highlighted distinctly by the two groups (countries of Western and Northern Europe, considered countries with developed economies, and countries of Central and Eastern Europe, considered countries with emerging/developing economies).

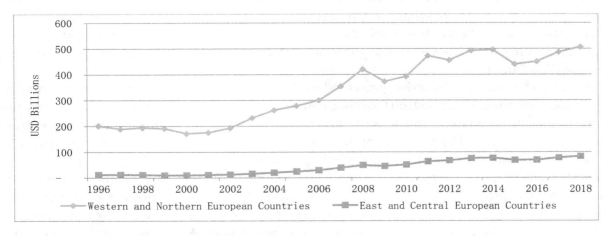

Figure 2. The evolution of total food exports in the countries included in the empirical analysis. Source: Own processing, data are sourced from the World Bank database [18].

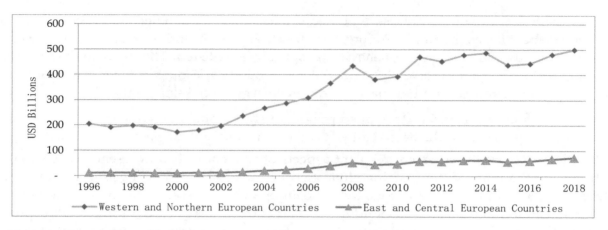

Figure 3. The evolution of total food imports in the countries included in the empirical analysis. Source: Own processing, data are sourced from the World Bank database [18].

Thus, it is observed that, in the case of countries with emerging/developing economies, the variations of exports and imports are more pronounced compared to those registered with the countries with developed economies, which can absorb the impacts of the influence factors. The economic literature analyzes the effects of imports and exports on private research and development expenditures in the food-processing sector. The empirical results [19] reflect that increasing the level of import intensity leads to reductions in private spending on research and development, while increasing the level of export intensity promotes higher private spending on research and development. These results imply that the effects of reducing the research and development activity of imports offset the effects of improving export research and development. Other studies examine the impact of EU enlargement on export performance of agri-food products in 12 new EU Member States and five new independent states on EU markets, covering the period 1999–2007 [20]. A longer duration for agri-food exports from the new EU member states was identified. The results confirm the gains from the eastern enlargement of the EU in terms of export growth and a longer duration for the export of specialized foods, with a higher added value for consumers and more competitive niche agri-food products [20].

3. Materials and Methods

This research applies scientific tests, uses specific estimators and statistical–econometric techniques, investigates data sets and collections, and assesses the most appropriate methods of investigation in order to provide accurate results. The activity of foreign trade in foodstuffs, transposed in the external balance of a country, can make a significant contribution to the economic (sustainable) growth of the respective country. Especially in the context of the transition to the green economy, a strategic vision must include the factors that achieve a significant influence. It is also necessary to integrate and study the behavioral evolution, habits/preferences and attitudes of consumption, and the degree of adoption and use of technologies along the value chain from plant culture/animal growth to food processing/distribution.

As regards the selected countries (presented in Table 2), on the one hand, the founding countries of the European Union were included; on the other, countries in Central and Eastern Europe, representative in terms of structural changes in the economy, were also included. The countries in Central and Eastern Europe are affected by processes of transition from a centralized economy to a market economy, or are even in the early stages of reforms, such as in Macedonia, Former Yugoslav Republic (FYR), a candidate country for EU accession. The authors opted for a split into two groups of countries based on the criteria: Geographical and economic development. The division into two groups of countries based on their level of development was made taking into consideration similar approaches to be found in the field's literature, such as the ones cited in this article. Additionally, a division into three groups would make the groups very unequal with respect to the volume of the sample, with advanced economies having much more representation than the developing or the emerging ones; consequently, the representativeness of such results would be far lower than in the present situation (lower accuracy). Similar divisions of the European countries are to be found in [21,22]. With respect to the criteria of economic development, in practice, international bodies also operate with the same classifications: 1. economically advanced countries and developing and emerging countries [23]; 2. developed markets, emerging markets, and border markets [24]; 3. developed economies and economies in transition [25].

Table 2. List of countries according to their grouping by level of development.

	Western and Northern European Countries Advanced Economies	East and Central European Countries Emerging and Developing Economies
1	Austria	Bulgaria
2	Belgium	Croatia
3	Denmark	Czech Republic
4	France	Estonia
5	Germany	Hungary
6	Italy	Latvia
7	Luxemburg	Lithuania
8	Holland	Macedonia, Former Yugoslav Republic (FYR)
9	Norway	Poland
10	Spain	Romania
11	Sweden	Slovak Republic
12	United Kingdom	Slovenia

Source: Authors' own processing.

In the empirical study, to determine the inter-correlation with the exogenous indicators of foreign food trade, six relevant endogenous indicators were selected, including from the perspective of the bioeconomy/sustainable development: Labor force, added value of agriculture, forestry and fisheries, research and development expenses, forest area, fuel consumption based on fossils, and renewable energy consumption:

- *Labor force* includes all persons who are available to provide labor for the production of goods and services over a certain period of time, whether they are employed or unemployed. In emerging countries, the problem of disguised unemployment (includes low productivity, poorly paid jobs) can be solved by transferring the labor force affected by this phenomenon to the industrial sector in order to support production and, implicitly, export development. [18]. The economic literature shows a statistically significant reduction in employment in the forestry sector and a simultaneous increase in labor productivity due to the increasing use of technological equipment [26]. The analysis focused on the Czech Republic, but the results can be applied to other European countries as well [26]. A significant decrease in employment leads to instability in the forestry sector.
- *Added value from agriculture, forestry, and fisheries.* This is represented by the net production of a sector after all of the results have been gathered and the intermediate inputs have decreased. It is calculated without making deductions for the depreciation of manufactured assets or the depletion and degradation of natural resources. [18];
- *Expenditure on research and development*, a notion aimed at a marketable application in practice of an invention or an integration of the invention into the economic–social practice. The usual method of measuring innovation is based on the use of indirect indicators: Data on research and development (that measure parts of the inputs to the innovation process, indicate resources

spent) and data on patents for invention (granted for inventive technologies with marketing prospects) [18].

- *The forest area,* which is represented by the land under natural trees or under planted trees, but excludes tree stands from agricultural production systems and trees from urban parks and gardens [18].

- *Fossil fuel energy consumption.* Fossil fuels are non-renewable resources from coal, oil, and oil and natural gas products; they take millions of years to form and reserves are depleted much faster than new ones are made [18].

- *Renewable energy consumption* is calculated as the share of renewable energy in the total final energy consumption [18]. Accordingly [26], the large-scale renewable energy implementation plans should include strategies for integrating renewable sources into coherent energy systems influenced by energy saving and efficiency measures. In the case of Denmark, the authors of [27] discussed the problems and prospects of converting the current energy systems into a 100% renewable energy system. Renewable energy sources are used in the context of further technological improvements; the renewable energy system can be implemented in a sustainable manner.

Through a quantitative mix of instruments, the nature of the inter-correlations between these indicators was studied in order to provide certain answers to the fundamental question of this research: Which of the indicators analyzed at the level of the 24 European economies, over a period of 22 years, has a positive impact on the determination/influence, in a relevant way, of the evolution of food exports and imports? To answer this question for the present analysis, a series of six working hypotheses has been constructed, which will be tested using the multiple regression model; the first of these is methodological in nature:

- *H1. The Pooled Mean Group (PMG) estimator is best suited for modeling the relationships between variables.* Judging from the fact that the countries included in the panel belong to the European area, it is expected to be characterized by a common long-term trend, but at the same time, to present short-term differences, considering the internal conditions specific to each state.

- *H2. The labor force directly impacts both exports and imports.* This hypothesis is because exports contribute to economic growth, and one of the most important sources of economic growth is the labor force. Thus, a direct relationship between the two variables is expected. The more the labor force is developed, the greater the ability of a country to export. As for imports, if the labor force grows, the availability of money on the market will increase, as well as its remuneration. Thus, the demand for products will increase, some of them being covered by imports; their value will also increase.

- *H3. The added value of the agricultural sector positively influences exports, but negatively influences imports.* A high added value in the agricultural sector implies a high productivity, which will increase the value of exports. Moreover, increasing productivity in the agricultural sector will lead to better coverage of domestic needs from their own production, which will decrease the value of food imports.

- *H4. Research and development expenditure leads to an increase in the value of exports, but also of imports.* Rising spending on research and development implies a development of this sector, which will increase the added value of the economy. This is because the final product of innovation is one with high gross added value, as it is intensive in innovation, knowledge, and capital. Thus, if part of the innovation process is carried out in the food sector, it will lead to an increase in exports of this sector. As in the case of the labor force, the increase of the expenditure with the research and development brings financial resources to the society, which positively impacts the imports.

- *H5. Energy consumption is higher in countries with higher food exports and lower in countries with higher imports.* Countries with large exports consume more energy, while importing countries will consume less energy;

- *H6. Renewable energy consumption is inversely correlated with food exports and directly correlated with imports.* Countries with important renewable energy sectors are more developed countries, which export products other than food, but mainly import this type of product because they do not have a highly developed agricultural sector.

The purpose of the present research is to link the foreign trade of foodstuffs, estimated both by exports (EXP) and by imports (IMP), with the following factors:

- labor force (LABOR),
- the gross added value of agriculture, forestry, and fisheries (AGRI),
- expenditure on research and development (RD).

In the preliminary stage of the actual modeling of the relation between the dependent variables and the main determinants considered in the present analysis, it is necessary to investigate the statistical properties of the series of variables. Following this examination, the most appropriate statistical–econometric techniques are decided to model the link between the variables included in the study. Moreover, before the start of the statistical analysis, all of the variables considered were respectively logarithmized for a possible normalization of their distribution for an easier interpretation of the associated coefficients in the form of elasticities. To determine whether the series of variables are stationary in the level or first difference, the Fisher–Phillips Perron unit root test developed by Choi [28] was applied. The main advantage of this test is that it can be applied to both balanced and unbalanced panel data. Thus, considering the series of our variables that sometimes have missing values, it was considered that the application of this test is the most appropriate. First, in the analysis of stationarity for the level of variables, it was included in the equation for both the constant and the trend. Considering that series of macroeconomic variables most often have a certain tendency, including the trend in the equation increases the accuracy of the results. Secondly, for the first difference of the series, only the constant in the equation was included, since the differentiation of the series leads, in most cases, to the elimination of a possible tendency. Moreover, to correct for potential data persistence, both equations are aggregated with a lag.

The results of the stationarity test presented in Table 3 suggest that both dependent variables, i.e., exports and imports, have a unit root (they are integrated with an order of one - I (1)); the p-value associated with the statistics calculated for the level of the variables is higher than the significance thresholds of 1% and 5%, thus leading to the acceptance of the null hypothesis that the series are characterized by the unit root. In contrast, for the first difference of the variables, the p-value associated with the calculated statistics is lower even than the significance thresholds of 1%, leading to the rejection of the null hypothesis and the acceptance of the alternative one, according to which the series are stationary.

Regarding the exogenous (independent) variables, the results are mixed, as the variables are both integrated by the first order and stationary at the level. Taking into account the characteristics of the series of variables included in this analysis—namely that the dependent variables are I(1), and the independent variables are both I(1) and I(0)—for modeling the relation between them, a dynamic model was considered, namely an ARDL (Autoregressive Distributed Lag) data panel. The mathematical form of the dynamic model for the ARDL panel data (p, q_1 ... q_k) [29] is as follows:

$$Y_{it} = \sum_{j=1}^{p} \partial_{ij} Y_{i,t-j} + \sum_{j=0}^{q} \gamma'_{ij} X_{i,t-j} + \mu_i + \varepsilon_{it}, \tag{1}$$

where $i = \overline{1,N}$ represents the countries analyzed, and $t = \overline{1,T}$ denotes the number of years included in the study (the period analyzed). Y_{it} is the dependent variable, and X_{it} $(k \times 1)$ represents the vector of explanatory variables with the vector of associated coefficients $\gamma_{ij}(k \times 1)$.

Table 3. Test of stationarity (unit root).

Test/ VAriable	Fisher-Type unit-Root Test			
	Level (Constant and Trend)		Δ (Constant)	
	Inverse Chi-Square	p-Value	Inverse Chi-Square	p-Value
Dependent variable				
EXP	22.9662	(0.9992)	260.7056 ***	(0.0000)
IMP	16.6480	(1.0000)	272.1349 ***	(0.0000)
Independent variable				
LABOR	31.0687	(0.9724)	356.7083 ***	(0.0000)
AGRI	87.9761 ***	(0.0004)	-	-
RD	10.4846	(1.0000)	250.6373 ***	(0.0000)
FOREST	67.7993 **	(0.0314)	-	-
ENG	70.6167 **	(0.0184)	-	-
RENEW	72.9023 **	(0.0117)	-	-

Source: own processing. Note: The null hypothesis (H0): All panels contain unit roots, and the alternative hypothesis (H1): At least one panel is stationary. For the stationary variables at a level of significance of 1% and 5%, the first difference of the series was not analyzed. *** and ** indicate statistical significance at a threshold of 1% and 5%. In the above table, the following abbreviations were used: Exports (EXP), imports (IMP), labor force (LABOR), gross added value of agriculture, forestry, and fisheries (AGRI), expenditure on research and development (RD), forest area (FOREST), fossil fuel energy consumption (ENG), and renewable energy consumption (RENEW).

In our case, the dependent variable is represented by the exports; the imports of food products, and as their main determinants, the labor force, the added value of agriculture, forestry and fisheries, and the expenditure for research and development were respectively considered. Moreover, μ_i and ε_{it} indicate the country-specific fixed effects and the error term, respectively.

The above Equation (1) can be rewritten in the form of a panel data error correction model if it is assumed that the variables are non-stationary and co-integrated. Thus, the equation incorporating both long-term and short-term coefficients, together with the error correction term Equation (2), has the following form:

$$\Delta Y_{it} = \phi_i\left(Y_{it-1} - \lambda_i' X_{it}\right) + \sum_{j=1}^{p-1} \partial_{ij}^* \Delta Y_{it-j} + \sum_{j=0}^{q-1} \gamma_{ij}'^* \Delta X_{it-j} + \mu_i + \varepsilon_{it} \quad (2)$$

where:

$$\phi_i = -(1 - \sum_{j=1}^{p} \partial_{ij}),$$
$$\lambda_i = \sum_{j=0}^{q} \gamma_{ij} / \left(1 - \sum_{k} \partial_{ik}\right),$$
$$\partial_{ij}^* = -\sum_{m=j+1}^{p} \partial_{im},$$
$$\gamma_{ij}^* = -\sum_{m=j+1}^{q} \gamma_{im},$$

and Δ represents the difference operator.

In order to confirm the long-term relationship between the variables, the coefficient associated with the error correction term, namely ϕ_i must be negative and statistically significant, and its values must be between [−1; 0]. Moreover, it helps us evaluate whether the model is specified correctly and to determine the speed of adjustment of the system to long-term equilibrium following an exogenous shock. First, it should be noted that one advantage of the ARDL technique on panel data is the accuracy (consistency) of the estimated coefficients when the dependent variable is I(1) and the independent variables have different integration orders of (I(0) and I(1)). Secondly, another advantage is given by the flexibility of the estimated coefficients, in the sense that it allows us to evaluate the influence of the independent variables on the dependency in both the long term and in the short term.

Considering all of the results related to the evaluation of the characteristics of the analyzed variables, an ARDL model (1.1) was estimated, including one lag for the dependent variable and, respectively, one lag for the independent ones. The decision to include a lag in the model is closely linked to the value of the Akaike Information Criterion (AIC) and the total number of panel-level observations. Considering that $N = 24$ and $T = 21$ ($N * T = 504$), the inclusion of several lags in the model significantly reduces the number of observations; the ARDL model is sensitive in this respect. It should be mentioned that the main analysis was started for a group of 24 countries in Europe (Austria, Belgium, Bulgaria, Croatia, Czech Republic, Denmark, Estonia, France, Germany, Hungary, Italy, Latvia, Lithuania, Luxembourg, North Macedonia, Netherlands, Norway, Poland, Romania, Slovak Republic, Slovenia, Spain, Sweden, and United Kingdom) for the period 1996–2017. The choice of the analysis period was strictly determined by the availability of data and, respectively, by the variable of expenditure on research and development, for which the values stop in 2017. The ARDL model (1.1) is estimated using three specific estimators, namely the Dynamic Fixed Effects (DFE), Pooled Mean Group (PMG), and Mean Group (MG). Then, the Hausman test helps us determine if the PMG estimator or the MG is best suited to model the evaluated data. It should be noted that the DFE estimator considers that both the short-term and long-term coefficients, together with the error correction term, are identical for all panel members (for the analyzed countries) only the constant is different, depending on the country. On the other hand, the MG estimator assumes the exact opposite (the short-term and long-term coefficients, together with the error correction term, are different for all panel members), and the PMG estimator is the intermediate version between the two, considering a common long-term trend for all countries, with a respective short-term heterogeneity between coefficients. The final step in the analysis was the validation of the final models by evaluating their robustness. For this purpose, three variables (related to both the agricultural sector and the size of sustainable development) were introduced in the analysis, taking into account the continuous discussions at the international level related to the problem of natural resources and their diminution. For the present analysis, the following variables were considered as control variables:

- forest area (FOREST),
- fossil fuel energy consumption (ENG), and
- renewable energy consumption (RENEW),

These were used to check if the relationships found in the main regressions remain stable in the presence of environmental and long-term sustainability factors, i.e., FOREST, ENG, and RENEW.

4. Results

Hypothesis H1 was first evaluated in order to determine the appropriate final model for the data sample used. In Tables 4 and 5 are presented the estimations of the ARDL model (1.1) through the three estimators for exports (EXP) and imports (IMP). In the basic vector, the variables of labor force (LABOR), the added value of agriculture, forestry, and fisheries (AGRI), and the expenditure of research and development (RD) were considered as factors influencing the exports/imports.

Table 4. Estimation of the Autoregressive Distributed Lag (ARDL) model (1.1) for exports.

	DFE	PMG	MG
The dependent variable: ΔEXP			
Long-term coefficients			
LABOR	−2.7674 ***	−1.6328 ***	−9.7459
	(0.7793)	(0.2483)	(10.1587)
AGRI	0.7217 ***	0.1591 ***	0.5002 **
	(0.2761)	(0.0881)	(0.2095)
RD	1.0087 ***	1.1345 ***	1.3584 **
	(0.1336)	(0.0448)	(0.6856)

Table 4. *Cont.*

	DFE	PMG	MG
Short-term coefficients			
ECT	−0.1297 ***	−0.1882 ***	−0.4012 ***
	(0.0189)	(0.0404)	(0.0506)
ΔLABOR	0.2795	−0.0688	−0.6799
	(0.3959)	(0.4027)	(0.4416)
ΔAGRI	0.2634 ***	0.2589 ***	0.0813
	(0.0392)	(0.0463)	(0.0719)
ΔRD	0.2762 ***	0.3132 ***	0.2191 ***
	(0.0459)	(0.0680)	(0.0754)
Constant	3.5616 **	3.7360 ***	−14.6986 **
	(1.4659)	(0.8157)	(6.3261)
Log Likelihood	-	637.0791	-
No. of countries	24	24	24
No. of observations	493	493	493

Source: own processing. Note: Hausman test: MG vs. PMG: Since the p-value = 0.4865 is greater than all significance thresholds, the null hypothesis that the PMG estimator is the preferred model to form the relationship between variables was accepted. *** and ** indicate statistical significance at a threshold of 1% and 5%. In the above table, the following abbreviation were used: Dynamic fixed effects (DFE), pooled mean group (PMG), mean group (MG), exports (EXP), labor force (LABOR), gross added value of agriculture, forestry, and fisheries (AGRI), expenditure on research and development (RD), and error correction term (ECT).

Table 5. Estimation of the ARDL model (1.1) for imports.

	DFE	PMG	MG
The dependent variable: ΔIMP			
Long-term coefficients			
LABOR	−0.6142	0.2027	1.9127
	(0.7179)	(0.1405)	(1.9953)
AGRI	0.4032	0.3155 ***	0.5320 **
	(0.2708)	(0.0366)	(0.2155)
RD	0.8464 ***	0.8605 ***	0.5569 ***
	(0.1303)	(0.0211)	(0.2047)
Short-term coefficients			
ECT	−0.1073 ***	−0.2529 ***	−0.4740 ***
	(0.0202)	(0.0547)	(0.0515)
ΔLABOR	0.3294	−0.0733	−0.4728
	(0.3281)	(0.3773)	(0.4393)
ΔAGRI	0.2559 ***	0.1876 ***	0.0570
	(0.0325)	(0.0485)	(0.0479)
ΔRD	0.3688 ***	0.3607 ***	0.2765 ***
	(0.0376)	(0.0701)	(0.0616)
Constant	0.5509	−1.5650 ***	−9.0339
	(1.2108)	(0.3585)	(5.4557)
Log Likelihood	-	702.2978	-
No. of countries	24	24	24
No. of observations	495	495	495

Source: own processing. Note: Hausman test: MG vs. PMG: Since the p-value = 0.7162 is greater than all significance thresholds, the null hypothesis that the PMG estimator is the preferred model to form the relationship between variables was accepted. *** and ** indicate statistical significance at a threshold of 1% and 5%. In the above table, the following abbreviation were used: Dynamic fixed effects (DFE), pooled mean group (PMG), mean group (MG), imports (IMP), labor force (LABOR), gross added value of agriculture, forestry and fisheries (AGRI), expenditure on research and development (RD), and error correction term (ECT).

First of all, it was noticed that the Error Correction Term (ECT) is negative and strongly significant for all models. Thus, the modeling technique is justified and the specification of the models is validated. Moreover, for the export equation, the associated coefficient varies between about 0.13 and 0.4, suggesting a low adjustment rate (similarly to the case for the import equation, where the adjustment rate varies between about 0.11 and 0.47). Second, the Hausman test, by which we discriminated between the estimator MG and PMG, suggests that the most appropriate of these is the PMG estimator for both the dependent variable exports and imports (p-value = 0.486 for exports, p-value = 0.716 for imports). Consequently, the working hypothesis H1 is accepted: The PMG estimator is the preferred one for modeling the relationships between variables. Countries have a common long-term tendency with respective short-term heterogeneities. The common long-term trend can be explained by the efforts of all states to increase trade openness, ultimately stimulating the economic growth. For example, at the European Union level, trade policies focus on coordinating states towards a common trajectory, which involves increasing the trade flow. In tandem, the heterogeneities in the short term can be given by the differences in the commercial structures of the countries, both in terms of exports and imports of food products, or respectively of the national macroeconomic policies. As stated in the methodological section, to evaluate the validity of the other working hypotheses, the results of the PMG estimator were analyzed. Regarding the export model, the following were observed:

1. The labor force has a negative long-term and strongly significant effect on the value of food exports in the countries analyzed. Thus, with an increase of 1% of the labor force, the value of exports decreases by approximately 1.6%. The second working hypothesis, H2, is invalidated for exports.

2. Both the added value of agriculture, forestry, and fisheries and the expenditure on research and development positively influence the value of exports in the long term. Moreover, as with the labor force case, the associated coefficients are strongly significant at a significance threshold of 1% (p-value < 1%). In addition, the magnitude of the coefficients suggests that an increase of 1% in the added value of agriculture, forestry, and fisheries and in expenditure on research and development causes an increase of approximately 0.16% and 1.13%, respectively, in the value of exports. The working hypotheses H3 and H4 are accepted for exports.

3. In the short term, the variable LABOR does not have statistical significance, and AGRI and RD contribute significantly to increasing the value of food exports in the analyzed countries.

The results of the import model show the following:

1. The labor force does not have a statistically significant impact on the value of food imports because of the lack of significance of the associated coefficient. This result invalidates H2 on the import side, as there is no relationship between the two variables.

2. On the other hand, as in the case of exports, the added value of agriculture, forestry, and fisheries and the expenditure on research and development help to stimulate the value of imports. The difference that appears is related to the magnitude of the coefficients associated with these variables compared to those of the export model. For example, the coefficient associated with the variable AGRI is approximately double for the model of imports, and suggests that, with a 1% increase in the added value of agriculture, forestry, and fisheries, the value of imports increases by about 0.32%. On the other hand, when the RD increases by 1%, the value of the imports of food products increases by about 0.86% (the magnitude of the coefficient being smaller in comparison with that of the expenditure for research and development in the case of the export model). The results invalidate H3 on the import side and accept H4.

3. In the short term, the coefficients associated with the variables AGRI and RD are statistically significant, having a positive impact on the value of food imports.

The robustness analysis of the basic model of exports and imports was made by introducing the three additional variables closely related to the concept of sustainable development (environmental quality); namely, the forest area (FOREST), the energy consumption of fossil fuels as a share of total energy (ENG), and the renewable energy consumption (RENEW). Considering that the ARDL technique provides consistent results for mixed independent variables (i.e., first order (I(1)), as well as stationary (I(0)), no prior analysis of the stationarity of these factors (additional ones included in the study) is required. However, it should be mentioned that due to the availability of data, the analysis period of the models that include the FOREST variable is 1996–2016, and for the other two variables, the analyzed period is 1996–2015. The results of the models estimated through the PMG estimator are presented in Tables 6 and 7. The additional independent variables were included one at a time in the vector of the basic variables, so that finally, in the last column (column (4)) of the two tables, all of the independent variables are considered. On the one hand, for the export model, it was observed that ECT is negative and strongly significant in all models, validating once again the chosen technique and specification of the models. Moreover, its magnitude is comparable from one model to another, also indicating a relatively low rate of adjustment to long-term equilibrium (see Table 6). Regarding the independent variables, overall, it can be observed that the statistical significance and the signs of the coefficients of the variables in the basic vector do not change with the inclusion of the additional factors. Moreover, if the last model, where all of the exogenous factors were included (see column (4) of Table 6), is considered, it is to be noticed that the additional factors have a positive impact on exports in the long term. In contrast, in the short term, it is to be mentioned that only the consumption of energy and renewable energy significantly influences exports, the first variable in the positive sense and the second in the negative sense. Hypothesis H5 is accepted for exports in both the short and long terms. Hypothesis H6 is invalidated at the level of the analyzed sample, highlighting a direct link between the renewable energy consumption and exports. On the other hand, for the model of imports, the associated coefficient of ECT is also strongly statistically and negatively significant, suggesting once again that the model is well specified and has a high accuracy. Contrary to the main model, it is observed that the inclusion of the variable of forest area or the inclusion of all variables in the equation (see column (1) and column (4) of Table 7) brings a significant gain to the long-term coefficient associated with the variable of labor force (the sign of the associated coefficient is positive). In addition, if the most complete model is analyzed (see column (4) in Table 7), it must be noted that the variables of the main vector (AGRI and RD) keep their positive sign for the associated coefficient and, respectively, the high statistical significance. In addition, the FOREST variable has a positive influence on imports, while RENEW has a negative impact on the value of imports. For the ENG variable, the associated coefficient does not have statistical significance. In the case of imports, both working assumptions related to sustainability factors are rejected (H5 and H6). The short-term coefficients are statistically significant for the variables AGRI, RD, and ENG, and the positive sign illustrates that all of these macroeconomic factors contribute to the increase of the value of imports. In total, for the models of both exports and imports, the inclusion of additional factors in the main equation does not significantly affect the results of the initial models, so it can be said that that they have a high robustness.

Table 6. Estimation of the ARDL model (1.1) for exports—robustness analysis.

The Dependent Variable: ΔEXP	(1)	(2)	(3)	(4)
Long-term coefficients				
LABOR	−2.6293 ***	−4.2200 ***	−2.3602 ***	−1.8319 ***
	(0.3809)	(0.5559)	(0.2952)	(0.3317)
AGRI	0.4649 ***	0.6135 ***	0.4354 ***	0.7151 ***
	(0.0988)	(0.0534)	(0.1098)	(0.0625)
RD	0.8934 ***	0.9876 ***	1.0177 ***	0.8004 ***
	(0.0574)	(0.0314)	(0.0611)	(0.0438)
FOREST	1.8069 ***			2.5942 ***
	(0.4867)			(0.6370)
ENG		0.9073 ***		1.3878 ***
		(0.1266)		(0.1590)
RENEW			0.0567	0.2954 ***
			(0.0417)	(0.0518)

Table 6. *Cont.*

The Dependent Variable: ΔEXP	(1)	(2)	(3)	(4)
Short-term coefficients				
ECT	−0.2266 ***	−0.1782 ***	−0.2105 ***	−0.2935 ***
	(0.0514)	(0.0443)	(0.0447)	(0.0554)
ΔLABOR	0.0867	0.1496	−0.2014	0.5162 ***
	(0.4107)	(0.4242)	(0.4082)	(0.3156)
ΔAGRI	0.1939 ***	0.1716 **	0.1965 ***	0.0851
	(0.0551)	(0.0683)	(0.0556)	(0.0705)
ΔRD	0.2766 ***	0.2951 ***	0.3284 ***	0.2001 **
	(0.0843)	(0.0984)	(0.0689)	(0.0851)
ΔFOREST	13.5043			3.0714
	(13.2621)			(13.6067)
ΔENG		0.3288 **		0.3317 **
		(0.1293)		(0.1574)
ΔRENEW			0.0250 ***	−0.0966 **
			(0.0515)	(0.0479)
Constant	3.2907 ***	7.1126 ***	5.6302	−8.7431 ***
	(0.7605)	(1.7842)	(1.2187)	(1.5790)
Log Likelihood	633.3564	606.3407	586.0194	656.261
No. of countries	24	24	24	24
No. of observations	468	441	447	439

Source: Own processing. Standard error in round brackets. *** and ** indicate statistical significance at a threshold of 1% and 5%. In the above table, the following abbreviation were used: Exports (EXP), labor force (LABOR), gross added value of agriculture, forestry, and fisheries (AGRI), expenditure on research and development (RD), forest area (FOREST), fossil fuel energy consumption (ENG) and renewable energy consumption (RENEW), and error correction term (ECT).

Table 7. Estimation of the ARDL model (1,1) for imports—robustness analysis.

The Dependent Variable: ΔIMP	(1)	(2)	(3)	(4)
Long-term coefficients				
LABOR	0.2607	−0.0293	−0.0791	0.6177 **
	(0.1406)	(0.2018)	(0.1831)	(0.2443)
AGRI	0.3826 ***	0.3961 ***	0.3964 ***	0.6869 ***
	(0.0472)	(0.0434)	(0.0438)	(0.0655)
RD	0.8105 ***	0.8556 ***	0.8561 ***	0.7026 ***
	(0.0277)	(0.0242)	(0.0246)	(0.0409)
FOREST	0.8985 **			5.7577 ***
	(0.3789)			(1.2851)
ENG		−0.1517 **		0.0036
		(0.0724)		(0.1189)
RENEW			0.0104	−0.2249 ***
			(0.0195)	(0.0626)
Short-term coefficients				
ECT	−0.3123 ***	−0.2799 ***	−0.2897 ***	−0.2591 ***
	(0.0566)	(0.0568)	(0.0515)	(0.0498)
ΔLABOR	−0.1696	−0.2139	−0.5791	−0.4609
	(0.4113)	(0.4697)	(0.4352)	(0.5869)
ΔAGRI	0.1445 ***	0.1438 ***	0.1394 ***	0.0861
	(0.0504)	(0.0518)	(0.0477)	(0.0461)
ΔRD	0.3153 ***	0.3636 ***	0.3853 ***	0.4164 ***
	(0.0737)	(0.0846)	(0.0714)	(0.0943)
ΔFOREST	−1.4857			−12.5173
	(13.4565)			(17.9804)
ΔENG		0.2821 ***		0.3724 ***
		(0.0911)		(0.1224)
ΔRENEW			0.0393	0.0239
			(0.0726)	(0.0701)
Constant	−5.1475 ***	−0.6655 ***	−1.0976 ***	−18.8824 ***
	(0.9681)	(0.1670)	(0.2242)	(3.7615)

Table 7. *Cont.*

The Dependent Variable: ΔIMP	(1)	(2)	(3)	(4)
Log Likelihood	695.4414	663.4577	660.5005	706.2197
No. of countries	24	24	24	24
No. of observations	470	443	449	441

Source: Own processing. Standard error in round brackets. *** and ** indicate statistical significance at a threshold of 1% and 5%. In the above table, the following abbreviation were used: Imports (IMP), labor force (LABOR), gross added value of agriculture, forestry, and fisheries (AGRI), expenditure on research and development (RD), forest area (FOREST), fossil fuel energy consumption (ENG) and renewable energy consumption (RENEW), and error correction term (ECT).

5. Discussion

The analysis identified and presented, mainly within the literature review section, a significant variety of research and studies related to the bioeconomy, reflecting the complexity of the transition process and its substantial transformations. Several studies approached a smaller number of countries for a shorter period of time. The focus within those studies was on one or two indicators. A study on bioeconomy [13] reflected the essential factors for the development of a sustainable bioeconomy, describing and linking economic factors, socio-economic factors, and ecological factors. The present analysis selects factors from each of the three mentioned categories, as per the study [13], identifies data, uses instruments to research the inter-linkages between the factors, tests hypotheses, and provides conclusions.

This study advances the reader's understanding of the research problem by considering for analysis a larger number of countries over a longer period of time. The focus of the study is on six indicators, significantly larger than in other related studies.

At the level of the sample of analyzed European countries, there is a tendency of long-term convergence in this area. The estimated models highlight the existence of a significant common long-term trend and an adjustment speed. These characteristics are due to the economic policies that exist not only at the level of the European Union, but also through the economic treaties it has with other European countries, which are meant to open up the trade and trade flows. Obviously, in the short term, heterogeneities are highlighted because they depend on the internal conditions and the structural specificity of the national economy of each country in the sample.

The study advances the reader's understanding by presenting the diversity of financial instruments, explaining the importance of financial mechanisms and financial resources as well as their risks and vulnerabilities, and the connection with bioeconomy and with the analyzed factors.

Considering the results of the empirical study, it is necessary to anchor them in the context of allocating/ensuring adequate financial support. In order to strategically orient the European economy towards the bioeconomy, there are needed financial resources, allocated through European programs, but also alternatives: Private and public financial resources to co-finance and support this transition process, the results of which will be reflected in the long term. The bioeconomy, by building a strategic vision and translating it at the regional level, can be the driving factor of societal changes. According to the European Commission [10], 67% of regions used European Structural and Investment Funds (ESIF) as a source of funding to support bioeconomic activities and access cooperation programs (for cooperation between regions: Joint Programming Initiatives; for other programs to promote the bioeconomy: Interreg, LIFE +, CIP/COSME, ERASMUS+, Intelligent Energy Europe) [10]. Specific mechanisms for granting ESIF funds to synergistic projects have been developed at the levels of several countries and regions (Italy, Czech Republic, Spain, France) together with public–private partnerships for the diversification of funding sources. The investment dimension is significantly larger than the capacity of the public sector. The European Commission estimates an additional annual investment requirement to reach the current targets set for 2030, in terms of climate and energy, of approximately 260 billion EUR, or, respectively, 1.5% of the GDP of 2018 [30]. It is also necessary to mobilize national budgets and private capital. In order to support the bioeconomic processes, it is appropriate to approach the issues related to sustainability, the sustainable financing of the European economy, and sustainable banking activities at European level in a convergent way.

Among the risks and uncertainties that may affect the bioeconomic processes, the following must be mentioned:

- the short-term orientation, the need for short-term financial results;
- non-integrated financial instruments/programs to provide the financing;
- the low degree of educational promotion/transparency regarding the sustainable impact of investments in bioeconomic activities, including for retail customers who could buy "green bonds" (EU Green Bonds) [31];
- the limited contributions of banks, insurance companies, institutional investors on the capital market, and entrepreneurs for re-orienting financial resources towards bioeconomic activities;
- the reduced promotion of scoring systems/standards accepted at the level of funders for better scoring of projects that support the bioeconomy;
- the low degree of connection of financial programs/resources offering specialized consultancy for the support of bioeconomic activities;
- the low level of education of the participants and lack of information and understanding about the impact of bioeconomic activities for each individual and for the European society as a whole.

The European Commission has set up a High-Level Expert Group (HLEG on Sustainable Finance) [32] with a mandate to propose changes to the investment chain in order to build a sustainable financing strategy for the European Union economy (to "achieve economic prosperity in the long term, increasing social inclusion and reducing dependence on the exploitation of finite resources and the natural environment" [10]). The objective proposed by the mandate is ambitious, but realistic: Transforming Europe into the main pole for low-carbon global investments, resource efficiency, and circular economy; in conjunction with this are the two strategic objectives: The adoption of the 2030 Agenda for Sustainable Development [33] and the Paris Agreement 2015/16, which regulates measures to reduce carbon dioxide emissions [34].

At the European banking level [35] a number of proposals have been formulated that will stimulate and contribute to the debate of European institutions, regulators, and banks on how to increase sustainable activities, mobilize and redirect private financial flows to support such activities, develop new tools, and increase the number of eligible projects. A specific element in the proposals made is the recommendation of the development of a "sustainable finance support factor" [33,36] as part of the legislation on bank capital requirements in the European Union. It has been proposed that the EBA (European Banking Authority) should explore the possibility of introducing a justification factor for certain assets that are classified as sustainable under the EU taxonomy. Europe has identified an annual financial gap of over 180 billion EUR to finance the policies and investments needed to maintain global temperatures [33] in line with the objectives of the Paris Agreement. It is more than obvious that, without the private sector, this funding gap cannot be closed. Because about two-thirds of the European economy is bank-financed, banks play—and will continue to play—a key role in the transition to a sustainable future, acting as investors, capital providers, and capital intermediaries for "green packaging" of some projects, in order to be eligible for funding through:

- Bank financing for "green projects" with the development of specific indicators or rating models. Due to the consistent role of the banking system in financing the European economy, in accordance with the results of the empirical study and with the theme of research—bioeconomy and foreign food trade—some relevant considerations from the perspective of sustainable banking activities were added. According to the report by the GABV (Global Alliance for Banking on Values) [36], sustainable banks consistently deliver products, services and social, environmental, and financial "profits" to support the real economy. The focus of sustainable banks is simultaneously on three components: People, environment, and prosperity. They are anchored in the communities in which they operate, establish long-term relationships with customers, and manage long-term risks.

The products and services of sustainable banks are mainly oriented toward supporting small- and medium-sized enterprises (SMEs) and microfinance, agriculture (food production, organic farms, rural, and agro-finance), financing energy efficiency (green energy, innovative technologies, alternative energy), financing eco-housing and social housing, and financing of educational and cultural programs (schools, kindergartens, theatres, museums). It is to be mentioned that the development and knowledge of activities/principles promoted by sustainable banks, together with other alternative financing instruments and mechanisms (crowdfunding platforms [37], specialized private energy investment funds, innovative institutional investors specializing in innovative products) could contribute synergistically to the developed programs of authorities and the orientation of financial and educational resources towards supporting bioeconomy at the European level.

- The technical support offered to obtain financing through capital markets, through the issuance of variable/fixed income securities for the financial support of "green projects".

Associations between international financial institutions, development banks, international banks, and local banks in the form of international networks—for example, the SBN (Sustainable Banking Network) operating in over 38 countries, with members of financial-banking institutions with assets of approximately 43 trillion USD [38]—lead to the promotion of sustainable financing as a global priority and to the transformation of the financial sectors/markets in which they operate. In addition, the efforts of coagulation and convergence in order to support the gradual transition to the bioeconomy were also made at the level of central banks, regulators, and supervisors. The Network for Greening the Financial Sector (NGFS) [39], which includes 46 members—central banks and financial supervisory authorities—emphasizes, at the level of the portfolios held, making socially responsible investments, managing the ESG risks (Environment, Social, Governance) related to the transition to the bioeconomy, and the development of scoring systems/new indicators that reflect the transition to a sustainable economy. The partnership between the bioeconomy and foreign trade with food products within the "green economy", together with the entire value chain that contributes to obtaining food, harmonized with the new technological context, must also be supported at the microeconomic level. Regional cooperation, especially in the agricultural field, between investment funds and small entrepreneurs, family associations, and authorized individuals contributes to the transformation of life and work, enhancing the connection between them, increasing the degree of innovation, and increasing social inclusion [40].

The future evolution of the bioeconomy will influence and will be influenced by public support and attitudes in the process of change [41]. The synergistic co-interest of all "actors" through harmonized, individual, and collective contributions can lead to the implementation and realization of this complex process of transition from the linear economy to the bioeconomy at the level of Europe.

This study adds to the existing literature by connecting the analysis of macroeconomic policies with the results of the empirical study and with potential guidelines for future policies. The study presents results and various influences between the analyzed factors in the short term, providing explanations and correlations for the long term trends as well.

This study advances the understanding in the complex topic of bioeconomy, providing an integrated approach for the reader. The approach explains and presents the relationships between the components of foreign trade in food products and macroeconomic variables, the various forms of financial support and mechanisms, and the newest established networks of financial institutions, focused on supporting the development of bioeconomy and its essential factors. Those correlations, together with the presentation of risks that may affect bioeconomic processes, offer a better understanding of the future mix of policies developed by authorities. The references included in the study offer a broad, up-to-date perspective with high relevance for the object of the research.

6. Conclusions

This research aimed to identify the factors with the highest capacity to stimulate the trade in food products, starting from a data panel over a limited time horizon (1996–2017), which included 24 European countries. The econometric results, statistically relevant, together with the qualitative aspects presented, highlighted that the bioeconomy and foreign trade in food products are in a sustainable partnership at the European level. All six independent variables analyzed act positively on the dependent variables; the main direct influencers of foreign food trade are the gross added value of the agricultural sector and the research and development expenses, both in the short and long term.

Sustainable bioeconomy can represent a strategic catalyst for economic growth at the European level and a beneficiary thereof; for all three visions of the bioeconomy included in the research, the main objectives are growth and sustainability. For future research, it is necessary to study the consumption habits, the behaviors regarding the food products, the modalities of their distribution at the regional level, and the connection with financing solutions, which will ensure the entire value chain necessary for the production, marketing, and promotion of the food products. A key point is to gather the cooperation and contributions of authorities, regulators, academic environment/researchers, investors, and financial-banking actors to harmonize the instruments in order to support the transition to the bioeconomy and make the partnership sustainable in the long term.

Author Contributions: D.C.N. and V.M. designed and implemented the research, analyzed the data and the results, and wrote and revised the manuscript together; both authors contributed equally (50%) for each task. All authors have read and agreed to the published version of the manuscript.

Abbreviations

The following abbreviations are used in this manuscript:

EU	European Union
WCED	World Commission on Environment and Development
OECD	European Organization for Cooperation and Development
FAO	Food and Agriculture Organization
UN	United Nations
ARDL	Autoregressive Distributed Lag
PMG	Pooled Mean Group
DFE	Dynamic Fixed Effects
MG	Mean Group
ECT	Error Correction Term
EXP	Food exports
IMP	Food imports
LABOR	Labor force, total
AGRI	Agriculture, forestry, and fishing, value-added
RD	Research and development expenditure
FOREST	Forest area
ENG	Fossil fuel energy consumption
RENEW	Renewable energy consumption GDP Gross Domestic Product
ESIF	European Structural and Investment Funds
HLEG	High Level Expert Group
EBA	European Banking Authority
GABV	Global Alliance for Banking on Values
SME	Small- and medium-sized enterprises
SBN	Sustainable Banking Network
NGFS	Network for Greening the Financial Sector
ESG	Environment, Social, Governance

References

1.   Brundtland at 25. Available online: https://greenblue.org/brundtland-at-25/ (accessed on 14 January 2020).
2.   Dinu, V. Food Security. *Amfiteatru Econ.* **2019**, *22*, 281–283.
3.   European Commission. What Is the Bioeconomy? Available online: https://ec.europa.eu/research/bioeconomy/index.cfm (accessed on 14 January 2020).
4.   European Commission. *Innovating for Sustainable Growth: A Bioeconomy for Europe*; Publication Office of the European Office: Luxembourg, 2012.
5.   Scarlat, N.; Dallemand, J.F.; Monforti-Ferrario, F.; Nita, V. The role of biomass and bioenergy in a future bioeconomy: Policies and Facts. *Environ. Dev.* **2015**, *15*, 3–34. [CrossRef]
6.   OECD. *The Bioeconomy to 2030. Designing a Policy Agenda*; Organisation for Economic Co-Operation and Development (OECD): Paris, France, 2009.
7.   Poltronieri, P. Alternative energies and fossil fuels in the bioeconomy era: What is needed in the next five years for real change. *Challenges* **2016**, *7*, 11. [CrossRef]
8.   Bugge, M.M.; Hansen, T.; Klitkou, A. What is the bioeconomy? A review of the literature. *Sustainability* **2016**, *8*, 691. [CrossRef]
9.   Staffas, L.; Gustavsson, M.; Mccormick, K. Strategies and policies for the bioeconomy and bio-based economy: An analysis of official national approaches. *Sustainability* **2013**, *5*, 2751–2769. [CrossRef]
10.   European Commission. *Research and Innovation Plans & Strategies for Smart Specialisation (RIS3) on Bioeconomy*; European Commission: Brussels, Belgium, 2017; p. 5.
11.   Porter, M. *The Competitive Advantage of Nations*; Free Press: New York, NY, USA, 1990.
12.   McCormick, K.; Kautto, N. The Bioeconomy in Europe: An overview. *Sustainability* **2013**, *5*, 2589–2608. [CrossRef]
13.   Food and Agriculture Organization of the United Nations. *Assessing the Contribution of Bioeconomy to Countries' Economy: A Brief Review of National Frameworks*; Food and Agriculture Organization of the United Nations: Rome, Italy, 2018.
14.   Urmetzer, S.; Pyka, A. *Varieties of Knowledge-Based Bioeconomies*; University of Hohenheim: Stuttgart, Germany, 2014; pp. 91–2014.
15.   Pașnicu, D.; Ghența, M.; Matei, A. Transition to Bioeconomy: Perceptions and behaviours in Central and Eastern Europe. *Amfiteatru Econ.* **2019**, *21*, 9–23.
16.   European Commission. Agri-Food Trade in 2018. Available online: https://ec.europa.eu/info/sites/info/files/food-farming-fisheries/news/documents/agri-food-trade-2018_en.pdf (accessed on 14 January 2020).
17.   European Commission. EU Agri-Food Trade Surplus Hits Record Levels in September 2019. Available online: https://ec.europa.eu/info/news/eu-agri-food-trade-surplus-hits-record-levels-september-2019-2019-dec-17_en (accessed on 15 January 2020).
18.   World Bank Data. Available online: https://data.worldbank.org/indicator?tab=all (accessed on 14 January 2020).
19.   Pascal, L. *Assessing the Effects of International Trade on Private R&D Expenditures in the Food-Processing Sector*; Department of Economics, University of Lethbridge: Lethbridge, AB, Canada, 2002; pp. 349–369.
20.   Bojnec, Š.; Fertő, I. Does EU Enlargement Increase Agro-Food Export Duration? *World Econ.* **2012**, *35*, 609–631. [CrossRef]
21.   Dragoș, S.; Mare, C.; Dragoș, C.M. Institutional drivers of life insurance consumption: A dynamic panel approach for European countries. *Geneva Pap. Risk Insur.-Issues Pract.* **2019**, *44*, 36–66. [CrossRef]
22.   Dragos, S.L.; Mare, C.; Dragota, I.M.; Dragos, C. M, Muresan G.M. The nexus between the demand for life insurance and institutional factors in Europe: New evidence from a panel data approach. *Econ. Res.-Ekon. Istraz.* **2017**, *30*, 1477–1496.
23.   International Monetary Fund (IFM). World Economic Outlook update January 2020. Available online: http://www.imf.org/en/publications/weo/issues/2020 (accessed on 21 January 2020).
24.   Morgans Stanley Capital International (MSCI). Market Classification. Available online: https://www.msci.com/market-classification (accessed on 19 January 2020).
25.   United Nations. World Economic Situation and Prospects 2020. Available online: https://www.un.org/development/desa/dpad/publication/world-economic-situation-and-prospects-2020 (accessed on 19 January 2020).

26.	Toth, D.; Maitah, M.; Maitah, K. Development and Forecast of Employment in Forestry in the Czech Republic. *Sustainability* **2019**, *11*, 6901. [CrossRef]

27.	Lund, H. *Elsevier Global Right*; Aalborg Universitet: Aalborg, Denmark, 2007; Volume 32, pp. 912–919.

28.	Choi, I. Unit root tests for panel data. *J. Int. Money Financ.* **2001**, *20*, 249–272. [CrossRef]

29.	Pesaran, H.M.; Shin, Y.; Smith, R.P. Pooled mean group estimation of dynamic heterogeneous panels. *J. Am. Stat. Assoc.* **1999**, *94*, 621–634. [CrossRef]

30.	European Commission. *The European Green Deal*; European Commission: Brussels, Belgium, 2019.

31.	Green Bonds. Mobilising the Debt Capital Markets for a Low-Carbon Transition. Bloomberg. Philantropies. 2015. Available online: https://www.oecd.org/environment/cc/Green%20bonds%20PP%20%5Bf3%5D%20%5Blr%5D.pdf (accessed on 5 February 2020).

32.	European Commission. *Final Report of the High-Level Expert Group on Sustainable Finance*; European Commission: Brussels, Belgium, 2018; Available online: https://ec.europa.eu/info/publications/180131-sustainable-finance-report_en (accessed on 19 January 2020).

33.	The 2030 Agenda for Sustainable Development. Available online: https://sustainabledevelopment.un.org/content/documents/21252030%20Agenda%20for%20Sustainable%20Development%20web.pdf (accessed on 15 January 2020).

34.	The Paris Agreement. Available online: https://unfccc.int/process-and-meetings/the-paris-agreement/the-paris-agreement (accessed on 15 January 2020).

35.	European Banking Federation—Press Release from 9 December 2019. Available online: https://www.ebf.eu/sustainable-finance/banks-present-proposals-to-scale-up-sustainable-finance/ (accessed on 12 December 2019).

36.	Global Alliance for Banking on Values (GABV). *Strong, Straightforward and Sustainable Banking. A Report on Financial Capital and Impact Metrics of Values Based Banking*; Global Alliance for Banking on Values: Zeist, The Netherlands, 2012.

37.	Crowdfunding the European Transition to Renewable Energy. 2016. Available online: http://www.crowdfundres.eu/index.html@p=695.html (accessed on 4 February 2020).

38.	Sustainable Banking Network, International Finance Corporation. *Global Progress Report of the Sustainable Banking Network*; Sustainable Banking Network, International Finance Corporation: Washington, DC, USA, 2019.

39.	Network for Greening the Financial System. *A Sustainable and Responsible Investment Guide for Central Banks' Portfolio Management*; Network for Greening the Financial System: Paris, France, 2019.

40.	European Fund for Southeast Europe. *Impact Report 2018. Cultivating Entrpreneurship*; European Fund for Southeast Europe: Luxembourg, 2019.

41.	Dinu, V. The Transition to Bioeconomy. *Amfiteatru Econ.* **2019**, *21*, 5–7.

9

Sustainable Brand Management of Alimentary Goods

Jana Majerova [1,*]**, Wlodzimierz Sroka** [2,3]**, Anna Krizanova** [1]**, Lubica Gajanova** [1]**,
George Lazaroiu** [4] **and Margareta Nadanyiova** [1]

[1]   Department of Economics, Faculty of Operation and Economics of Transport and Communications,
University of Zilina, Zilina 026 01, Slovakia; anna.krizanova@fpedas.uniza.sk (A.K.);
lubica.gajanova@fpedas.uniza.sk (L.G.); margareta.nadanyiova@fpedas.uniza.sk (M.N.)

[2]   Management Department, WSB University, Dąbrowa Gornicza 41-300, Poland; WSroka@wsb.edu.pl

[3]   North-West University, Potchefstroom 2351, South Africa

[4]   Department of Economic Sciences, Spiru Haret University, Bucharest 030045, Romania;
phd_lazaroiu@yahoo.com

*   Correspondence: jana.majerova@fpedas.uniza.sk

Abstract: Sustainability of food production and consumption has become one of the most discussed topics of sustainable development in global context. Thus, traditional managerial patterns have to be revised according to the social request. The revisions that have been done so far are based on relevant specifics of production and have mostly general character. Moreover, traditional managerial postulates do not change; only their way of implementation is modified. These two facts are possible reason of the practical fails in sustainable management of alimentary goods. One of these traditional managerial concepts is brand. Within this context, it has been considered as a facilitator of CSR (corporate social responsibility) activities. But the situation has changed, and the suspicion that brand loyalty is not a facilitator but an obstacle to the sustainable management is high. Thus, the importance of research of brand loyalty in scope of sustainable management of alimentary goods is indisputable. According to the above mentioned, the main goal of the contribution is to identify relevant brand value sources of loyalty in scope of sustainable brand management of alimentary goods. To achieve this, the factor analysis has been applied to provide statistical evaluation of data obtained from our own questionnaire survey. We have found out that components of brand value sources do not vary when comparing brands and those without loyal consumers. Based on this, appropriate recommendations for the theory and practice of sustainable brand management of alimentary goods have been formulated.

Keywords: brand; branding; brand management; sustainability; alimentary goods; brand loyalty

JEL Classification: Q01; M11; M31

1. Introduction and Current Situation Insight

Contemporary scientific literature highlights the importance of corporate social responsibility in the sustainable management process, not only in general, but also regarding the sectoral specifics of production [1–3]. Thus, according to this trend, producers of alimentary goods should put moral pressure on their suppliers, and consumers should be more involved in their buying decision processes [4]. In order to achieve these changes in so far functioning stereotypes, the issue of brand and brand management should be highlighted. But the fact that alimentary goods are characterized by traditional habitual buying behavior creates a specific obstacle to such an approach, because the reason is that, in this type of buying behavior, consumer's loyalty is the leading motivation for purchase whether or not the brand is socially responsible on the market [5,6]. The above-mentioned fact can be considered as a real problem when seeking to achieve sustainability in scope of food production and consumption. The main reason for this is that brand managers of alimentary goods in case of brands subjectively perceived as valuable are not motivated enough to act responsibly towards society, and consumers are not motivated to play active role in the process of information searching [7]. In this case, socially responsible activities are not directly connected with desired effect, even if they are applied in accordance with contemporary state of knowledge. While the main theoretical attention is generally paid to the issue of facilitators of sustainable management, practice shows the need of application of an opposite approach—that is, a focus on possible obstacles to optimal implementation of managerial patterns. Thus, it is vital to change the approach and to focus not on the facilitators but on the contrary, also on the obstacles to sustainable development. Only when these barriers are identified will it be possible to manage them and to eliminate their negative impacts on the effectiveness of sustainable management. Thus, the irony is, on the one hand, the phenomenon of consumer loyalty can be considered as a stimulus, and on the other hand, as an obstacle to sustainable performance of brands on alimentary goods market. According to this fact, it is vital to analyze value sources of brands characterized by presence vs. absence of consumer loyalty and to apply a conscious and responsible approach to the consumer´s loyalty as one of the leading buying behavior motivations.

Thus, this article focuses on the meaning of brand loyalty research in the scope of sustainable management of alimentary goods. The first part of the article analyses interactions between loyalty and sustainable brand management in general, as well as in case of alimentary goods. This is followed by a methodological part of the article where the main postulates of the article as well as factor analysis and relevant statistical tests are described. The next part synthetizes obtained results formulated on the basis of detected discrepancies in value sources of brands characterized by presence vs. absence of consumer's loyalty. The discussion is also an immanent part of the article and it is included in the same part as results of the research itself. The last part contains the summary of the main results and outcomes of research with implications for the future.

The conceptualization of the research is as follows: (1) literature review and current situation summary; (2) formulation of original presumptions of authors; (3) creation of the model of research based on questionnaire survey; (4) realization of the research itself; (5) statistical testing and evaluation of obtained data; (6) formulation of conclusions and managerial implications, and (7) critical consideration of framework conditions of applicability of the research outcomes, obstacles and limits of the research, and possible ways of further research in scope of sustainable brand management of alimentary goods. Doing all the mentioned activities while respecting the contemporary state of knowledge in the area of brand management, corporate social responsibility, strategic management, and statistics.

2. Literature Review and Theoretical Background of Research Itself

In recent decades, many researches in scope of the complex issue of corporate social responsibility have been carried out. However, it is still necessary to continue investigating its benefits and causalities in scope of marketing [8]. Thus, nowadays, it is crucial that the attention is paid to the impact of contemporary global trends on the practical aspects of managerial challenges stimulated by incorporation of the concept of corporate social responsibility. The reason is the assumption that it affects significantly the success of each company and its competitiveness by modifying traditional managerial patterns according to the postulates of corporate social responsibility [9]. Until now, the research has investigated this issue separately, without any deeper interactions with overall managerial framework. The representative example of such an approach is a huge scientific school focused on the analysis of supply chain management and its importance in sustainable management. It has been concluded that such an approach is insufficient, and formulated advices for managers should be revised according to the wider consequences of implementation of sustainable management [10]. Despite this fact, the importance of sustainable supply chain management as an immanent part of corporate social responsibility with significant impact on consumer buying behavior is indisputable. The reason is that consumers adapt supply chain practice to the real socially responsible behavior of the company and they perceive their own value across the prism of the market performance of the company and its transformation into real corporate citizenship [11]. So it is obvious that brands strongly affect overall sustainable development of the society as they can be considered as a powerful tool to create and maintain public opinion [12]. According to the above mentioned, brands are becoming parts of people's lives all over the world. They form an immanent part of human beings' reality and they co-act in the process of own value creation as they have strong interpretation power about their consumers [13]. Moreover, brands have become a very effective tool of communication. By these, consumers inform other members of their social group who they are and if they really belong to a specific social group. It is a way to communicate one's own social status and life values. Thus, once the brand starts to be perceived as valuable from the point of view of selected group of consumers, it is very likely that this group starts to be identified with this brand and all the group members will transform into loyal consumers who consider such a brand as subjectively valuable for them. A logical consequence of this situation is that there will be many benefits, like less price sensitivity, lower need of communication activities of such a brand, less critical approach towards quality, and so forth. Thus, a loyal consumer is a dream of each brand manager because only by personalizing a consumer with a brand, can brand management be considered as really effective. However, there is not only a one-way influence between brand and consumers. While, at the beginning, the brand is created by consumers, very shortly after creating a loyal consumer platform, consumers start to be modified by the brand. It means that the brand has potential to change social attitudes and life values and it is willing to be the real tool of social change leading to the sustainable development [14]. Thus, brands characterized by their environmental conformity can be transformed into a strong tool to learn who their consumers are. On the one hand, there is a huge amount of brands who are primarily focused on already environmentally oriented consumers, but on the other hand, there are many more brands (traditionally perceived as valuable and significant for selected social groups) that can change the environmental orientation of their consumers [15]. In scope of the above mentioned, it is vital to pay attention to the detailed segmentation of consumers. The reason is that the segment of socially responsible consumers can vary internally according to the highlighted green attitudes connected to brands. It means that there is at least a double construct of a socially responsible consumer: (1) a consumer who is really internally environmentally oriented and who prefers socially responsible brands when making a buying decision and (2) a consumer whose environmental attitudes are only derived and narrowly connected with the essence of his/her favorite brand. The main difference between these two types of consumers lies in the fact that, while in the first case, socially responsible orientation of the brand is the key attribute of its brand value building, in the second case, such a consumer does not prefer socially responsible brands automatically as it is only consequence of long-term subjectively perceived brand value and the main motive for brand value substitution would not be its replacement by a more environmentally friendly brand. Thus, it is vital to identify real internal attitudes of consumers and to discover if the subjectively perceived brand value is reason or consequence of its socially responsible market behavior [16]. So the main task for managers of not only formally but really sustainably manageable brands is to identify internal motives and brand value sources of their consumers and to co-act in the process of market education as one of the prerequisites of socially sustainable development [17].

One of possible tools to achieve this state is to apply a conscious and responsible approach to the consumer's loyalty and its creation through brand management [18]. The reason of such a postulate is that the traditional educational model of consumers (push model) has failed and thus, it is necessary to apply the opposite one (pull model). The main idea of such a model lies in identification of appropriate brand value sources of brands characterized by loyal and nonloyal attitudes and in subsequent usage of relevant brand loyalty sources (brand value sources in case of loyal brands) for purposes of sustainable management. Thus, subconscious consumers' education would be applied—that is, brands will affect consumer's environmental attitudes through existing brand value sources, and this change in attitudes will affect their general attitudes towards sustainable development of society. Unless this is done, the consumer's loyalty can be, in specific product categories, considered as a significant obstacle to the development of sustainable management and corporate social responsibility. That is why companies who are devoting significant resources to socially responsible activities, insights into the optimal formulation, implementation, and effectiveness estimation of socially responsible strategies are nowadays still on the crossroad [19].

Although the scientific literature clearly stated strong positive correlation between corporate social responsibility implementation into strategic management of the company and its positive image, the individual processes and mechanisms which are relevant from the point of view of this positive effect creation have not been analyzed in details so far [20]. Although the possible significant impact of psychographic characteristics of consumers on the process of a brand's positive image creation has been highlighted, there is a scientific gap lying in the need of the mechanisms of behavioral economics investigation [21]. In scope of the above mentioned, Song et al. focused on the identification of a relationship between selected structural associations and brand loyalty. They provided a case study on the example of coffee shop brands. As a result of the scientific effort of Song et al., it has been discovered that the phenomenon of so-called love brands is extremely effective in the process of brand loyalty creation and management [22]. Other researches have proven the significance of brand satisfaction and brand trust in the process of brand loyalty creation and maintenance [23,24]. On the other hand, not only the impact of selected subjectively perceived brand value sources on the brand loyalty has been investigated, but also the impact of brand loyalty and its value sources on the effectiveness of brand social responsibility has been analyzed. As a result, it has been concluded that brand loyalty is essential in the process of sustainable brand management implementation via systematic manipulation with (1) attitudinal loyalty, (2) expenditure level, and (3) intention to buy and recommend [25].

Since the very beginning of brand loyalty research, authors have examined separately purchase loyalty and attitudinal loyalty as two main aspects of brand loyalty. Recently, the first of them has been connected with factual brand performance, while the second one has been described as only a hypothetical construct of brand loyalty without significant impact on brand performance on the market [26]. This research rejected the original concept of dual structure of brand loyalty, which was constructed on the presumption of the meaning of attitudinal loyalty as a key issue in the pricing fences setting [27].

In addition to this dual approach to the brand loyalty research, also another one has been applied—brand as a way to build loyalty and loyalty as a way to build a brand. Bhattacharya and Sen have determined the conditions under which consumers enter into an emotional relationship with brands. Based on this, they have detected loyalty as one of the presumptions of subjectively perceived brand value. Thus, they have applied the opposite concept to the so far implemented. According to this, Bhattacharya and Sen doubt subjectively perceived brand value as an antecedent of brand loyalty creation [28]. Following this approach, Stocchi and Fuller identified brand loyalty with the main brand equity source, discussing different segments of consumers and different markets [29]. In their approach, we can see another dimension of brand loyalty research because they draw from the general approach and they also apply a diversified approach. They conclude that difference in ranking of individual brand value sources (not only its quantity but also quality) perceived by loyal and nonloyal consumers exists.

Thus, also the relationship between the sources of sustainable brand value management and brand loyalty is analyzed [30]. It has been found that there is a positive correlation between (1) sustainable brand management and brand attitude; (2) brand attitude and brand loyalty, and (3) brand loyalty and sustainable brand management [31]. This conclusion is essential as so far, the authors have mainly stated that sustainable management has a positive impact on perceived brand value. The key importance of brand loyalty in this process has not been detected until now. Moreover, it has been also highlighted that the concept of brand value and its patterns could vary across socio-cultural specifics of consumers. It means that the need of focusing on the consumers at the regional basis has been stated. According to this trend, contemporary research not only highlights the importance of consumer loyalty in the process of brand management in general, but authors also focus on identification of sources of brand loyalty across markets (in both product and regional prospective) [32,33].

Chatzipanagiotou et al. have applied a cross-cultural approach to the analysis of brand value sources [34]. They state that most of the so far created models of brand value building and managing are very simple and they do not take into account the complexity of relevant factors. As such a factor, it is identified also the consumer behavior and its individuality and difficult predictability due to the rejection of the traditional neoclassical concept of so-called *Homo Oeconomicus*. So, they have constructed regression model with significant factors affecting the final subjectively perceived brand value as well as they have identified critical points of this model implementation when applying it to cross-cultural environment. Regional specifics in perception of brand value sources with implications to brand loyalty have been discussed by Sukalova et al., Tamuliene and Pilipavicius, Rozgina, Jain and Zaman, and Christodoulides et al. These authors have verified the effectiveness of traditional Aaker's quadratic model of brand value sources in the wider perspective of unified European single market as well as individual national markets, formulating advices for the practice of international brand value management [35–39].

Not only regional but also sectoral specifics are relevant for the research of interactions between brand loyalty and sustainable management. Rather et al. focused in their research on sectoral specifics of brand loyalty using factor analysis [40]. They have developed an integrated model of brand loyalty building based on the consumer's perception of identified key brand value sources (brand commitment, brand trust and brand satisfaction).

Emotional attributes of brand loyalty in general (not respecting product or regional prospective but focusing on the pure nature of brand value sources in scope of consumer's characteristics) have been analyzed by Poushneh and Vasquez-Parraga [41]. On the one hand, they have removed the traditional heterogeneous approach, but on the other hand, they have incorporated another selective criterion—consumer typology. This approach follows the research which examined the changes in brand loyalty over time, a case study of plenty of product categories worldwide [42].

Since then, product categories have been analyzed mainly separately, focusing on these product categories where the loyalty can be considered as main motivation of buying behavior due to the capital demanding of the purchase [43]. Previously, customer satisfaction and image were priority proven as the main attributes of brand loyalty creation across product categorization. Unfortunately, only a few scientists have focused on their synergic effect and complex research of the mechanisms of causalities and correlations between customer satisfaction and image in the process of brand loyalty creation in the light and shadow of sustainable brand management [44]. One of these studies has formed the main premise for future research aimed to tourist services, as a significant subcategory of services where the decision-making process is based on rational pillars [45].

Similarly, the importance of brand loyalty in scope of corporate social responsibility has been analyzed in the banking sector. According to the results of this study, it has been proven that sustainable brand management directly influences brand image and brand value subjectively perceived by consumers. Moreover, dominant importance of brand loyalty in the process of sustainable brand management has been identified [46,47]. IT sector is another field where consumer's loyalty has been discussed as one of the pillars of CSR (corporate social responsibility) effectiveness [48]. Not only stating but also investigating the significance of consumer's loyalty has been the aim of the study provided in scope of luxury brands [49]. Generally, all above-mentioned researches focused on the identification of brand value sources relevant in scope of brand loyalty creation in a specific product category. The comparison of brand value sources in case of presence vs. absence of brand loyalty has not been done so far. However, contemporary trend in brand loyalty research indicate such an ideological change. On the contrary, research becomes more complex, trying to find common mechanisms in wider groups of brands.

Such a wider group of brands is also the category of so-called private label brands. Contemporary research of brand loyalty in case of this category escalates turbulently nowadays [50]. One of the trends identified in this field is the investigation of seasonality in brand loyalty, which has been detected by the practice of private label brands. Mainly in case of private label brands of alimentary goods has the observation of seasonality in consumer's preferences been really obvious [51]. This can be considered as a possible obstacle to sustainable brand management based on brand loyalty. This is because the variability in brand loyalty has not described in details so far, and thus, complex analysis of this phenomenon needs to be carried out [52]. Moreover, this fact partially rejects theoretical constructs based on presumption of positive effect of brand loyalty on the overall effectiveness of sustainable brand management. This is because such fluctuations in brand loyalty could affect negatively sustainable brand management activities. Thus, although many approaches to brand loyalty and its importance in sustainable management have been applied so far, there is still a scientific gap lying in the fact that brand loyalty is traditionally considered as a facilitator of corporate social responsibility while there are various indicators of the opposite—especially in the category of products characterized by habitual buying behavior.

In scope of the above mentioned, these research questions have been set:

(1)   How are brand value sources of alimentary goods?
(2)   What is the order of importance of these brand value sources and their components?
(3)   Do these sources differ in case of brand loyalty absence vs. brand loyalty presence?

3. Methodological Background

According to the literature review above, the main aim of the article is to identify brand value sources of loyalty which are relevant to sustainable brand management of alimentary goods. To achieve this aim, we have used the data from our own research provided on the socio-demographically representative sample of 2000 respondents (sample without outliers and incompatible units was 697) during the first half of 2019. We conducted this research via a questionnaire survey in the form of computer-assisted web interviewing respecting the ICC/ESOMAR (International code on Market, Opinion and Social Research and Data Analytics). The questionnaire was administered in Slovak Republic among its inhabitants over 15 years of age who were asked to complete the questionnaire because of their legal working subjectivity. Thus, the main presumption of autonomous buying decision-making has been fulfilled. On the other hand, one of the limitations of general applicability of the research outcomes has been caused by this fact—that is, territorial applicability of the recommendations done on the basis of research outcomes only in scope of Slovak consumer´s preferences. Thus, possible implementations of statements which result from research itself are applicable only in case of alimentary goods brands addressed to Slovak consumer (domestic or foreign). The questionnaire consisted of three parts with the following reasoning: (1) the first part covered the general socio-demographic profile of respondents; (2) the second part covered questions about perception of brand value sources generally, and (3) the third part covered questions about perception of brand value sources in details across the traditional typology of buying behavior and representative product categories.

To provide research of brand value sources in scope of buying behavior typology, the traditional quadratic typology of buying behavior has been used, where on the basis of the degree of engagement and differentiation, we can identify the following categories: (1) complex buying behavior (high involvement/significant differences between brands); (2) variety seeking behavior (low involvement/significant differences between brands); (3) dissonance-reducing buying behavior (high involvement/few differences between brands), and (4) habitual buying behavior (low involvement/few differences between brands) [53]. The last mentioned category is the category which is relevant for purposes of research of sustainable brand management of alimentary goods. Brand value sources are analyzed in their traditional structure defined by Aaker—that is, (1) imageries; (2) attitudes; (3) attributes, and (4) benefits. The components of brand value sources are set in accordance with provided literature review and with relevance to so far identified specifics of psychographic profile of Slovak consumers [54].

The model of brand value sources identified by Aaker was used in accordance with the provided literature review due to its general applicability regardless specifics of product categories formulated on the principle of typology of buying behavior. The reason is that the presented article is only a partial outcome of complex research aimed to verify the internal diversification in brand value sources, ranking in case of brand value presence vs. absence across four basic product categories. Brand value sources and their relevant components which have been, through the realized questionnaire survey, tested in scope of their importance across product categories relevant for the types of buying behavior are summarized in the Table 1 below.

Table 1. Coding of brand value sources and their components relevant to further research evaluation.

Brand Value Sources	Components of Brand Value Sources	Code	
		Brand Loyalty Absence	Brand Loyalty Presence
imageries	happiness	2	4
	expectations	3	5
	satisfaction	1	1
	certainty	5	2
	positive associations	4	3
attitudes	I aim to buy branded products	12	13
	I am interested in branded products on a regular basis	13	12
	branded products attract my attention because I consider them better	11	11
	branded products attract my attention because I consider them more prestigious	14	14
attributes	quality	19	19
	creativity of ad	16	16
	popularity	15	15
	availability	17	17
	innovativeness	18	18
benefits	branded product makes me happier	10	10
	branded product increases my social status	6	8
	branded product makes it easier for me to get friends	7	6
	branded product attracts the attention of others	8	7
	branded product belongs to my lifestyle	9	9

Source: Authors' own research, 2019.

Factor analysis has been chosen as the main statistical tool for evaluation of the consumer's perception of brand value sources in case of brand loyalty absence vs. brand loyalty presence. This analysis is one of the group of multidimensional statistical methods which are used to create so-called factors (previously unobservable variables) to reduce the amount of originally set attributes without losing the relevant information obtained inside the data set [55,56]. Recently, this statistical tool has been used with higher frequency in the social sciences due to the boom in information technology development and the need of reducing subjectivity. The definition of the relevant statistical model as well as the identification of rational assumptions is the base of this analysis. In the process of identification of relevant factors, it is primarily important to identify and test the dependence between originally defined variables through the correlation matrix. The basic presumption for the data reduction is the correlation of these variables verified by the correlation matrix creation as well as the fulfilment of the assumption that identified correlation exists as a consequence of less undetected hidden variables (factors). Based on this, it is possible to diversify originally defined variables into partial groups. In these groups, there are unified factors which internally correlate more inside the group than in comparison with other groups.

We assume that x is a p-dimensional random vector of the considered variables with a vector of mean values μ, a covariance matrix $C(X) = \Sigma$, and a correlation matrix of simple correlation coefficients $P(X) = P$. One of the basic assumptions of factor analysis is the existence of R common background factors $F_1, F_2, \ldots, F_R$; trying to have them as little as possible, preferably less than p. The P-dimensional random vector consists of the j-observable random variables $x_j, j = 1, 2, \ldots, p$; which can be expressed by Equation (1) as

$$X_j = \mu_j + \gamma_{j1}F_1 + \gamma_{j2}F_2 + \ldots + \gamma_{jR}F_R + \varepsilon_j, \tag{1}$$

where $\varepsilon_1, \varepsilon_2, \ldots, \varepsilon_p$ are p stochastic error terms referred to as specific factors. If we write this in matrix, we get the Equation (2):

$$x = \mu + \Gamma f + \varepsilon, \tag{2}$$

where Γ is a matrix of factors loadings type p R; f is R-member vector of common factors, and ε is p-member vector of specific factors. Factors loadings can be considered as regression coefficients p of observed variables on R nonobservable factors, and when certain conditions of solution are met, they are also covariance between the original and the new variables. Factors loadings can be interpreted as the contribution of the r-factor of the j-specified variable, when the same units of measurement are used. To determine the adequacy of the statistical sample, we use the KMO (Kaiser–Meyer–Olkin) Test Equation (3):

$$KMO = \frac{\sum_{j\neq j'}^{p} \sum_{j\neq j'}^{p} r^2(x_j, x_{j'})}{\sum_{j\neq j'}^{p} \sum_{j\neq j'}^{p} r^2(x_j, x_{j'}) + \sum_{j\neq j'}^{p} \sum_{j\neq j'}^{p} r^2(x_j, x_{j'} . \text{ other } x)} \tag{3}$$

where $r^2 (x_j, x_{j'})$ are simple correlation coefficients and $r^2 (x_j, x_{j'} \cdot \text{ other } x)$ are partial correlation coefficients under the condition of statically constant remaining p-2 variables. $(x_1, x_2, \ldots, x_{j-1}, x_{j+1}, \ldots, x_{j'-1}, x_{j'+1}, x_p)$.

Required value of KMO test should be higher than 0.6. By acquiring it, the adequacy of statistical sample is proved [57]. Required value of Barlett's test of sphericity should be lower than 0.05. By acquiring it, the dependence between variables is proved [58]. Required value of Cronbach's Alpha should be higher than 0.8. By acquiring it, the intrinsic consistency of the factors is proved [59]. Detection of the optimal values of these tests forms appropriate basis to the identification of the order of brand value sources in case of loyalty absence vs. loyalty presence. Thus, a set of advices formulated on the basis of factors identification and comparison of obtained results can be submitted to the practice of sustainable brand value building and managing of alimentary goods.

4. Results and Discussion

Provided KMO test (Kaiser–Meyer–Olkin test) has indicated the adequacy of the used statistical sample in case of brands with consumer's loyalty absence as well as in case of brands with consumer's loyalty presence (>0.6). When testing the brand value sources in case of brand loyalty absence, the value of 0.902 has been reached, and in the case of brand loyalty presence, the value of 0.931 has been reached. Barlett's test of sphericity has proved the existence of dependence between variables by acquiring the resulting value at 0.00 in case of brands with consumer's loyalty absence as well as in case of brands with consumer's loyalty presence (<0.05). We have also detected statistical relevance of four relevant factors in both cases.

The testimonial value of factor analysis in case of brand value sources research when brand loyalty is absent has reached a value of 76.552%. (See Table 2)

Table 2. Total variance explained—brand loyalty absence.

Code	Initial Eigenvalues			Extraction Sums of Squared Loadings			Rotation Sums of Squared Loadings		
	Total	% of Variance	Cumulative %	Total	% of Variance	Cumulative %	Total	% of Variance	Cumulative %
1	9.368	49.303	49.303	9.368	49.303	49.303	4.312	22.694	22.694
2	2.595	13.660	62.963	2.595	13.660	62.963	3.871	20.374	43.068
3	1.380	7.264	70.228	1.380	7.264	70.228	3.233	17.015	60.083
4	1.202	6.325	76.552	1.202	6.325	76.552	3.129	16.469	76.552
5	0.691	3.637	80.189						
6	0.494	2.598	82.787						
7	0.446	2.347	85.134						
8	0.374	1.967	87.101						
9	0.347	1.824	88.925						
10	0.307	1.614	90.539						
11	0.289	1.522	92.061						
12	0.280	1.472	93.534						
13	0.247	1.300	94.834						
14	0.206	1.082	95.916						
15	0.194	1.020	96.936						
16	0.184	0.967	97.903						
17	0.163	0.857	98.760						
18	0.132	0.693	99.454						
19	0.104	0.546	100.000						

Source: Authors' own research, 2019.

In case of brand loyalty absence, the existence of four relevant factors with significant components has been proved. These factors are (1) imageries with five components where the value of Cronbach's Alpha has been 0.813; (2) benefits with five components where the value of Cronbach's Alpha has been 0.842; (3) attitudes with four components where the value of Cronbach's Alpha value has been 0.849, and (4) attributes with five components where the value of Cronbach's Alpha has been 0.813. (See Table 3)

Table 3. Rotated component matrix—brand loyalty absence.

Code	Brand Value Source			
	Imageries	Benefits	Attitudes	Attributes
1	0.855			
2	0.825			
3	0.790			
4	0.790			
5	0.784			
6		0.908		
7		0.899		
8		0.859		
9		0.579	0.406	
10	0.442	0.551		
11			0.798	
12			0.785	
13		0.414	0.733	
14			0.709	
15				0.798
16				0.784
17				0.739
18				0.664
19	0.415		0.421	0.594

Source: Authors' own research, 2019.

The creation of a rotated component matrix has allowed to rank the brand value sources in case of brand loyalty absence according to their priority in the impact on consumer's perception as follows: (1) imageries; (2) benefits; (3) attitudes; (4) attributes. (See Table 4)

Table 4. Brand value sources—brand loyalty absence.

Factors	F1 Imageries	F2 Benefits	F3 Attitudes	F4 Attributes
N of Items	5	5	4	5
Cronbach's Alpha	0.813	0.842	0.849	0.813
% of Variance	49.303	13.660	7.264	6.325

Source: Authors' own research, 2019.

The testimonial value of factor analysis in case of brand value sources research when brand loyalty is present has reached a value of 74.614%. (See Table 5)

Table 5. Total variance explained—brand loyalty presence.

Code	Initial Eigenvalues			Extraction Sums of Squared Loadings			Rotation Sums of Squared Loadings		
	Total	% of Variance	Cumulative %	Total	% of Variance	Cumulative %	Total	% of Variance	Cumulative %
1	9.500	50.002	50.002	9.500	50.002	50.002	4.177	21.982	21.982
2	2.080	10.949	60.951	2.080	10.949	60.951	3.941	20.741	42.723
3	1.456	7.665	68.616	1.456	7.665	68.616	3.125	16.446	59.170
4	1.140	6.001	74.617	1.140	6.001	74.617	2.935	15.448	74.617
5	0.685	3.606	78.223						
6	0.480	2.527	80.751						
7	0.425	2.235	82.986						
8	0.413	2.171	85.157						
9	0.392	2.061	87.218						
10	0.359	1.889	89.106						
11	0.309	1.625	90.732						
12	0.294	1.547	92.279						
13	0.280	1.473	93.752						
14	0.261	1.372	95.124						
15	0.247	1.301	96.426						
16	0.222	1.169	97.595						
17	0.171	0.899	98.494						
18	0.151	0.797	99.291						
19	0.135	0.709	100.000						

Source: Authors' own research, 2019.

In case of brand loyalty presence, the existence of four relevant factors with significant components has been proved. These factors are (1) imageries with five components where the value of Cronbach's Alpha has been 0.854; (2) benefits with five components where the value of Cronbach's Alpha has been 0.837; (3) attitudes with four components where the value of Cronbach's Alpha value has been 0.841 and (4) attributes with five components where the value of Cronbach's Alpha has been 0.869. (See Table 6)

Table 6. Rotated component matrix—brand loyalty presence.

Code	Brand Value Source			
	Imageries	Benefits	Attitudes	Attributes
1	0.800			
2	0.781			
3	0.776			
4	0.767			
5	0.751			
6		0.880		
7		0.860		
8		0.850		
9		0.604		
10	0.521	0.531		
11			0.812	
12			0.784	
13			0.762	
14		0.437	0.643	
15				0.784
16				0.752
17				0.747
18				0.586
19	0.417		0.411	0.559

Source: Authors' own research, 2019.

The creation of a rotated component matrix has allowed ranking the brand value sources in case of brand loyalty presence according to their priority in the impact on consumer's perception as follows: (1) imageries; (2) benefits; (3) attitudes; (4) attributes. (See Table 7)

Table 7. Brand value sources—brand loyalty presence.

Factors	F1 Imageries	F2 Benefits	F3 Attitudes	F4 Attributes
N of Items	5	5	4	5
Cronbach's Alpha	0.854	0.837	0.841	0.869
% of Variance	50.002	10.949	7.665	6.001

Source: Authors' own research, 2019.

Thus, it is possible to make the conclusion that importance of factors does not vary across analyzed categories of brands of alimentary goods (i.e., brand loyalty absence vs. presence). For detailed information, see Table 8.

Table 8. Ranking of groups of components in analyzed categories.

Rank	Brands	
	Brand Loyalty Absence	Brand Loyalty Presence
1	Imageries	Imageries
2	Benefits	Benefits
3	Attitudes	Attitudes
4	Attributes	Attributes

Source: Authors' own research, 2019.

As it is obvious, the brand value sources ranking created on the basis of their priority in the impact on consumer's perception in case of brands with consumer's loyalty absence is the same as in case of brands with consumer's loyalty presence. However, when analyzing groups of components deeply, we can see that differences exist. The internal ranking inside identified groups of components is equal only in case of the less important group of brand value sources—in case of attributes. All others brand value sources are internally different from the point of view of relevance of individual components of these groups of brand value sources. The most visible example can be seen in scope of imageries, where only one component of brand value sources has the same ranking in case of brand loyalty absence and in case of brand loyalty presence. Thus, in case of brand loyalty absence, the order is the following: (1) satisfaction; (2) happiness; (3) expectations; (4) positive associations, and (5) certainty, while in case of brand loyalty presence, the order is the following: (1) satisfaction; (2) certainty; (3) positive associations; (4) happiness, and (5) expectations. This finding indicates the need of a selective approach to brand value sources and implementation of so far defined patterns in the practice of brand management.

While in both cases satisfaction is considered a main component in case of brands of alimentary goods, complementary components should be used differently. In case of brand value absence (similarly in phase of brand value building), it is happiness and expectations which should be mainly used, while in case of brand value presence (similarly in phase of brad value managing), it is certainty and positive associations. In case of benefits, the order is also mixed - in case of brand loyalty absence, the most important component is the ability to increase social status, while in case of brand loyalty presence, it is the ability to make it easier to get friends. Based on these findings, it is crucial to unify the consumer's satisfaction (as a main component of the most important brand value source) with corporate social responsibility and to implement sustainable managerial tools focused on stimulation of the consumer's socially conformal behavior, mainly on the basis of this brand value source. For detailed information about internal order of components inside identified brand value sources of alimentary goods, see Table 9 as a modification of Table 1.

Table 9. Brand value sources and components.

Brand Value Sources	Components of Brand Value Sources	
	Brand Loyalty Absence	Brand Loyalty Presence
imageries	satisfaction happiness expectations positive associations certainty	satisfaction certainty positive associations happiness expectations
benefits	branded product increases my social status branded product makes it easier for me to get friends branded product attracts the attention of others branded product belongs to my lifestyle branded product makes me happier	branded product makes it easier for me to get friends branded product attracts the attention of others branded product increases my social status branded product belongs to my lifestyle branded product makes me happier
attitudes	branded products attract my attention because I consider it better I aim to buy branded products I am interested in branded products on a regular basis branded products attract my attention because I consider them more prestigious	branded products attract my attention because I consider it better I am interested in branded product on a regular basis I aim to buy branded products branded products attract my attention because I consider them more prestigious
attributes	popularity creativity of ad availability innovativeness quality	popularity creativity of ad availability innovativeness quality

Source: Authors' own research, 2019.

Another important dimension of these findings lies in the differences between the order of brand value sources which have been identified generally and in case of alimentary goods. In case of brand loyalty absence, both categories are characterized by imageries as a main brand value source, while in case of brand loyalty presence, imageries are the most important only for brands of alimentary goods. Generally, imageries have been replaced by benefits. Thus, the need of a selective approach to sustainable brand management across product categories which has been so far only assumed, has been definitely proved [45–50,52]. For detailed information, see Table 10.

Table 10. Comparative ranking of grouped brand value components (in general/alimentary goods).

Rank	Brand Value Sources			
	Brand Loyalty Absence		Brand Loyalty Presence	
	In General	Alimentary Goods	In General	Alimentary Goods
1	Imageries	Imageries	Benefits	Imageries
2	Attitudes	Benefits	Attributes	Benefits
3	Benefits	Attitudes	Imageries	Attitudes
4	Attributes	Attributes	Attitudes	Attributes

Source: Kliestikova et al. [60].

Practical implications of these results indicate that imageries are the leading brand value source in case of brand loyalty absence regardless of the category of product (i.e., in general or in case of alimentary goods). On the contrary, in case of brand loyalty presence, the leading brand value source with significant impact on brand value subjectively perceived by consumer is benefits, while in case of alimentary goods, imageries remain to be the most relevant brand value source. According to this fact, we can observe these main findings: (1) in general, brand value sources vary due to the phase of brand management, while in case of alimentary goods, brand value sources remain identical; (2) the position of brand loyalty is ambivalent when applying point of view of stimuli vs. point of view of obstacle to prospective sustainable brand management, and (3) traditional general patterns of sustainable brand management are inapplicable in case of brands of alimentary goods.

Thus, we can state that the process of brand value building and management in case of alimentary goods does not have to be selective if it is connected with the phase of brand value building or brand value management (taking brand value sources into account and not their components). This coherent approach facilitates the managerial practice of brands of alimentary goods, where it is enough to identify relevant components of imageries as the most important brand value sources (happiness, expectations, satisfaction, certainty, and positive associations) at the very beginning of the process of brand value building, and these components can be subsequently used during all the brand life cycle. A representative product which declares the applicability of this approach is Coca-Cola, which is systematically built on the basis of joy and happiness as the leading brand value pillars.

On the other hand, as it has not been identified the difference between brand value sources of alimentary goods regarding to the presence vs. absence of brand loyalty, we cannot conclude that its existence is vital in the process of sustainable brand management. It means that managers of brands of alimentary goods should not expect bigger receptivity of sustainable brand management activities by the customers in case of brand loyalty presence. In other words, if the transition to sustainable brand management is done and the consumer identifies that it is not in accordance with brand value source accented so far, the brand value could be harmed.

The fact that we have identified the difference between brand value sources in case of brand loyalty presence vs. brand loyalty absence among categories verifies the presumption that universal sustainable brand management patterns should not be applied, as individual product categories are specific, and the modification of formulated models and processes of sustainable brand management should be supported by further market, opinion, and social research to arrange optimal applicability and effective goal fulfilment in scope of sustainable brand management.

We have confirmed that the main task for managers of not only formally but really sustainably manageable brands is to identify internal motives and brand value sources of their consumers and to co-act in the process of market education as one of the prerequisites of socially sustainable development [17,61]. Similarly, we verified the importance of satisfaction-affected trust and brand loyalty in the category of alimentary goods [23,24]. On the other hand, we have rejected the theory which highlights the importance of emotional sources of brand value. Thus, the scientific gap lying in the need of the mechanisms of behavioral economics investigation has been disputed [21]. In scope of these facts, it is also disputable the phenomenon of so-called love brands and its effectiveness in the process of brand loyalty creation and management [22]. When analyzing these findings, it is possible to apply specific point of view based on the regional psychographic specifics of consumers [34–39]. It is because we have accepted theories from authors who investigated regionally closer markets, while the theories of other authors have been rejected. Such a specific attitude of Slovak consumers to the brands of alimentary goods can be the reason of the phenomenon of double quality of food, which is typical for Slovak market in comparison with other markets of neighbor countries (mainly Austria). Surprisingly, although double quality has been clearly proven by independent tests, Slovak consumers do not change their attitudes towards brands and they follow their buying habits. This fact is extremely dangerous in scope of sustainable brand management concept implementation because brands of alimentary goods which are subjectively perceived as valuable are not motivated enough to behave responsibly towards society, and consumers do not have motivation to be active in the process of information searching. Thus, it is extremely important to analyze value sources of brands characterized by presence vs. absence of consumer loyalty and to apply a conscious and responsible approach to the consumer´s loyalty as to the one of the leading buying behavior motives. However, there are still possibilities for further research that should be focused in more detail on the specifics of consumer segmentation. A possible way to obtain brand management benefits in this case is the application of the generation approach. It is because we can suppose that the ranking of brand value sources and their components in case of alimentary goods varies if analyzing Generations X, Y, and Z.

5. Conclusions

Until now, the phenomenon of brand loyalty has not been analyzed in details connected with possible negative impact on sustainable development. Thus, the main aim of the article is to identify brand value sources of loyalty which are relevant to sustainable brand management of alimentary goods. To achieve this aim, we have used the data from our own research provided on the socio-demographically representative sample of 2000 respondents (sample without outliers, and incompatible units was 697) during the first half of the year 2019. We have realized this research via a questionnaire survey in the form of computer-assisted web interviewing. The questionnaire was administered in Slovak Republic among its inhabitants aged over 15 years who have been asked to fulfil the questionnaire because of their legal labor subjectivity. Thus, the main presumption of autonomous buying decision-making has been fulfilled. To provide research of brand value sources in scope of buying behavior typology, traditional quadratic typology of buying behavior has been used, where based on the degree of engagement and differentiation, obtained data were statistically evaluated by the factor analysis supported by relevant statistical tests (KMO Test, Barlett's test of sphericity, and calculation of Cronbach's Alpha). Based on this, it has been possible to identify relevant brand value sources of alimentary goods in case of brand loyalty absence as well as brand loyalty presence. It has not been proved the existence of significant difference between brand value sources ranking according to their priority in the impact on the consumer's perception in case of loyal and nonloyal consumers. The order of the brand value sources has been in both cases following (1) imageries; (2) benefits; (3) attitudes, and (4) attributes. However, when analyzing groups of components deeply, we can see that differences exist. The internal ranking inside identified groups of components is equal only in case of the less important group of brand value sources—in case of attributes. All others brand value sources are internally different from the point of view of relevance of individual components of these groups of brand value sources. The most visible example can be seen in scope of imageries, where only one component of brand value sources has the same ranking in case of brand loyalty absence and in case of brand loyalty presence. From a managerial point of view, these findings are even more important as they provide more details potentially used in scope of sustainable brand management of alimentary goods. Even though the conclusions formulated on the basis of provided research obtain useful information for the practice of sustainable brand management, there have been identified various relevant limitations of the research. The most important is the territorial validity of the research. These findings are fully applicable only in case of Slovak consumer, meaning that in case of entering Slovak company on foreign market, these findings have to be critically re-evaluated in scope of specifics of selected market. When respecting this fact, managers have to their disposal a very wide portfolio of information usable in all the portfolio of sustainable brand management implications. Not only valuable introspection into the previous practical successes and fails of brands is provided, but also the platform for optimal managerial decision-making in the future is created. In scope of the above mentioned, provided research offers the information relevant to appropriate setting of the content communicated with the consumers according to their identified preferences, demands, and expectations. The main managerial recommendation consists of the fact that "imageries" have been detected as a most valuable source of brand value from the consumer´s point of view. This source consists of happiness, expectations, satisfaction, certainty, and positive associations as its relevant components. That is to say, these are the basic pillars of subjectively perceived brand value which should be systematically used in the process of sustainable brand management of alimentary goods in all its complexity. It means that there is no need to distinguish between the process of brand value building and the process of brand value management as the main brand value source does not change.

The outcomes of the research and subsequently formulated conclusions provide the understanding of overall complexity of internal and external factors which motivates consumers to be interested in strong and functional interaction with brand. These findings have already been partially outlined by various authors, but no clear and uniform statement has been formulated thus far in the scope of sustainable brand value management of alimentary goods and the individual brand value sources. We have mainly verified the importance of satisfaction affected by trust, and brand loyalty in the category of alimentary goods, but on the other hand, we have rejected existing theory which highlights the importance of emotional sources of brand value.

Regardless of the declared importance and usability of the research results, there are still many points of view which could enrich its managerial applicability. One of them is the consideration of generational stratification and critical discussion of specifics of the consumer's brand value perception in the light and shadow of sustainable brand management optimal implementation.

Author Contributions: J.M. and A.K. designed the experiments. M.N. and L.G. analysed the data. G.L. contributed analysis tools. J.K. and W.S. wrote the paper. All authors have read and agreed to the published version of the manuscript.

References

1. Oh, W.-Y.; Choi, K.J.; Chang, Y.K.; Jeon, M.-K. MNEs' Corporate Social Responsibility: An Optimal Investment Decision Model. *Eur. J. Int. Manag.* **2019**, *13*, 307–327. [CrossRef]

2. Popadic, I.; Borocki, J.; Radisic, M.; Stefanic, I.; Duspara, L. The Challenges While Measuring Enterprise Innovative Activities-The Case from a Developing Country. *Teh. Vjesn. Tech. Gaz.* **2018**, *25*, 452–459. [CrossRef]

3. Zvirgzdina, R.; Linina, I.; Vevere, V. Efficient Consumer Response (ECR) Principles and Their Application in Retail Trade Enterprises in Latvia. *Eur. Integr. Stud.* **2015**, *9*, 257–264. [CrossRef]

4. Gonzalez-Ramos, M.I.; Donate, M.J.; Guadamillas, F. An Empirical Study on the Link Between Corporate Social Responsibility and Innovation in Environmentally Sensitive Industries. *Eur. J. Int. Manag.* **2018**, *12*, 402–422. [CrossRef]

5. Bellucci, M.; Bini, L.; Giunta, F. Implementing Environmental Sustainability Engagement into Business: Sustainability Management, Innovation, and Sustainable Business Models. *Innov. Strateg. Environ. Sci.* **2020**, 107–143. [CrossRef]

6. Demir, M.; Lenhart, S. Optimal sustainable fishery management of the Black Sea anchovy with food chain modeling framework. *Nat. Resour. Modeling* **2019**, e12253. [CrossRef]

7.   Vafaei, S.; Bazrkar, A.; Hajimohammadi, M. The Investigation of the Relationship Between Sustainable Supply Chain Management and Sustainable Competitive Advantage According to the Mediating Role of Innovation and Sustainable Process Management. *Braz. J. Oper. Prod. Manag.* **2019**, *16*, 572–580. [CrossRef]

8.   De los Salmones, M.D.G.; Crespo, A.H.; del Bosque, I.R. Influence of Corporate Social Responsibility on Loyalty and Valuation of Services. *J. Bus. Ethics* **2005**, *61*, 369–385. [CrossRef]

9.   Ganushchak-Efimenko, L.; Shcherbak, V.; Nifatova, O. Assessing the Effects of Socially Responsible Strategic Partnerships on Building Brand Equity of Integrated Business Structures in Ukraine. *Oeconomia Copernic.* **2018**, *9*, 715–730. [CrossRef]

10.   Zhulega, I.A.; Gagulina, N.L.; Samoylov, A.V.; Novikov, A.V. Problems of Corporate Economics and Sustainable Development in the Context of the Sanction World Order: Living Standards and Live Quality. *Ekon. Manaz. Spektrum* **2019**, *13*, 83–95. [CrossRef]

11.   Gillespie, B.; Rogers, M.M. Sustainable Supply Chain Management and the End User: Understanding the Impact of Socially and Environmentally Responsible Firm Behaviors on Consumers' Brand Evaluations and Purchase Intentions. *J. Mark. Channels* **2016**, *23*, 34–46. [CrossRef]

12.   Dumitriu, D.; Militaru, G.; Deselnicu, D.C.; Niculescu, A.; Popescu, M.A.M. A Perspective Over Modern SMEs: Managing Brand Equity, Growth and Sustainability Through Digital Marketing Tools and Techniques. *Sustainability* **2019**, *11*, 2111. [CrossRef]

13.   Hu, D.; Wang, Y.D.; Yang, X. Trading Your Diversification Strategy for a Green One: How Do Firms in Emerging Economies Get on the Green Train? *Organ. Environ.* **2019**, *32*, 391–415. [CrossRef]

14.   Lu, J.T.; Ren, L.C.; He, Y.F.; Lin, W.F.; Streimikis, J. Linking Corporate Social Responsibility with Reputation and Brand of the Firm. *Amfiteatru Econ.* **2019**, *21*. [CrossRef]

15.   Drugau-Constantin, A. Emotional and Cognitive Reactions to Marketing Stimuli: Mechanisms Underlying Judgments and Decision Making in Behavioral and Consumer Neuroscience. *Econ. Manag. Financ. Mark.* **2018**, *13*, 46–50. [CrossRef]

16.   Torres, A.; Bijmolt, T.H.A.; Tribo, J.A.; Verhoef, P. Generating Global Brand Equity through Corporate Social Responsibility to Key Stakeholders. *Int. J. Res. Mark.* **2012**, *29*, 13–24. [CrossRef]

17.   Grubor, A.; Milovanov, O. Brand Strategies in the Era of Sustainability. *Interdiscip. Descr. Complex. Syst.* **2017**, *15*, 78–88. [CrossRef]

18.   Riera, M.; Iborra, M. Corporate Social Irresponsibility: Review and Conceptual Boundaries. *Eur. J. Manag. Bus. Econ.* **2017**, *26*, 146–162. [CrossRef]

19.   Bhattacharya, C.B.; Sen, S. Doing Better at Dong Good: When, Why, and How Consumers Respond to Corporate Social Initiatives. *Calif. Manag. Rev.* **2004**, *47*, 9–24. [CrossRef]

20.   Hur, W.M.; Kim, H.; Woo, J. How CSR Leads to Corporate Brand Equity: Mediating Mechanisms of Corporate Brand Credibility and Reputation. *J. Bus. Ethics* **2014**, *125*, 75–86. [CrossRef]

21.   Castro-Gonzalez, S.; Bande, B.; Fernandez-Ferrin, P.; Kimura, T. Corporate Social Responsibility and Consumer Advocacy Behaviors: The Importance of Emotions and Moral Virtues. *J. Clean. Prod.* **2019**, *231*, 846–855. [CrossRef]

22.   Song, H.; Wang, J.; Han, H. Effect of Image, Satisfaction, Trust, Love, and Respect on Loyalty Formation for Name-brand Coffee Shops. *Int. J. Hosp. Manag.* **2019**, *79*, 50–59. [CrossRef]

23.   Olah, J.; Karmazin, G.Y.; Farkasne Fekete, M.; Popp, J. An Examination of Trust as a Strategical Factor of Success in Logistical Firms. *Bus. Theory Pract.* **2017**, *18*, 171–177. [CrossRef]

24.   Olah, J.; Sadaf, R.; Mate, D.; Popp, J. The Influence of the Management Success Factors of Logistics Service Providers on Firms' Competitiveness. *Pol. Jounal Manag. Stud.* **2018**, *17*, 175–193. [CrossRef]

25.   Rivera, J.J.; Bigne, J.; Curras-Perez, R. Effects of Corporate Social Responsibility on Consumer Brand Loyalty. *Rbgn-Rev. Bras. De Gest. De Neg.* **2019**, *21*, 395–415. [CrossRef]

26.   Vevere, V.; Sannikova, A. Developing Intercultural Negotiations Skills to Meet Current Challenges of Diverse EU Business Environment as Part of University Social Responsibility. *Eur. Integr. Stud.* **2018**, *12*, 8–18. [CrossRef]

27.   Chadhuri, A.; Holbrook, M.B. The Chain of Effects from Brand Trust and Brand Affect to Brand Performance: The Role of Brand Loyalty. *J. Mark.* **2001**, *65*, 81–93. [CrossRef]

28.   Bhattacharya, C.B.; Sen, S. Consumer-company Identification: A Framework for Understanding Consumers' Relationships with Companies. *J. Mark.* **2003**, *67*, 76–88. [CrossRef]

29. Stocchi, L.; Fuller, R. A Comparison of Brand Equity Strength Across Consumer Segments and Markets. *J. Prod. Brand Manag.* **2017**, *26*, 453–468. [CrossRef]

30. Curras-Perez, R.; Dolz-Dolz, C.; Sanchez-Garcia, I. How Social, Environmental, and Economic CSR Affects Consumer-Perceived Value: Does Perceived Consumer Effectiveness Make a Difference? *Corp. Soc. Responsib. Environ. Manag.* **2018**, *25*, 733–747. [CrossRef]

31. Yang, C.Y.; Yang, C.H. The Impact of Sustainable Environmental Management in the Food and Beverage Industry on Customer Loyalty: A View of Brand Attitude. *Ekoloji* **2019**, *28*, 965–972.

32. Abdullah, M.I.; Sarfraz, M.; Arif, A.; Azam, A. An Extension of the Theory of Planned Behavior Towards Brand Equity and Premium Price. *Pol. J. Manag. Stud.* **2018**, *18*, 20–32. [CrossRef]

33. Popp, J.; Olah, J.; Kiss, A.; Lakner, Z. Food Security Perspectives in Sub-Saharan Africa. *Amfiteatru Econ.* **2019**, *21*, 361–376. [CrossRef]

34. Chatzipanagiotou, K.; Christodoulides, G.; Veloutsou, C. Managing the Consumer-based Brand Equity Process: A Cross-cultural Perspective. *Int. Bus. Rev.* **2019**, *28*, 328–343. [CrossRef]

35. Sukalova, V.; Ceniga, P.; Janotova, H. Harmonization of Work and Family Life in Company Management in Slovakia. *Procedia Econ. Financ.* **2015**, *26*, 152–159. [CrossRef]

36. Tamuliene, V.; Pilipavicius, V. Research in Customer Preferences Selecting Insurance Services: A Case Study of Lithuania. *Forum Sci. Oeconomia* **2017**, *5*, 49–58. [CrossRef]

37. Rozgina, L. The Latvian Audit Services Market: Current Issues and Challenges. *Forum Sci. Oeconomia* **2018**, *6*, 7–21. [CrossRef]

38. Jain, T.; Zaman, R. When Boards Matter: The Case of Corporate Social Irresponsibility. *Br. J. Manag.* **2019**. [CrossRef]

39. Christodoulides, G.; Cadogan, J.W.; Veloutsou, C. Consumer-based Brand Equity Measurement: Lessons Learned from an International Study. *Int. Mark. Rev.* **2015**, *32*, 307–328. [CrossRef]

40. Rather, R.A.; Tehseen, S.; Itoo, M.H.; Parrey, S.H. Customer Brand Identification, Affective Commitment, Customer Satisfaction, and Brand Trust as Antecedents of Customer Behavioral Intention of Loyalty: An Empirical Study in the Hospitality Sector. *J. Glob. Sch. Mark. Sci.* **2019**, *29*, 196–217. [CrossRef]

41. Poushneh, A.; Vasquez-Parraga, A.Z. Emotional Bonds with Technology: The Impact of Customer Readiness on Upgrade Intention, Brand Loyalty, and Affective Commitment through Mediation Impact of Customer Value. *J. Theor. Appl. Electron. Commer. Res.* **2019**, *14*, 90–105. [CrossRef]

42. Dawes, J.; Meyer-Waarden, L.; Driesener, C. Has Brand Loyalty Declined? A Longitudinal Analysis of Repeat Purchase Behavior in the UK and the USA. *J. Bus. Res.* **2015**, *68*, 425–432. [CrossRef]

43. Smith, A.; Stirling, A. Innovation, Sustainability and Democracy: An Analysis of Grassroots Contributions. *J. Self-Gov. Manag. Econ.* **2018**, *6*, 64–97. [CrossRef]

44. Brunner, T.A.; Stocklin, M.; Opwis, K. Satisfaction, Image and Loyalty: New Versus Experienced Customers. *Eur. J. Mark.* **2008**, *42*, 1095–1105. [CrossRef]

45. Alvarado-Herrera, A.; Bigne, E.; Aldas-Manzano, J.; Curras-Perez, R. A Scale for Measuring Consumer Perceptions of Corporate Social Responsibility Following the Sustainable Development Paradigm. *J. Bus. Ethics* **2017**, *140*, 243–262. [CrossRef]

46. Hafez, M. Measuring the Impact of Corporate Social Responsibility Practices on Brand Equity in the Banking Industry in Bangladesh the Mediating Effect of Corporate Image and Brand Awareness. *Int. J. Bank Mark.* **2018**, *36*. [CrossRef]

47. Long, C.Z.; Lin, J. The Impact of Corporate Environmental Responsibility Strategy on Brand Sustainability: An Empirical Study Based on Chinese Listed Companies. *Nakai Bus. Rev. Int.* **2018**, *9*, 366–394. [CrossRef]

48. Alcaide, M.A.; De La Poza, E.; Guadalajara, N. Assessing the Sustainability of High-Value Brands in the IT Sector. *Sustainability* **2019**, *11*, 1598. [CrossRef]

49. Pinto, D.C.; Herter, M.M.; Goncalves, D.; Savin, E. Can Luxury Brands Be Ethical? Reducing the Sophistication Liability of Luxury Brands. *J. Clean. Prod.* **2019**, *233*, 1366–1376. [CrossRef]

50. Tofighi, M.; Bodur, H.O. Social Responsibility and its Differential Effects on the Retailers' Portfolio of Private Label Brands. *Int. J. Retail. Distrib. Manag.* **2015**, *43*, 301–313. [CrossRef]

51. Lizbetinova, L. The Quality of Communication in the Context of Regional Development. *Deturope-Cent. Eur. J. Reg. Dev. Tour.* **2014**, *6*, 22–38.

52. Casteran, G.; Chrysochou, P.; Meyer-Waarden, L. Brand loyalty evolution and the impact of category characteristics. *Mark. Lett.* **2019**, *30*, 57–73. [CrossRef]

53. Sukalova, V.; Ceniga, P. Customer Protection in the Field of Life Insurance. *Lect. Notes Manag. Sci.* **2017**, *73*, 17–22. [CrossRef]

54. Kolnhofer Derecskei, A. Relations between Risk Attitudes, Culture and the Endowment Effect. *Eng. Manag. Prod. Serv.* **2018**, *10*, 7–20. [CrossRef]

55. Berzakova, V.; Bartosova, V.; Kicova, E. Modification of EVA in Value Based Management. *Procedia Econ. Financ.* **2015**, *26*, 317–324. [CrossRef]

56. Lizbetinova, L.; Starchon, P.; Lorincova, S.; Weberova, D.; Prusa, P. Application of Cluster Analysis in Marketing Communications in Small and Medium-Sized Enterprises: An Empirical Study in the Slovak Republic. *Sustainability* **2019**, *11*, 2302. [CrossRef]

57. Tuffnell, C.; Kral, P.; Durana, P.; Krulicky, T. Industry 4.0-based Manufacturing Systems: Smart Production, Sustainable Supply Chain Networks, and Real-time Process Monitoring. *J. Self-Gov. Manag. Econ.* **2019**, *7*, 7–12. [CrossRef]

58. Popescu Ljungholm, D. Employee–employer Relationships in the Gig Economy: Harmonizing and Consolidating Labor Regulations and Safety Nets. *Contemp. Read. Law Soc. Justice* **2018**, *10*, 144–150. [CrossRef]

59. Svabova, L.; Kramarova, K.; Durica, M. Prediction Model of Firm´S Financial Distress. *Ekon. Manaz. Spektrum* **2018**, *12*, 16–29. [CrossRef]

60. Kliestikova, J.; Kovacova, M.; Krizanova, A.; Durana, P.; Nica, E. Quo Vadis Brand Loyalty? Comparative Study of Perceived Brand Value Sources. *Pol. J. Manag. Stud.* **2019**, *19*, 190–203. [CrossRef]

61. Krylov, S. Strategic Customer Analysis Based on Balanced Scorecard. *Ekon. Manaz. Spektrum* **2019**, *13*, 12–25. [CrossRef]

10

Crisis Response and Supervision System for Food Security: A Comparative Analysis between Mainland China and Taiwan

Chun-Chieh Ma [1], Han-Shen Chen [2,3] and Hsiao-Ping Chang [2,3,*]

[1]   Department of Public Administration and Management, National University of Tainan, No.33, Sec. 2, Shu-Lin St., Tainan 70005, Taiwan; ccma@mail.nutn.edu.tw

[2]   Department of Health Diet and Industry Management, Chung Shan Medical University, No.110, Sec. 1, Jianguo N. Rd., Taichung City 40201, Taiwan; allen975@csmu.edu.tw

[3]   Department of Medical Management, Chung Shan Medical University Hospital, No. 110, Sec. 1, Jianguo N. Rd., Taichung City 40201, Taiwan

*   Correspondence: pamela22@csmu.edu.tw

Abstract: In Mainland China, major food security incidents have occurred with high frequency, of which the number and degree of harm are both increasing. At the same time, Taiwan's food security crisis has also been spreading. For these reasons, this article makes a comparative analysis of food security issues between Mainland China and Taiwan from a legal point of view and identifies the blind spots of the legal system and supervision using official documents and research papers regarding the most typical incidents in the period of 2008–2019. The results indicate that, compared with Mainland China, Taiwan has a better food security supervision system, and its experience with the supervision system, specific rules, social supervision, and responsibility is worth investigating. However, while there are loopholes in criminal law in Mainland China, which has not formed a complete system, criminal law in Taiwan is also weak in terms of regulation of food security incidents. Based on the results, this article puts forward suggestions with the expectation that, in the face of an increasingly severe food security crisis, Mainland China and Taiwan will strengthen their cooperation in constructing legal systems for food security supervision and inspection, exchange experience, cooperate in inspection, and share food security information to avoid rumors of food insecurity circulating in popular science. It is expected that the results and suggestions of this study will be helpful in the crisis response, as well as in the supervision systems in Mainland China and Taiwan for guarding food security. Although the comparative analysis is specific to the two regions, its characteristics are typical of food security globally, especially in Asia.

Keywords: crisis response; supervision system; food security; Mainland China; Taiwan

1. Introduction

With the rapid development of the social economy under globalization, various kinds of public emergencies in traditional and nontraditional fields have become increasingly prominent, which has become an important hidden threat to national security and people's property and life security [1]. Frequent food security incidents have resulted in the repeated questioning of the credibility of the government [2]. Researchers have found that an effective emergency system can reduce accident losses to less than 6% compared to a situation without an emergency system [3]. In view of this, in the process of dealing with crisis events, the appropriateness of the emergency response system has become the focus of researchers.

In recent years, major food security incidents have occurred frequently in Mainland China, the number and degree of harm of which are both increasing. Food security has attracted more attention from all walks of life than ever before, resulting in a crisis of public confidence in food security [4]. In terms of recent food security incidents, poisonous milk powder incidents broke out in March 2008, but due to the preparation for the Olympic Games, the local government of Hebei did not deal with these in time. The poisonous milk powder incident continued until the New Zealand government reported the incident to China in September. This incident affected China's foreign trade and seriously damaged China's international image [5]. This incident caused social unrest because the melamine mixed in the infant milk powder caused kidney lesions in infants [5]. In view of the importance of food security issues, the Food Security Law of Mainland China regulates scientific and effective supervision measures with a clear and systematic division of responsibilities for food security supervision, which provides a legal guarantee to resolve the current prominent food security issues in a proper and orderly manner. However, due to the unique and complex market environment of Mainland China, food security supervision will still face multiple dilemmas with respect to the system and technology [6].

Taiwan's food security crisis has also been spreading [7]. From the "plasticizer" to the "Datong Tainted Oil" incidents, food security issues have also been emerging in Taiwan one after the other. The emergence of the poisonous starch incident in 2013 in Taiwan set off a panic in Taiwanese society. The official poisonous compound in the starch was maleic anhydride. It was originally used as a food packaging material and adhesive, but some practitioners mixed it into edible starch to increase the viscosity and elasticity of the starch. This substance was illegally added and may cause kidney damage. In April 2013, seven county and municipal health bureaus, such as New Taipei City, began to investigate the manufacturers using the poisonous starch closely. In May, the New Taipei City Health Bureau found that Mingji tofu contained the toxic starch elements, and its raw material was procured from Sanjin Powder Trading Company, located in Tainan; therefore, the Health Bureau of Tainan City continued to investigate the incident. In order to eliminate the circulation of the toxic starch in the market and require the raw material manufacturers of the starch to start providing security certificates from the downstream manufacturers, pitchmen, and food practitioners starting in June, the Health Department launched the "0527 Food Security Project". The "0527 Food Security Project" was a project for food safety control initiated by the resolution of the Conference on Research and Strategies organized by the Ministry of Health and Welfare, which invited experts from food management, science and technology, and public health and medical fields in order to solve the food safety problem caused by the malicious addition of maleic anhydride to starch. It was launched by the Health Department to check all starch manufacturers in Taiwan. Because the poisonous starch incident was a significant issue, it brought food security issues to the public's attention, as well as governmental actions on food security issues [4].

Mainland China and Taiwan are facing a major impact from food security issues. In the face of many major food security incidents in Mainland China and Taiwan over the years and in view of the fact that food security affects people's lives and health, this article performs a comparative study of food security in the two regions, which should help governments and nongovernmental organizations to learn from each other's progress at the levels of food supervision and the legal system and to establish an effective food supervision mechanism and rigorous legal system to protect consumers' rights and interests. Taking advantage of the promotion of the "Food Security Law" in Mainland China, this study explores the legal system of food security supervision, which should have theoretical and practical significance to solve food security problems effectively [8]. A comparative analysis of the current situation of food security in Taiwan could be helpful in implementing a cross-strait food security agreement, enhancing food security communication and mutual trust in Mainland China and Taiwan, thereby ensuring the security and health of local citizens [9].

2. Literature Review

2.1. Concept of Food Security

The precise concept of food security was first proposed by the Food and Agriculture Organization of the United Nations (FAO) at the World Food Conference in November 1974. The FAO believes that food security is a fundamental right of people that includes their subsistence rights. At this conference, the FAO defined food security as "guaranteeing that everyone can get what they need for survival and health everywhere."

In the 1980s, the Food and Agriculture Organization of the United Nations gave a new definition of food security in response to food shortages in developing countries: "The ultimate goal of food security is to ensure that all people can buy and afford any food they need at any time." After the 1990s, with the improvement of the economy and the increase in material consumption, the quality and nutrition of food became the theme of food security. In 1991, the International Conference on Nutrition gave a new definition of food security: "Everyone can get safe and nutritious food at any time to maintain a healthy and active life" [10].

2.2. Crisis Response and Supervision System for Food Security

Crises often occur in our daily lives. Especially in the modern era of rapid technological change, the impact of a crisis becomes more and more important. Fink described a crisis as a watershed, "turning point", or "deterioration" of an event, which is a key time point or opportunity for decision making. The time and situation during the occurrence of the incident are not stable, thus demanding immediate decisions [11]. Coombs suggested that no organization can be free from crisis [12], so when a crisis occurs, if policymakers can respond appropriately, implement corresponding prevention/mitigation measures for the crisis incident in stages, prepare for response, advanced recovery, and other management behaviors, and cooperate with all relevant stakeholders with good communication channels, they can defuse the crisis and transform the organization from being in crisis to becoming a model for crisis management [13]. Therefore, in the modern era of academic debate, how to respond to a crisis has become an important issue [14].

Among the numerous crises from the past to the present, food security incidents have emerged one after the other in recent years and have attracted great attention from society. With regard to food security, Yeung and Morris [15] found that the main influencing factor of consumer behavior was consumers' own perception of the food security crisis, rather than the risk of food security incidents. Therefore, previous literature has argued that the potential harms caused by food security incidents may be far less than the exaggerated crisis [16]. Smith [17] further pointed out that the public would first use their usual information channels to understand the situation in the face of the crisis. At the beginning of the event, they would be more sensitive to the information they first received and would more easily believe the information. Therefore, at the moment of the crisis, the first response is very important in influencing the public's perception.

With the improvement of the economy and the standard of living, the public attach more and more importance to food security issues. In a multirisk society, when facing a potential food security crisis, the public collects information about new crises mainly through the media. After receiving such information, people internalize their awareness of the crisis and its risks. With modern science and technology, the circulation of information becomes easier, and the public can have a deeper and more confident understanding of the crisis [18]. According to the analysis of a study conducted by Palenchar and Heath [18], in the face of a crisis, community residents can achieve self-protection through the relevant norms of crisis response and are willing to adopt other contingency measures to make them feel more in control of the crises. Jacob et al. [19] also found that uncertainty in crisis information increases people's fear and cognitive bias. It is important to inform consumers about the risks so that they can choose the right strategy that reflects their demand for information, their morals, and their values [20].

From those studies, the public must deal with an abundance of information in order to manage their own risk profile connected to nourishment and possibly place trust in various expert systems delivering the information in order to overcome anxiety in approaching food-related crises [21,22]. Since different groups of actors—governments or institutions on the one hand and private commercial enterprises on the other—influence decision making about food, often conflicting with each other and enhancing the anxiety in food choices [23], the public's fear or aversion about food security could be contained through both information provided by expert systems and adequate communication to trigger positive emotional responses [24].

Moreover, most public opinion concerns derive from problems with food safety and government supervision, coupled with a slow or lacking action by the government, resulting in aggravated negative public sentiment and a decrease in the credibility of the government. Therefore, the government must start from the source, from research experiments to market entry, and further improve the system at each level to strengthen approval and supervision. For these reasons, it is important to be open and transparent in policy formulation and to give people an opportunity to express their demands while using the media and other forces to increase information disclosure, actively address popular science rumors, guide the public in rational discussions, effectively protect the public's rights, and enhance the credibility of the government [25].

Accordingly, because of the increase in the risk to food security, the governments of various countries should rethink their current response programs to establish suitable food security management systems and adequate communication features. Under such a situation, priority should be given to establishing adequate norms for the crisis response to lead the execution of the related orders, as well as implementing supervision systems to reduce damage during the crisis [26]. Based on the studies, suitable norms for the crisis response can improve the public's sense of control over a crisis of food security, and a better supervision system, specific rules, social supervision, and responsibility investigation will also be helpful to prevent as well as mitigate crises in food security.

3. Methods: Comparative Analysis Approach and Qualitative Research Method

Comparative political and social research is generally defined in two ways: Either on the basis of its supposed core subject, which is almost always defined at the level of political and social systems [27], or by means of descriptive features that claim to enhance knowledge about politics and society as a process [28]. These descriptions are generally considered to differentiate the comparative approach from other approaches within political and social sciences.

Since the food security crises depicted in this study were the first major emergencies to occur in Mainland China and Taiwan after the enforcement of food security systems comprehensively adopted at that time, by adopting a comparative analysis approach, the observation of the crisis responses in these two regions could show the descriptive features of two systems so as to enhance knowledge regarding food security policies, the food security situation, and supervision issues.

Additionally, based on qualitative research, this study performs a holistic exploration of social phenomena using various data collection methods in the natural environment. The practical induction method was used to analyze data and formation theory. This is an activity to gain an explanatory understanding of the behavior and meaning construction of the research object through interaction with the research object [29]. Hence, this study also chose a qualitative research method for exploration and collected data in the period of 2008–2019, both in Mainland China and Taiwan, so as to avoid any subjective judgment and to make more objective judgments and a comparative study.

This paper takes the legal norms of food security in Mainland China and Taiwan as the research scope and identifies the blind spots of the legal system and supervision through comparative analysis using official documents and research papers regarding the two regions. It also puts forward suggestions with the expectation that Mainland China and Taiwan will learn from each other and make improvements in their responses to food security crises.

4. Food Security Situation and Supervision Issues in Mainland China

4.1. Situation of Food Security in Mainland China

In recent years, many major food security incidents have occurred frequently in Mainland China, as shown in Table 1. For this reason, food security has attracted more attention from all domains than ever before, resulting in a crisis of public confidence in food security [4].

Table 1. Situation of food security in Mainland China.

Situation	Description
The security situation of agricultural products and poultry products is worrying.	Harmful residues, such as fertilizers and pesticides, antibiotics, and other harmful substances, excessive heavy metals, etc., exist in agricultural products.
There are potential security hazards in the production, processing, and marketing of agricultural products, which can easily cause secondary food poisoning and foodborne diseases.	Due to inadequate supervision, production, processing, and marketing places often become the transmission centers of infectious diseases.
Food poisoning and foodborne diseases occur from time to time.	1. The "Sanlu poisonous milk powder" incident in 2008 that resulted in 14 newborns suffering from urolithiasis; 2. The "Shuanghui lean meat powder" incident in 2011; 3. The "Vitamin C Yinqiao tablets of Guangzhou Pharmaceuticals containing highly toxic substances" case in 2013. 4. According to the statistics of the National Health and Family Planning Commission, the frequent occurrence of food poisoning and foodborne diseases poses a serious threat to food security in 2019.
The technological achievements and technical reserves of food security are inadequate.	The key detection technologies of foodborne hazards and food security control technologies in Mainland China are lagging behind those in developed areas such as Taiwan.

4.2. Issues for Food Security Supervision in Mainland China

Mainland China currently implements a multisectoral and phased supervision system [30]. In detail, before the reform, the highest coordinating body for food security in Mainland China was the Commission on Food Security of the State Council, which served as a high-level deliberative and coordinating body for food security work under the State Council. It comprised the Commission Office on Food Security of the State Council (referred to as the Food Security Office of the State Council) as the body handling daily affairs of the Commission. According to the institutional reform and functional transformation plan of the State Council approved in March 2013, the relevant organizations and affiliations will be adjusted and merged into the National Health and Family Planning Commission and the State Food and Drug Administration.

Continuing the above, the food security supervision system in Mainland China can be visualized as in Figure 1. This supervision system seems to be comprehensive, but its biggest drawback is that it is prone to unclear powers and responsibilities, overlapping functions, inefficiency, and shirking of responsibilities by the food supervision departments. Moreover, more than 10 supervision departments, including agriculture, industry and commerce, quality inspection, health, and food and medicine, are involved in various links closely related to production, processing, circulation, and consumption. If a certain link is broken and the connection is not good, the whole food supervision system will have problems [31,32].

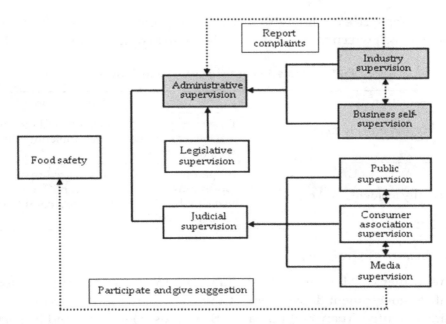

Figure 1. Food security supervision system in Mainland China.

Apart from the power and responsibility problems, in the actual operation process, the supervision system also has the problem of inadequate supervision [33,34]. Consumers often abandon attempts at solving food security problems through judicial procedures because of the high cost of safeguarding rights, which leads to the failure of mass supervision. Media supervision and consumer association supervision mainly rely on the reports of those whose rights and interests have been damaged, but because of consumers' apprehension, the effect of supervision will be greatly weakened. Enterprise self-supervision and industry supervision mainly depend on the morality and conscience of the enterprise itself, and once in conflict with economic interests, food manufacturers are likely to choose to pursue maximum benefits and market share. These two kinds of supervision are also difficult to achieve.

As in administrative supervision, information about the food supply chain is not open and transparent, so consumers can only passively believe the information issued by third-party government departments when they want to buy products manufactured by enterprises. Not only are they unable to respond to the government's supervision, but they also have little information exchange with food enterprises. For certain reasons, the government may not provide all of the real information to consumers [35]. This kind of one-way government trust very easily causes crises. Although government departments will invest much money in monitoring food security risks, as long as food security incidents occur, "government supervision is not in place" must be one of the reasons for the occurrence of food security incidents [36,37].

At the technical level, the food security standards in Mainland China are lagging behind, and the means of inspection and detection are relatively undeveloped. Therefore, it is difficult to guarantee the scientific rigor and authority of the test results [38]. Current food security supervision urgently needs to solve the technical problems such as the immature means of the evaluation and detection of food security risks [39].

5. Food Security Situation and Supervision Issues in Taiwan

5.1. Situation of Food Security in Taiwan

Taiwan attaches great importance to food security. It not only formulates relatively complete laws and regulations, but also has a complete process supervision and guarantee system from farm to dining table, as shown in Table 2. In 2011, Taiwan responded decisively and quickly to the plasticizer incident, which took only three months to subside. However, in 2013, a series of food security incidents broke out, such as "organic rice mixing with low-price imported rice", "poisonous starch", "tainted oil", and "poisonous milk", which also shook the Taiwanese people's confidence in their food security.

Table 2. Situation of food security in Taiwan.

Situation	Description
Responded decisively and quickly to the plasticizer incident.	Taiwan responded decisively and quickly to the plasticizer incident, which took only three months to subside in 2011.
A series of food security incidents broke out.	"Organic rice mixing low-price imported rice", "poisonous starch", "tainted oil", and "poisonous milk" also shook Taiwanese people's confidence in their food security in 2013.

5.2. Issues for Food Security Supervision in Taiwan

Compared with that of Mainland China, Taiwan's food security supervision is mainly the responsibility of three government departments: Council of Agriculture, Executive Yuan, Ministry of Health and Welfare, Executive Yuan, and Bureau of Standards, Metrology, and Inspection, Ministry of Economic Affairs. Among them, the Council of Agriculture, Executive Yuan is mainly responsible for the supervision of the production of raw food materials, including raw material management for agricultural products, animal husbandry, and aquatic products, and assisting in the work of the Ministry of Health and Welfare, Executive Yuan. The Ministry of Health and Welfare, Executive Yuan and its subordinate health agencies are responsible for the supervision of food market circulation under the leadership of Executive Yuan. The Bureau of Standards, Metrology, and Inspection, Ministry of Economic Affairs (MOEA) is responsible for entry and exit food inspection and entrusted inspection. Only food inspected up to the standard can be circulated in the market [40]. Finally, food security and quality certification based on the Food Security and Health Management Law is the Good Manufacturing Practice (GMP) Smile Mark promoted by the Industrial Development Bureau, MOEA, and the Certified Agricultural Standards (CAS) Mark of the Council of Agriculture, Executive Yuan [41]. Taiwan's food security supervision system is shown in Figure 2.

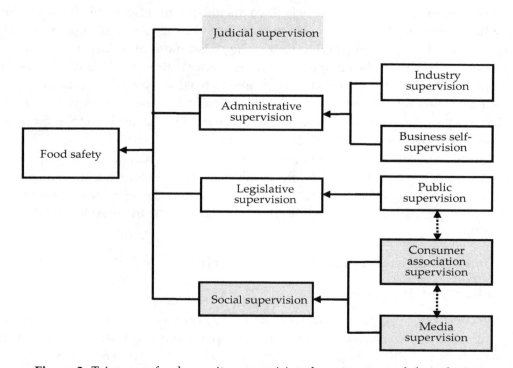

Figure 2. Taiwanese food security supervision departments and their duties.

6. Comparison of Food Security Status and Supervision Issues in Mainland China and Taiwan

From the food security supervision system in Mainland China and Taiwan, we can deduce the food security responses relative to the system, as shown in Table 3. The legal construction on food security issues in Mainland China has not only such problems as a lack of system, integrity, and deterrence in supervision laws and regulations, duplicated multidepartment enforcement, repeated enforcement, conflict of laws and regulations, a discrepancy of technical standards, and imperfect enforcement measures [42], but also supervision blind spots, such as imperfect legislation, content lags, discoordination between laws, the insufficient force of legal penalties, and even the weak legal awareness of the general public.

Table 3. Comparisons of food security issues in Mainland China and Taiwan.

Difference	Mainland China	Taiwan
Time point of food security accidents	The problem was found only when patients appeared. Food security accidents resulted in bad effects and serious consequences.	Poisonous and harmful food was reported by the public before it caused serious casualties.
Penalty intensity	The criminal law of the mainland is more stringent and the penalties heavier, but for the administrative penalty, enterprise food security information is not disclosed.	Taiwan's criminal law has not been amended and does not meet the requirements of the times. The level of penalty is too light, and there is collusion between the government and businesses.
Public response	It is relatively difficult for public opinion to create pressure.	The public response is intense and can generate strong public opinion pressure, but social network and social media rumors spread.
Scope of problem	Since the beginning of the 21st century, food security issues have been widespread, with numerous categories and a high frequency of outbreaks.	Food security incidents have occurred frequently in the last decade, but the severity is much lower than in Mainland China.
Social and economic conditions	Mainland China is currently a developing region, with economic development as the central task, so it is inevitable that some manufacturers with low social morality will only care about their interests, regardless of risk to life.	Taiwan is relatively developed with a relatively better quality of life and fewer food security incidents than Mainland China.
Social system	As it implements the socialist system with Chinese characteristics, it is vulnerable to the impact of the market economy, which leads to money-oriented food producers and processors.	With the implementation of the capitalist social system, utilitarianism is rampant, and companies are easily driven by profits.
Legislative branch	The legislative branch is the National People's Congress (NPC). The establishment of new laws and the revision of old laws take a long time, which hinders the thorough eradication of food security issues.	The legislative branch is the Legislative Yuan; the law-passing rate and number have greatly increased recently, but the law on food security still needs to be amended.
Inspection and testing measures	Food security testing systems are duplicated, the testing system is not perfect, and the testing standards are confusing (provincial, prefectural, and municipal standards are different).	CAS Good Food System, GMP Certification System, Processed Food Traceability System, Food Consumption Traffic Light System, Food Recycling System.
Attitudes towards informers	Informers do not receive corresponding personal security protection.	The government helps informers of illegal acts by keeping their identities secret, protecting their personal security through legislation, and through an incentive system to encourage consumers to report substandard food actively.

Compared with that of Mainland China, Taiwan's food security management is under the jurisdiction of the "Ministry of Health and Welfare, Executive Yuan" and the "Industrial Development Bureau, MOEA". Relevant laws and regulations are also incorporated in the food hygiene management law, food hygiene standards, and regulations to promote industrial upgrading. This model has its own unique features, but it also has many unsatisfactory defects. The more noticeable ones are: (1) Political mafias are often involved in food security incidents, and some public opinion representatives arbitrarily carry out "lobbying" in judicial cases. There is a suspicion of collusion between officials and businessmen, resulting in difficult judgment in major cases and obstacles for law enforcement and inspection, which makes food practitioners act with impunity. (2) Taiwan's criminal law has a weak effect in the punishment of food security felonies to achieve its proper purpose. (3) Taiwan's administrative punishment is too light, which may lead to disguised incentives for economic crime and social network and social media rumors.

Comprehensive research and analysis has shown that the fundamental reason for the frequent occurrence of food security accidents in China lies in the more humanistic supervision mode, under which most departmental managers often act on their own will, regardless of the regulations. To change this dependence on the bad paths of free-riding, weak government rules, abuse of power, and buck-passing, we must change the supervision mode of food security from the rule-of-man mode of supervision to supervision based on rules and regulations [43]. The inherent reason for the frequent occurrence of food security accidents lies in the unequal interests among all levels of the regulatory bodies and food enterprises caused by information asymmetry. Based on incentive compatibility theory, improving food security supervision in China should start with clarifying the responsibility of food security supervision at all levels and increasing the punishment of enterprises violating regulations [44]. In contrast, Taiwan should promote the basic law on food security, starting from reviewing and revising its existing cumbersome and decentralized food security regulations [45]. The primary task is to strengthen the promotion of third-party inspection, the "production resumé" system, food security knowledge advocacy, moral education, information provision, etc. [20].

7. Discussion and Suggestions

As food security is the focus of the whole society, it is of great importance. The legal system of supervision is the source of food security problems and thus is the most important of aspect of food security. Based on the comparative study of Mainland China and Taiwan, it was found that there were similarities and differences in the legal system of food security supervision between Mainland China and Taiwan. In view of this, the researchers put forward the following opinions and suggestions about the current legal system of food security supervision in Mainland China and Taiwan.

7.1. Suggestions for Food Security in Mainland China

In view of the current food security problems in Mainland China, this paper considers that it should establish a scientific, rational, advanced, and sound food security supervision system, learn from Taiwan's "product-management-oriented, stage-management-supplemented" food security supervision model, and change the current situation of unclear powers and responsibilities of various departments, different divisions of labor, and low efficiency of supervision [45–47].

(1) One suggestion is to deepen and refine relevant food security laws, such as clarifying the relevant unclear legal interpretation in the Food Security Law.

(2) In order to inform people, it could be a good idea to realize a specific international system for the rigorous evaluation of human health and environmental consequences [21]. This entails standardizing food security certification activities, improving the food recall system, establishing a food security liability insurance system, and formulating strict food security standards, which could be strictly implemented with reference to Taiwanese or European and American standards.

(3) Another suggestion is to foster diversified food security intermediary organizations with powerful supervision, improve their legal status, strengthen their supervision and restraint mechanisms,

improve their role of social supervision, and form a four-in-one supervision system involving government, enterprises, intermediaries, and individuals.

(4) Another suggestion is to formulate strict punishment measures, increase the intensity of punishment, and combine fines and corporal punishment, with multiple and more severe penalties for more serious offenses.

(5) Another suggestion is to establish an incentive system to encourage consumers to report illegal acts, bring out public enthusiasm to participate in food security, and give full play to their supervisory role.

(6) Civil education cultivated appropriately would enforce civil awareness, as well as public cognition of crisis prevention [20,48], providing moral education to food companies and raw material suppliers and popularizing and advocating food security knowledge among consumers to solve food security problems in a two-pronged way [21].

7.2. Suggestions for Food Security in Taiwan

In view of the abovementioned food security problems in Taiwan and the analysis results, this paper considers that the key point of response to these problems is to formulate the basic laws for the promotion of food security, review and revise Taiwan's existing cumbersome and decentralized food security regulations, and integrate them into a set of transparent, consistent, and comprehensive food security regulations [45,49,50].

(1) Food security legislation must cover every stage of food production and should be extended to the scope of animal feed. This is because only by ensuring feed security can it guarantee the security of livestock and poultry products, and also indirectly guarantee the security of consumers. Regulations should clearly specify that during the process of food production and marketing, suppliers of each production and marketing supply chain should bear the responsibility for the security of the food or raw materials they provide. Illegal food suppliers should be severely penalized and exposed to severe criminal liability.

(2) A special administrative body should be set up to coordinate all food security businesses so as to take effective food security management measures to safeguard public health and restore consumers' confidence in the security of food sold in Taiwan; for example, further integrating food security and hygiene regulatory bodies to centralize the power of food security and hygiene supervision, establishing the Food and Drug Administration, and reclaiming the business of imported food inspection from the Label Inspection Authority.

(3) Introduce third-party inspection: Inspection fees of all of the food on the shelves can be paid by consumers themselves to transfer the inspection fee to consumers by means of increasing the prices of commodities by the inspection cost, so that consumers can buy assurance and producers cannot counterfeit it.

(4) Improve and strictly control the food "production resumé" policy: The so-called "production resumé" system refers to the establishment of a "traceability system" to track the information of raw material ingredients, production, processing, circulation, sale, date, and other stages of food and to mark related information on the products. After the establishment of this system, on the one hand, consumers can understand the relevant ingredients and service life of the food; on the other hand, once the food has problems, we can immediately identify the crux of the problem from relevant information and deal with it properly, which will play an important role in ensuring food security in Taiwan.

(5) The government should strengthen food knowledge regarding security education advocacy and moral education, help to teach food producers at basic levels to follow good farming practices, help food processors to abide by good operation practices and instructions, and help food cooks to follow good hygienic practices and habits [20]. Food security knowledge, education, and training should be provided to producers at each stage of the food supply chain in order to prevent the generation of harmful residues during the food production process and avoid harm to consumers' health and the environment [21,22].

7.3. Implications

In recent years, food problems have been occurring in Mainland China and Taiwan. In order to ensure food security and considering the fact that food security is directly related to people's lives, the public's rights, and the credibility of the government, the relevant departments in Mainland China and Taiwan have realized the importance of responding to the food security crisis and signed the preliminary "Agreement on Food Security in Mainland China and Taiwan" to safeguard the rights and interests of consumers in Mainland China and Taiwan [25]. Furthermore, in the face of increasingly severe food security crises, governments shall strengthen cooperation in constructing legal systems on food security supervision and inspection, exchange experience, and cooperate in inspection (for example, when dealing with the international food security problem of imported poisonous milk powder, Mainland China can refer to strict inspection practices, such as "Food GMP Certification" and the high import standard requirements and policy means implemented in Taiwan over the years, starting from the source management of food security, so as to prevent the entry of problematic foreign food). Finally, because, in general, the more unfamiliar things are, the more they are perceived as risky, and avoidance and fear are the consequences [21,22], to avoid the occurrence of hygienic hazards of food, governments should establish adequate communication, be open and transparent in policy formulation, provide consumers with more information about how to fill the gap in food security knowledge, and share food security information to avoid rumors of food insecurity circulating in popular science [20,24,25]. It is expected that the results and suggestions of this study will be helpful in crisis response, as well as to the supervision system in Mainland China and Taiwan to guard food security. Although the comparative analysis was specific to the two regions, its characteristics are typical of food security globally, especially in Asia.

Author Contributions: Writing–review, analysis & interpretation of data, C.-C.M.; Writing – review & editing, H.-S.C.; Writing–review, editing & Project administration, H.-P.C. All authors have read and agreed to the published version of the manuscript.

References

1. Wang, H.T. Crisis Communication of the Government: A Case Study of the Ba-zhang Creek Event. Master's Thesis, University of Shih Hsin, Taipei, Taiwan, 2010, unpublished. Available online: https://hdl.handle.net/11296/6xnbhw (accessed on 23 November 2019).
2. Luo, S.T. The Media Agenda and Government Agenda in China's Food Safety Crisis. Master's Thesis, Southwest University of Political Science and Law, Chongqing, China, 2012, unpublished.
3. Tan, Y.H. Effectiveness evaluation of emergency response plan. *Sci. Technol. Manag. Res.* **2010**, *24*, 56–59.
4. Li, Y.Z. Harmony for Chaos, Food Safety, No Small Matter. Available online: http://shipin.people.com.cn/GB/86164/363431/index.html (accessed on 20 May 2013).
5. Chiou, W.W. The Development and Challenge of the Food Safety Institution in the PRC: On the Case of China's Poisoned-Milk Scandal. Master's Thesis, Nanhua University, Chiayi, Taiwan, 2010, unpublished.
6. China Economic Net. 2012 Food Safety Confidence Index Research Report. Available online: http://www.ce.cn/cysc/ztpd/12/xcz/xxzs/201206/15/t20120615_21179342.shtml (accessed on 15 June 2012).
7. China Food Safety Rule Network. Food Safety and Hygiene in Taiwan. Available online: http://foodlaw.cn/lawhtml/ywjy/109.shtml (accessed on 18 June 2010).
8. Zhang, F. The development and improvement of China's modern food safety supervision legal system. *Polit. Law* **2007**, *5*, 18–23.
9. Xia, Y.J. Food safety supervision system in Taiwan and its enlightenment. *J. Wuhan Univ. Technol. (Soc. Sci. Ed.)* **2013**, *26*, 163–168.
10. Wang, K.Y. Non-conventional security and reforms of Taiwan's military strategy. *Taiwan Int. Stud. Q.* **2010**, *6*, 1–43.
11. Fink, S. *Crisis Management: Planning for the Inevitable*; AMACOM: New York, NY, USA, 1986.
12. Coombs, W.T. *Ongoing Crisis Communication*; Sage: London, UK, 1999.
13. Howitt, A.M.; Leonard, H.B.; Giles, D. (Eds.) *Managing Crises*; Sage: New York, NY, USA, 2009.

14. Wang, H.Z. Comparative study on the background of Chinese and American food safety legislation. *Reform Open.* **2010**, *8*, 13–14.
15. Yeung, R.W.; Morris, J. Consumer perception of food risk in chicken meat. *Nutr. Food Sci.* **2001**, *31*, 270–279. [CrossRef]
16. Chen, K.L. Food Safety Issues and Risk Perception of Public: A Situational Theory of Problem Solving. Master's Thesis, University of Shih Hsin, Taipei, Taiwan, 2012, unpublished. Available online: https://hdl.handle.net/11296/4s75j9 (accessed on 12 October 2019).
17. Smith, R.D. Responding to global infectious disease outbreaks. *Public Relat. Rev.* **2006**, *33*, 120–129.
18. Palencher, M.J.; Heath, R.L. Strategic risk communication: Adding value to society. *Public Relat. Rev.* **2007**, *33*, 120–129. [CrossRef]
19. Jacob, C.; Mathiasen, L.; Powell, D. Designing effective messages for microbial food safety hazards. *Food Control* **2010**, *21*, 1–6. [CrossRef]
20. Xu, R.; Wub, Y.; Luan, J. Consumer-perceived risks of genetically modified food in China. *Appetite* **2020**, *147*, 104520. [CrossRef]
21. Boccia, F.; Covino, D.; Sarnacchiaro, P. Genetically modified food versus knowledge and fear: A Noumenic approach for consumer behaviour. *Food Res. Int.* **2018**, *111*, 682–688. [CrossRef] [PubMed]
22. Boccia, F. Fear of Genetically Modified Food and the Role of Knowledge for Consumer Behavior: The New Noumenic Approach. *EC Nutr.* **2019**, *14*, 205–207.
23. Baker, G.A. Food safety and fear: Factors affecting consumer response to food safety risk. *Int. Food Agribus. Manag. Rev.* **2003**, *6*.
24. Eiser, J.R.; Miles, S.; Frewer, L.J. Trust, perceived risk, and attitudes toward food technologies 1. *J. Appl. Soc. Psychol.* **2002**, *32*, 2423–2433. [CrossRef]
25. Li, Y.; Gao, X.; Du, M.; He, R.; Yang, S.; Xiong, J. What Causes Different Sentiment Classification on Social Network Services? Evidence from Weibo with Genetically Modified Food in China. *Sustainability* **2020**, *12*, 1345. [CrossRef]
26. MA, C.C. Responding in Crises: A Comparative Analysis of Disaster Responses between Mainland China and Taiwan. *J. Homel. Secur. Emerg.* **2012**, *9*, 1. [CrossRef]
27. Keman, H. Comparative research methods. *Comp. Polit.* **2014**, *3*, 47–59.
28. Pennings, P.; Keman, H.; Kleinnijenhuis, J. The comparative approach: Theory and method. In *Meaning and Use of the Comparative Method: Research Design*; Sage Publications: London, UK; Thousand Oaks, CA, USA, 1999.
29. Chen, X.M. *Qualitative Research Methods and Social Science Research*; Educational Science Publishing House: Beijing, China, 2000.
30. Overseas Agricultural Product Regulations. Mainland China Food Regulations Food Additives. Available online: http://foodadd.cas.org.tw/ (accessed on 2 November 2017).
31. Zhang, H. On Perfecting the Food Safety Legal System. *Port Health Control* **2006**, *3*, 4–5.
32. Luo, J.; Ren, D.P.; Yang, Y.X. The loopholes and improvement of food security regulation system in China. *Food Sci.* **2006**, *7*, 250–253.
33. Xie, S.F. A study on product liability insurance system in food industry-Inspiration from China's Taiwan food industry compulsory insurance legislation. *Econ. Manag.* **2011**, *25*, 83–87.
34. Jiang, X.F. On the crime of producing and selling toxic and harmful food. *Leg. Syst. Soc.* **2012**, *36*, 278–279.
35. Zhou, H.A. Comparative Study of Criminal Law Protection of Food Safety across the Taiwan Straits. Master's Thesis, Nanjing Normal University, Nanjing, China, 2010, unpublished. Available online: https://max.book118.com/html/2014/0329/7056505.shtm (accessed on 8 November 2019).
36. Kai, Z.; Fanfan, H.; Kui, Y.; Xueqiong, R.; Si, C.; Xiaojing, L.; Lixia, G. Current situation, problems, challenges and countermeasures of food safety risk communication in China. *Chin. J. Food Hygiene* **2012**, *6*, 19.
37. Fan, S.; Brzeska, J. Feeding more people on an increasingly fragile planet: China's food and nutrition security in a national and global context. *J. Integr. Agric.* **2014**, *13*, 1193–1205. [CrossRef]
38. Marucheck, A.; Greis, N.; Mena, C.; Cai, L. Product safety and security in the global supply chain: Issues, challenges and research opportunities. *J. Oper. Manag.* **2011**, *29*, 707–720. [CrossRef]
39. Pei, X.; Tandon, A.; Alldrick, A.; Giorgi, L.; Huang, W.; Yang, R. The China melamine milk scandal and its implications for food safety regulation. *Food Policy* **2011**, *36*, 412–420. [CrossRef]
40. Bi, B. Taiwan's food safety supervision legal system and its reference significance. *J. Anhui Agric. Sci.* **2013**,

41, 2261–2265.

41. Guo, W.; Zhou, W.; Wang, W.Z.; Nie, X.M.; Li, L.; Chu, X.G. Introduction to Taiwan food safety supervision system and quality certification. *Chin. Agric. Sci. B* **2009**, *25*, 79–83.

42. Chen, L.F.; Zhao, F.J. Analysis on the construction of China's food safety legal system. *J. Anhui Agric. Sci.* **2008**, *10*, 4275–4277.

43. Liu, P. Tracing and periodizing China's food safety regulation: A study on China's food safety regime change. *Regul. Gov.* **2010**, *4*, 244–260. [CrossRef]

44. Zhou, Y.H.; Song, Y.L.; Yan, B.J. Compatible-incentive Mechanism Design of Food Safety Supervision in China. *Commer. Res.* **2013**, *1*, 4–11.

45. Zhou, J. A Study on Punitive Compensation System of Food Safety Law. Master's Thesis, Tianjin Polytechnic University, Tianjin, China, 2010, unpublished.

46. Yu-li, S.H.A.N. The Motives, Measures, Achievements and Enlightenment of Three Rural Land Reforms in Taiwan. *Taiwan Res. J.* **2010**, *3*, 8.

47. Lu, J.H. The status quo and enlightenment of Taiwan agricultural product safety retrospect. *Straits Sci.* **2012**, *9*, 33–35.

48. Useem, M.; Kunreuther, H.; Michel-Kerjan, E. *Leadership Dispatches: Chile's Extraordinary Comeback from Disaster*; Stanford University Press: Palo Alto, CA, USA, 2015.

49. Wei, G.; Yu, Z.; Wenzhi, W.; Xuemei, N.; Li, L.; Xiaogang, C. Introduction of Food Safety Supervision System and Quality Certification in Taiwan. *Chinese Agric. Sci. B.* **2010**, 18.

50. Chang, C.F. Constructing the Process Model for the Traceability of Rice Producing and Distribution. Master's Thesis, Tatung University, Taipei, Taiwan, 2007, unpublished. Available online: https://hdl.handle.net/11296/ucb5w4 (accessed on 10 April 2020).

11

The Challenge of Feeding the World

Dániel Fróna [1,2,*], János Szenderák [1,2] and Mónika Harangi-Rákos [2]

[1] Karoly Ihrig Doctoral School of Management and Business, University of Debrecen, 4032 Debrecen, Hungary; szenderak.janos@econ.unideb.hu

[2] Institute of Sectoral Economics and Methodology, University of Debrecen, 4032 Debrecen, Hungary; rakos.monika@econ.unideb.hu

* Correspondence: frona.daniel@econ.unideb.hu

Abstract: The aim of the present research is to provide a comprehensive review about the current challenges related to food security and hidden hunger. Issues are presented according to major factors, such as growing population, changing dietary habits, water efficiency, climate change and volatile food prices. These factors were compiled from reports of major international organizations and from relevant scientific articles on the subject. Collecting the results and presenting them in an accessible manner may provide new insight for interested parties. Accessibility of data is extremely important, since food security and its drivers form a closely interconnected but extremely complex network, which requires coordinated problem solving to resolve issues. According to the results, the demand for growing agricultural products has been partly met by increasing cultivated land in recent decades. At the same time, there is serious competition for existing agricultural areas, which further limits the extension of agricultural land in addition to the natural constraints of land availability. Agricultural production needs to expand faster than population growth without further damage to the environment. The driving force behind development is sustainable intensive farming, which means the more effective utilization of agricultural land and water resources. Current global trends in food consumption are unsustainable, analyzed in terms of either public health, environmental impacts or socio-economic costs. The growing population should strive for sustainable food consumption, as social, environmental and health impacts are very important in this respect as well. To this end, the benefits of consuming foods that are less harmful to the environment during production are also to be emphasized in the scope of consumption policy and education related to nutrition as opposed to other food types, the production of which causes a major demand for raw materials.

Keywords: nutrition; agriculture; food security; hidden hunger

1. Introduction

Currently, one of the most important challenges to achieve food security is the intensification of global food production. Most surveys and research efforts in agriculture focus on crop production. However, these analyses do not take into account the instability of yield over time or the variability and reliability of cereal production over the years [1]. As the global population continues to grow, agricultural production must also keep pace with it. Over the upcoming 40 years, agricultural emissions will increase by approximately 60% so that humanity can be supplied with food in appropriate quantity and quality. Various studies predict strong population growth within 30 years [2]. According to Röös et al. (2017) that number will be approximately 9–11 billion by 2050 [3] but the number is disproportionate in terms of territorial distribution as it is mostly based on urbanized environments [4]. Concerns about food production are not unfounded. Scientific and technological innovations beat Malthus' predictions in 1798 over the long run and increasing food production has met with the increasing food demand of the growing population. To continue to prove Malthus wrong in the future will require serious efforts, especially in terms of agricultural livestock production [5]. If current global processes continue and population growth tendencies remain unchanged, another 2.4 billion people will live in developing countries by 2050 (in South Asia and Sub-Saharan Africa, the population is expected to grow steadily). The size of urbanized areas is expected to increase threefold between 2000 and 2030 [6]. In these regions, agriculture is of outstanding national economic importance. In total, 75% of the world's poorest people live in rural areas, where agriculture is their most important foundation of subsistence [7]. Nevertheless, on average, over 20% of the population living in rural areas is suffering from food supply security problems [8]. Satisfying the demand requires increased productivity, structural changes in the livestock sector and the need to increase animal products [9]. According to forecasts, the average daily intake per capita is projected to exceed 3000 kcal globally by 2050 to reach 3500 kcal in developed countries and to exceed 2500 kcal even in the poorest sub-Saharan areas [10].

The demand for food, feed and crops with high fiber content is constantly increasing. So, there is increasing pressure on the already "impoverished" arable land and freshwater resources. The size and proportion of land used to produce food and feed depend largely on the evolution of global eating habits and the achievable average yields. The production of raw materials for the Western diet (involving high meat, dairy and egg consumption), which is becoming more widespread in the world, poses serious environmental challenges [11]. In addition to the competition between food and feed production, the increasing utilization of biomass also has a significant impact on land use and water management. The global food sector is heavily dependent on fossil fuels. Therefore, the volatility of energy markets might have a significant direct impact on food prices and an indirect impact on the security of food supply [7].

The issues presented above have been under intensive research for several decades, surrounded by disputes in many cases. Different drives of food supply and security form a complex network, with strongly interconnected factors. The complexity of this network poses a major challenge for interested parties and requires close cooperation between parties to resolve the issues. Despite the overwhelming scientific results, some of the related areas are discussed based on emotion and by taking a subjective approach. Synthesizing scientific results and presenting them in an accessible manner may provide novel insight for related parties. This research is a comprehensive review about these issues and the possible solutions.

2. Materials and Methods

The overall objective of this paper is to provide comprehensive research about the topic, with the processing of international and relevant literature in a literature study. Food security, nutrition and livelihood security is connected at the global and national level as well. Thereby, they are affected with the risk of so-called "shocks" such as climate disasters (drought, flood, etc.), human conflict (such as war, radical protests, etc.), pests and diseases (such as invasive species, etc.). Multi-sectoral cooperation is essential because most of the studies have shown that only a few countries have achieved fast economic growth without preceding agricultural growth. The development of food production systems is based on agricultural diversification, the conservation of water sources and efficient land usage while biodiversity is being preserved.

Qualitative research is suitable for exploring results and situations from previous relevant research and comparing with our research. However, methodological examination regarding the data analysis process is limited, but there are no systematic rules for analyzing qualitative data. Of course, there must be a logical structure or a framework behind the analysis. Computer-assisted search engines make qualitative data analysis more efficient and faster. Qualitative research often provides results, insights and concepts rather than data analysis methods to assess hypotheses or theories. The economic impact of food security is analyzed by various, relevant studies, but there is a close relationship between the environmental and social effects (like water and soil management, climate change, energy security). In addition, upon preparing the study a combination of the following terms was applied during the search for relevant studies: food security, agriculture, population growth, food and environmental safety, food demand, yield trends, change in land use, biofuels, sustainability requirements, and mitigation of climate change. These relevant studies were mostly analyzed from Google Scholar, AgEconSearch, EconBiz.de and Scopus. The literature review is based on recent, relevant studies between 2005 and 2019. In every case, the latest database of the World Bank [12] and Food and Agriculture Organization of the United Nations (FAO) [13] was analyzed during the creation of this study. In some cases, the databases have been merged, e.g., for the exchange of currencies, the harmonization of units of measurement, the frequency of communication. These circumstances were harmonized with each other, so that there was no need to "beautify" the database. Based on these databases, we covered some of the major related results to gain insight into the complexity of these processes. Graphic representations can help readers to better understand the results. Data from the previously mentioned databases were analyzed by R Studio. This program supports the graphic appearance of the analyzed data. Comparing the results is difficult, since most of the results can be viewed as crude approximations of the future developments. We rather focus on the trends related to food security.

3. Results and Discussion

3.1. Land Use

In recent decades, the demand for growing agricultural products has been partly met by increasing cultivated land [14]. However, in the future, the efficiency of agricultural production and specific yields must be increased, since there is serious competition for existing agricultural areas. Various relevant studies outline an increasingly gloomy prospect, namely that increasing yields will not be able to meet the demands for raw materials [15–17]. Nowadays, the increase in agricultural performance is mainly due to the cultivation of new areas, which is hardly sustainable in the long term. Consequently, new areas must be incorporated into agricultural production. Agricultural production needs to expand faster than population growth and this objective needs to be achieved in a sustainable way without further damage to the environment. The driving force behind development is sustainable intensive farming, which means the even more effective utilization of agricultural land and water resources [18].

Urbanization takes away an increasing amount of agricultural land and puts pressure on current land use and biodiversity as well [19]. According to estimations, the growing food demand will require approximately 320–850 million hectares of agricultural land additionally by 2050. The demand for additional agricultural land is limited by the changes in future dietary habits, which will mainly be influenced by socio-economic developments in developing countries. Depending on these changes, consumption will be shifted towards food sources of animal origin by 2050 due to improving welfare.

Moreover, this will cause a major change in land use as well, since the demand for feed crops will increase [20]. In addition, future yields based partly on the introduction of new plant varieties and improved agronomic practices will determine how much arable land will be needed. The requirement to reduce greenhouse gas emissions originating from agriculture will inevitably limit further land allocations [21]. At the same time, it should also be taken into account that the reduction of greenhouse gas emissions in agriculture depends largely on the attitude of producers as they are the ones who are directly affected by the effects of climate change. Farmers who believe in climate change and its anthropogenic or man-made nature are much more open to reducing greenhouse gas emissions but, at the same time, farmers are often more easily able to adapt to changing circumstances than to reduce emissions of harmful substances [22].

3.2. Population

It is beyond dispute that population growth is among the main drivers of global changes (notice in Figure 1). In approximately 10,000 B.C., agriculture began to develop with a global population of approximately 2.4 million people [23]. At the beginning of our chronology, Earth's population was 188 million. As a result of the industrial revolution and the parallel development of health care and medicine, a major change occurred. By the end of the 1800s, the global population reached or already exceeded one billion people [24]. Currently, China alone has a population of 1.4 billion people [25]. The next major period was the 1930s when global population exceeded 2 billion people (when maize hybrids began to spread) (Figure 1). Due to the achievements of the Green Revolution, the global population doubled to over 3 billion (1960). It has been established that the global population grew from 1.65 billion to 6 billion during the 20th century. In 1970, there were nearly half as many people in the world as today [26]. By the middle of the 20th century, annual global population growth rose to 2.1% (1962), which is the highest annual growth rate in history. Nowadays, the growth rate has fallen to 1.2%, which is less than 80 million people annually. According to forecasts, the annual growth rate will decline to 0.1% by 2100 [27].

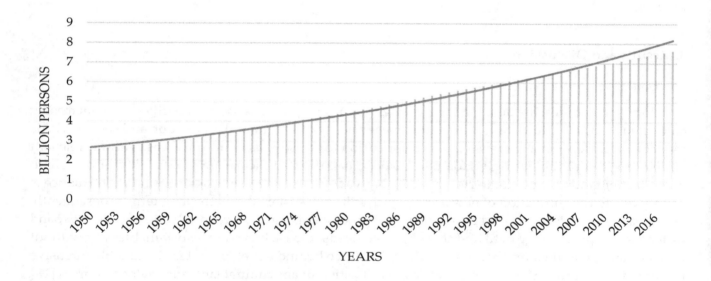

Figure 1. Global population growth. (Source: Own calculation and editing based on the database of FAO, 2019) [13].

Population growth in itself does not completely explain the changes in food consumption. While the volume of food consumption is dependent on the size of the population, quality of the consumed food is dependent on the average household income. According to Figure 2, there is no apparent connection between the (log of the) population and the (log of) GDP per capita, which means that independently from the population of the given countries, the GDP per capita may vary freely. At the same time, a higher GDP is more likely to be associated with a low share of agricultural added value. At the bottom of the graph, African countries possess a very high agricultural share in the GDP, while Europe at the top of the graph is very much the opposite with a high GDP per capita and a low share of agriculture (Figure 2).

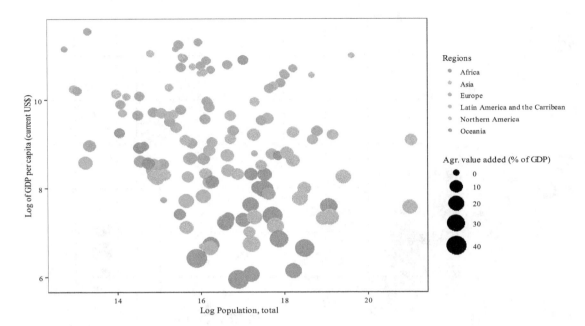

Figure 2. The connection among the population—GDP per capita and agriculture, 2017a. (Source: Own calculation and editing based on the database of the World Bank, 2019) [12].

Figure 3 shows the connection between the share of agriculture (as value added in % of GDP) and the GDP per capita. GDP per capita is measured in current US$, plotted in logarithmic form. As the GDP per capita increases in a given country of the world, the share of agriculture decreases quickly. Generally, African countries (red color) have the highest share of agriculture in GDP among the regions, with over 30% or even more in some cases. However, at the same time, the value of GDP per capita is very moderate. Asian and African countries have a relatively low GDP per capita but high share of agriculture in GDP. The values of Asian countries are extremely diverse. At the same time, European and North American countries typically have a high GDP per capita, while the share of agriculture is only a few percent. The share of agriculture is much lower in Latin America, Oceania, Northern America and especially in Europe, mostly under 5% of the GDP. At the same time, GDP per capita is the highest among the countries. The graph indicates that African and Asian countries are still very much dependent on agriculture, as they take a high share of the GDP (denoted by red and orange).

In Figure 4, a similar methodology-based editing can be noticed. It deals with the connections between energy consumption, agriculture and the GDP per capita. With these elements, the latest database is from 2013. It also indicates the regions with different colors and the GDP per capita value in US dollars. The "X" axis represents the energy usage (in kg of oil equivalent per capita) and the "Y" axis represents the share of agriculture in the GDP. It can be read that energy use per capita is very low in the case of Africa—the value is well under 2500 kg of oil equivalent per capita. A correlation also can be discovered between the agricultural share in the GDP and energy usage. In regions, where the share of agriculture in GDP is high (over 20%), there is a low value energy use per capita (under 2500 kg of oil equivalent per capita). In general, as a country became more industrialized, energy demand increased rapidly and, at the same time, the share of agriculture in the GDP quickly took a downturn. It can be concluded that countries with a high consumption of energy have much higher living standards in terms of the GDP per capita compared to countries with low energy consumption.

It is also worth noticing that the relationship is not linear. Below the 2500 kg of oil equivalent per capita consumption, a small increase in the energy use comes with a rapid decrease in agricultural share in the GDP. Above this level, no further change is expected. According to the regional distribution, Africa shows a lower level of energy consumption and the share of Asia varies between the lowest levels and the highest levels of oil consumption.

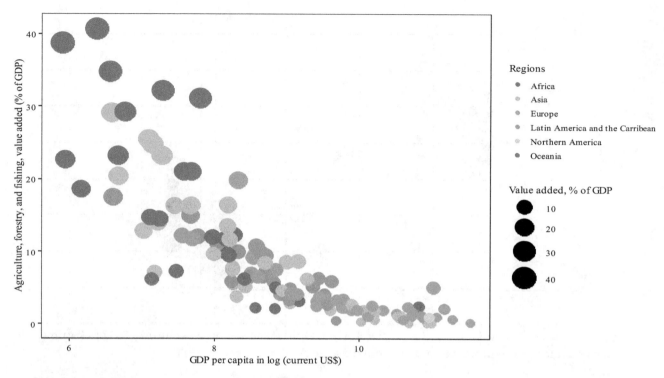

Figure 3. The relationship between GDP per capita and the share of agriculture in GDP, 2016. (Source: Own calculation and editing based on the database of the World Bank, 2019) [12].

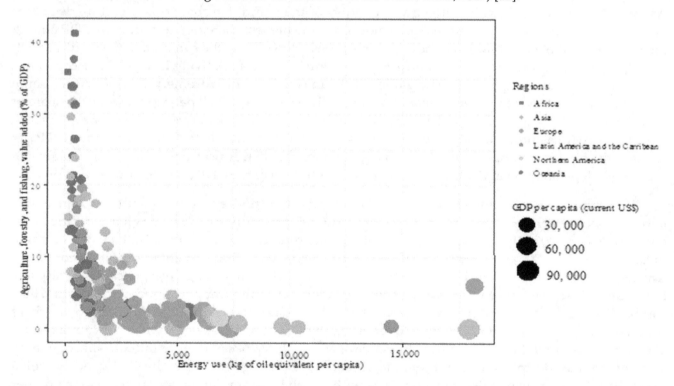

Figure 4. The main connections between energy consumption, agriculture and the GDP per capita, 2013. (Source: Own calculation and editing based on the database of the World Bank, 2019) [12].

3.3. Crop Biology

Crop production and harvest research have traditionally been limited to studies on the physiology and genetics of plants, the creation of new plant varieties, the development of new agricultural chemicals and the development of better agronomic methods [28]. Such research is necessary and there is an increasing number of global initiatives, which are aimed at the achievement of higher cereal yields [29]. Similarly, reducing the yield gap is also a keen research topic and objective, since it is becoming more and more justifiable for many crops to increase yields. Genetic enhancements are likely to be the potential solutions for achieving maximum yields for plants of key importance. There is considerable potential for improvement in yields and flexibility in the so-called "orphan crops". These crops have not been genetically modified yet and they are not traded on an international level. Consequently, less attention is focused on them in terms of their agricultural utilization. As they receive less research attention, breeding technology of "orphan crops" is lagging behind modern technology (e.g., millet, cowpea, manioc, etc.) [30]. On the other hand, agricultural research is increasingly driven by problems of a wider scope, such as the expected decline in yields due to climate change and severe weather events [31]. In addition to problems elicited by weather, the focus of research is on the emission of greenhouse gases and the pollution of water associated with the production of nitrogen-based fertilizers [32–34].

Similarly, research on pests and diseases is important as they are also major risk factors in the case of yield differentials and, due to the effects of climate change and efforts to conserve biodiversity, they are considered as urgent factors [35]. Due to the concerns about soil degradation discussed above, all agricultural practices related to conservation should be applied for the sake of yield improvements, such as tillage and other measures such as the conservation of crop residues and the application of crop rotations [36]. In order for agriculture to meet the emerging challenges it faces, new scientific discoveries should be adapted into practice as soon as possible. In addition, closer cooperation between farmers and scientists is required to integrate new developments appropriately into developments that complement agricultural practices [37]. Nevertheless, the use of genetically modified plants still sets off contradictions among researchers. The debate is mainly present between representatives of natural science and social science; however, it must be resolved in order for the reasons of aversions towards technologies to become understood [38]. Especially, as the latest technological advancements, for example, the application of genome editing in agriculture—and indirectly in food production—might exceed the significance of current GMO crop production [39]. GMO crops can play a radically different role in certain markets: while they have been present in the US since the mid-1990s [40], their distribution in other countries is strictly prohibited.

3.4. Reasons of Changing Eating Habits

Consumption patterns are constantly shifting towards products of animal origin and dairy products that contain higher value added, which results in the increasing demand for the production of feed crops. This process is already typical as, between 1960 and 2010, global arable land per capita decreased from 0.45 to 0.25 hectares and by 2050, it is expected to shrink to less than 0.20 hectares [41]. Approximately 66% of agricultural land is currently used by livestock farming in the European Union as well. This ratio is 40% on a global scale and is expected to rise further by 2050 [41]. According to the data above, dietary change will have a more prominent impact on land use than population growth. The problems mentioned above could be addressed by putting emphasis on wider cultural changes, which focuses on the necessity for coordinated actions of government and political activities, industry, communities, family and society. Recognizing the social needs and attitudes of consumer behavior, a number of research studies analyzing dietary changes are published, which increasingly reveal the routine nature of consumer habits and the institutions and infrastructure supporting them [42–44]. Initiatives aimed at the promotion of healthier and more sustainable patterns of consumption should address the social and technical systems that are able to respond to changing consumer habits. According to certain research activities, the decision-making process for choosing a diet might force consumers to face various ethical challenges [45]. For example, consumer preferences for organic food (with respect to health or sustainability) or the need for locally produced food (minimizing the so-called "food miles"). Obviously, these preferences greatly influence the decision-making process of consumers in relation to the choice of diet [42]. In addition, it should be emphasized that the sustainability of the food supply could be significantly improved even by the reduction of food losses [46].

The growing population should strive for sustainable food consumption, as social, environmental and health impacts are very important in this respect as well. To this end, the benefits of consuming foods that are less harmful to the environment during production are also to be emphasized in the scope of consumption policy and education related to nutrition as opposed to other food types, the production of which causes a major raw material demand [47]. In several countries—at primary schools—lunch break is a basic place of the learning process, where students learn about hygiene, healthy eating habits and/or recycling waste. Acquiring knowledge about healthy eating and recycling waste is fundamental at a young age [48]. Current global trends in food consumption are unsustainable, analyzed in terms of either public health, environmental impacts or socio-economic costs [49]. On different geographic scales, there are clear correlations between the socio-economic situation and the intake of high-quality food and the resulting health outcomes. The change in production structure is caused by the increase in the number of people with higher incomes in low- and medium-income countries. Primarily, this induces a change in consumption habits through the increasing consumption of meat, fruit and vegetables compared to different kinds of cereals [50]. The fact that the seasonal consumption of fruit and vegetables completely disappears is a particularly interesting development. From this point of view, transportation can be a critical factor of environmental impact. Currently, a person eats an average of 42 kg of meat annually, which is expected to rise to 52 kg by 2050, and 1.5 billion new consumers will appear on the market [27]. The growing share of poultry meat among other kinds of meat should be mentioned here. Due to changing eating habits, more and more people consume chicken meat. It can be produced relatively quickly, it is relatively cheaper, and it is not prohibited by religions.

The focus of research is increasingly shifted towards the relationship between nutrition and food production, especially the problems caused by climate change, increasing population and urbanization. As an example, many studies on Africa have been published [51,52], which have pointed out that there is a need for intervention at a social level to modify nutrition habits and to avoid malnutrition. Areas that are different in terms of public health so far are likely to become even more diversified, as low-income countries in particular find it more difficult to adapt to the consequences of climate change, food shortages and water shortage, as well as to the associated socio-demographic changes and the resulting dietary modifications [53]. Subsequent research activities and their practical implementation should address the impact of dietary changes on the natural environment and the impact of environmental changes on all components of food safety [20].

The integrated approach of agri-food research draws attention to the impact of social and political conflicts on health and malnutrition. Changes occurring in the environment might aggravate malnutrition by limiting the ability to produce food products. Extreme weather events (for example drought and floods) might contribute to the volatile change in food prices, which in extreme cases might result in serious problems, in the form of riots or the further increase in the proportion of famine [54].

3.5. Links between Nutrients and "Hidden Hunger"

There is a detectable positive change in the reduction process of global famine. However, despite progress, the world is still far from a sustainable food safety system. Obesity is a phenomenon that exists nearly in parallel with famine and malnutrition. Nearly 800 million people are chronically underfed in terms of energy intake, while 2 billion people suffer from micronutrient deficit, but at the same time 1.9 billion people are overweight or obese [55,56].

People suffering from hidden hunger typically consume food items with high calorie but low nutrient content, which can easily lead to obesity (although not necessarily). This also proves that famine and obesity, as well as under- and overnutrition occur in parallel at a global scale. This means the inadequate consumption of sufficient vitamins, minerals and trace elements. Therefore, it is interesting that overnutrition (in calories) may be associated with malnutrition (micronutrient). It will be a great challenge for the future to produce food of not only sufficient quantity but quality as well. As a summary of the above, three phenomena appear as contradictions but parallel to each other: malnutrition, overnutrition, and hidden hunger. These three forms of nutritional problems are also referred to as the "triple burden" of malnutrition [57]. This triple effect contributes to the reduction of physical and cognitive human development, the loss of productivity, sensitivity to infectious and chronic diseases and aging [58].

Micronutrient-deficient nutrition is a global phenomenon that may affect certain social groups, such as those over the age of 65 even in the most advanced countries [60]. Reduction of the various forms of malnutrition requires better food policy and targeted nutrition-related interventions. In Africa and Asia, urban populations are growing at a high rate, which may lead to a further decrease in per capita nutrients (an average reduction of 36% in Africa, 30% in Asia) (Figure 5). A possible solution for slowing down the process might be nutrient reuse. In contrast, average per capita amount of nutrients in Europe will decrease by 10%, but a steady decline in population numbers is also expected here [61]. Obviously, these analyses are limited by certain factors as they do not take into account, for example, the size of the city or changes that have occurred in terms of land use. By 2030, urban expansion will require an additional 2% of the available global land, but local effects might be more significant in the life of individual cities, affecting reuse opportunities and making adaptive decisions [8].

Figure 5. Map of hidden hunger. (Source: Muthayya et. al. (2013) [59]).

3.6. Climate Change and Water

According to estimations, climate change has already reduced global crop yields of maize and wheat by 3.8% and 5.5% respectively and researchers are warning that further decline in productivity is expected as temperature changes exceed critical physiological thresholds [62]. The progressively extreme climate change increases production risk and puts an increasing burden on the subsistence of agricultural producers. Climate change also poses a threat to the food supply of both rural and urban populations. Extreme climatic events have a long-term negative impact, since exposure to risk and increasing uncertainty affect the introduction of effective economic innovations. Consequently, the number of low-risk but low-yield activities begins to increase [31]. Agricultural activity also contributes to warming the planet. Total carbon dioxide emissions from agriculture in 2010 were equal to 5.2–5.8 gigatons of CO_2 equivalent annually, representing approximately 10%–12% of global anthropogenic emissions [63]. Agricultural categories with the highest level of emission are fermentation, manure, synthetic fertilizer and biomass combustion. Considering that there will be a need for further increases in agricultural production, the emission of harmful substances is also expected to increase. The main source of planned emission growth is the application of conventional agricultural techniques (as opposed to precision farming) that will result in the further, severe damage of the ecosystem, such as further water and soil pollution [64]. Some recent publications discuss the impact of climate change on yields, especially for the most important crops, such as wheat, maize, rice and soybean [65–69], which means that scientific processing of the topic is ongoing.

Currently, 97.5% of Earth's water resources are saltwater and only 2.5% is freshwater, 69% of which are glaciers and persistent snow, 30.7% groundwater, and 0.3% in the form of lakes and rivers [41]. There is some similarity between freshwater resources and land in terms of their availability. If we look at both factors on a global level, they are available in sufficient quantities, but the distribution is very uneven. This is also illustrated by the fact that there are huge differences between countries in the same regions, but even within countries. Demand for water is expected to increase by 100% by 2050, which can be attributed to population growth, urbanization and the effects of climate change [70]. As the urban population grows, household and industrial water consumption are expected to double. Climate change implies a greater chance of more extreme weather phenomena, because of which water consumption of crop production might increase considerably [70].

Humanity consumes the most water in the course of food production and global production of cereals. Due to increasing food production, water resources from the rivers and groundwater are primarily used for the irrigation of cultivated crops. Most irrigation systems usually provide more water to plants than they actually require [71]. Improving living standards, changing food preferences and the increasing demand for goods require a higher amount of water consumption. At the same time, more than 650 million people—especially south of the Sahara—have no access to drinking water of adequate quality. The current situation is further exasperated by the fact that 2.4 billion people do not have modern wastewater management [72]. The United Nations Organization puts special emphasis on the issue of sewage disposal.

This is also well illustrated by the fact that the 6th element of the Sustainable Development Goals is clean and sanitary water. Ensuring appropriate management and sustainable treatment of water resources is essential for our future.

Climate change is a global phenomenon, but developing countries are in greater danger. In addition, the problems posed by urbanization, increasing water shortage and technological backwardness are the most important challenges to be addressed. Rural areas should have access to the fundamental services of the 21st century, such as public utilities, health care, electrification, education, etc. This is important for the improvement in the living conditions of the population living here [63].

3.7. China's Food Supply

Inputs (fertilizer, water) and their impact on the environment are vital elements of food production (mostly cereal). According to surveys, the global center of nitrogen fertilizer utilization was in Western Europe and the US in the 1960s but was relocated to East Asia, especially to China by the beginning of the 21st century [73]. In the last century, China faced a number of food shortages. In the course of one of them, a quota system was introduced (1955–1993) followed by a land contract reform (1981) that was implemented. Total cereal production increased by 74%, from 354 million tons in 1982 to 618 million tons by 2017, which exceeded the rate of population growth [74]. Currently, China feeds 20% of the global population on 7% of the total agricultural land. In order to maintain this performance, China has paid a high price. The use of chemical fertilizers has tripled in the last three decades. Excessive and inefficient use achieved 32% efficiency compared to a global average of 55%. China's water supply is in a similar situation since, apart from low-efficiency utilization and poor-quality quantitative distribution across the country, it is also uneven. China's available water supply per person is only 2050 m^3, which is 25% of the global average. In North China, where only a low amount of water is available, a large volume of underground water is used for agricultural purposes. Therefore, it is of utmost importance for China to proactively investigate how food security can be achieved through the balance of resource management, environmental protection and sustainable agricultural development [74–76]. In 2015, the one-child policy was abolished; families are now allowed to have a second child. However, many people choose not to have more children because they cannot afford the high costs of their upbringing. Thus, according to demographic estimations, the two-child policy will result in only 2–4 million additional people in China annually for the next 10 years. The accelerated growth of China in terms of urban population as compared to rural areas continues to affect food consumption [77]. In 2016, China's urbanization rate rose to 57% and it might increase to 65% by 2025 and to 80% by 2050. However, it should be noted that in the east (China, India), a significant part of the population is concentrated because it is often impossible to live outside these areas (e.g., deserts, high mountains, and jungles). In light of these statistical data, they need to find a solution for further safe and healthy food supply [78].

3.8. Food Prices and Food Security

Changes in food prices fundamentally affect the quantity and quality of food available to an individual. In developing countries, where a high proportion of household income is spent on food, changes in food prices are a critical factor. In these areas, relatively moderated price changes can also have a significant impact on food security. The past few decades have been marked by rising food prices and rising price volatility. These market events require the collective cooperation of the countries concerned in order to mitigate the adverse effects of price changes. Swinnen and Squicciarini (2012) drew attention to the contradictory messages being transmitted by the parties involved in the food safety debate. These messages do not always correctly convey the true effect of high or even low food prices [79].

While the food price boom dates back at least to the 1970s, rising food prices (in nominal terms) in 2007/08 renewed the attention of the policy makers and market analysts to the so-called "commodity boom" again. Not only the price levels, but the higher variability became a concern as well [80]. As Baffes and Haniotis (2016) noted, that the reversal of the downward trend in food prices seen until 2000 has already had consequences for food security in developing countries [81]. The main sources of the price boom between 2000 and 2007 was the increased commodity demand induced by the global economic growth, the dollar depreciation and the changes in the stock to use ratio, according to Timmer (2008) [82]. However, these sources provided an inadequate explanation for the sudden increase in the prices. Additional factors were the growing demand for biofuels (where food crops are the input materials, especially maize), unfavorable weather events, plant diseases and the changes in trade policy. In some cases, panic and hoarding and further speculation has some effect as well [80]. According to the United Nations Conference on Trade and Development (UNCTAD) (2011) study, agricultural prices are more vulnerable to fluctuations by their nature. These effects will require a more efficient risk distribution mechanism among the markets, which would strengthen the safety net related to food price changes. Increased price fluctuation has an adverse effect on developing countries, since there is a high share of rural households with low household income, that often rely heavily on self-produced agriculture commodity products [83].

Oil price changes (and in general, energy price changes) became a crucial factor as well. The effect of oil prices is twofold. Firstly, high oil prices would increase the demand for alternative energy sources, such as biofuel. These changes, in turn, will increase the demand for input materials, which can change the allocation between food, feed and fuel. Second, higher oil prices lead to higher production costs, which decreases the supply of food in the long run [84]. In general, the cost of energy is approximately 10 per cent of the agricultural production, according to the World Bank (2016) estimates, which means that agriculture and its related sectors are highly energy intensive. In developing countries, production technologies and transportation are inefficient. Thus, energy price changes can have serious effects. A significant number of studies have shown a stronger impact of energy prices on agricultural prices and a closer integration of the two markets [85–88]. These studies have found a stronger connection between the energy and the agricultural market after the global economic and financial crisis. Among the results, there was apparent support for possible non-linear effects, increased spillover mechanisms and long-term relations (cointegration). At the same time, the root of the price developments is the fundamental market mechanism, as supply and demand. As Timmer (2008) noted, the long-term question is whether supply can keep up with demand generated by rapid economic growth. While the possibility existed in recent decades and supply could keep up with demand, this time, it is compounded by the scarcity of high-quality, accessible agricultural area, stagnation in yields seen over decades, and rising costs of basic inputs. As research results are often lagging behind in this field, the only possibility is to increase yields until new agricultural technologies emerge. The most effective solution to high food prices is therefore to stimulate an increase in agricultural output. Combining the effect of climate change and water scarcity, the problem requires a quick and efficient solution.

In the wealthiest countries, the concentration of retail trade and the increasing complexity of food businesses, as well as the extended impact of supply chains, play a role of key importance. In poorer countries, many of the listed effects can be overcome. However, according to researchers, cooperation along the supply chain is less effective [89,90]. In addition to cooperation between chain members, traceability is also very important in modern agriculture. The implementation of technological innovations is essential in food supply chains from farm to plate [89].

Cooperation along the supply chain is particularly important in developed countries because food retail is highly concentrated and, in many countries, there are numerous companies with a very strong bargaining position with suppliers and they therefore often push down purchasing prices. Lower profit ratios and higher volumes from more limited suppliers encourage lower prices and increase the number of sales, creating a vicious circle of addiction [91].

Additional income generated by the rising prices of agricultural products and food, therefore, does not reach producers in most of the cases, who are consequently able to introduce production-related innovation only from fewer resources [92]. The share of supermarket-type stores in food retail has become increasingly significant on a global scale in recent decades. Companies dealing with food retail often employ suppliers to ensure a continuous supply of certain product types, which may further increase the exposure of producers in the supply chain [93].

4. Conclusions

Growing population and changing dietary habits, with the intensifying demand for food with higher value added in developing countries are expected to increase food demand by 60% by 2050. In addition, unprecedented developments are taking place, especially in areas where the demand for fossil resources has traditionally been very low. Agricultural production can only be intensified with the increasing use of fertilizers. Thus, the efficiency of fertilizer usage needs to be improved. Almost all developed and developing countries have accepted the need to increase agricultural productivity and efficiency. The sustainable production of more food for human consumption requires technology that makes better application of limited resources, including land, water and fertilizer. Traditional agricultural production is not sustainable economically or environmentally. The question is whether the existing knowledge on agro-ecological practices is able to achieve the rate of yield that is required to feed the growing population. Without answering the question, a substantial investment is needed in research and innovation. In addition to food security, food stability is also important, and the most important issue here is predictability.

Food production requires a fundamental transformation in order to preserve the ecological conditions of the planet and to avoid the associated health risks. The key to the solution is so diverse that it is essential to integrate and renew the relevant branches of science. This includes, for example, molecular and taxonomic biology, food science and medicine, agronomy, ecology, earth science, computer science and nature biology. Long-term, interdisciplinary human health studies need to be further integrated in order to achieve a higher standard and compatibility of sustainable food production. Globally, sustainable development goals require an industrial and scientific revolution. Food production, affected human population growth and the global ecological challenges it generates, will play a crucial role in the future of the Earth.

Climate change and extreme negative weather conditions are key drivers of global famine and food insecurity. They have a negative impact on livelihood and all aspects of food security (accessibility, stability, etc.) and contribute to other malnutrition related to childcare and nutrition. Due to the growing energy and food demand, it has become evident that greenhouse gas emissions, especially carbon dioxide, have an impact on the global climate. There is a growing demand for suitable land, where food production, feed production, energy crops and urbanization are in competition. These problems are further exacerbated by the gradual change in soil productivity caused by climate change (erosion, water stress, increasing soil salinity, etc.). The health of the soil is also crucial during agricultural production because healthy crops can only be produced on soils in good conditions. Producing crops that meet the high criteria of healthy foods requires soil in good conditions. That means the farmers have to pay attention to the health status of the soil during agricultural production and plant seeds or use fertilizers which do not harm the soil. However, the change in indirect land use may also increase greenhouse gas emissions. Precision plant breeding is a good solution to increase crop production and yields. Farmers always have to pay attention to saving biodiversity. Increasing yields by starting agricultural production on new lands cannot be a solution anymore in order to save the available natural resources. This is due to the fact that crop production has shifted to previously unused land, which can lead to the transformation of forests and savannah. Such land use change will damage biodiversity and increase greenhouse gas emissions. The science of global climate change indicates that, as a result of the increasing level of greenhouse gases, the Earth as a whole has a general warming trend. While natural resources have an impact on greenhouse gas concentrations over time, global scientific consensus indicates that human resources for greenhouse gases also contribute to global climate change. The risk of food insecurity and malnutrition is greater today, especially in low-income regions, which are more exposed and sensitive to climate change.

Technological innovations may allow mankind to increase food production in a sustainable way to meet the reasonable needs. The use of smart devices including smartphones, other IT tools and different applications of precision and automatized agriculture can help farmers to increase the efficiency of agriculture. The spread of smart IT devices can help the spread of precision and automatized agriculture as well as more agriculture employees will have knowledge about these technological solutions. When professional agricultural users start to introduce new smart solutions in the operation of agricultural companies, they can count on the IT knowledge of their workforce; however, during the self-evaluation of employee knowledge, managers always have to pay attention to the Dunning–Kruger effect [94]. The above effect means that less educated workforce usually overestimates their knowledge—and this circumstance is totally typical in the case of IT knowledge in the agricultural sector [95]. Ultimately, the issue of food security applies to people as well as to finite resources. There is no simple or easy solution to sustainably feed nine billion people, especially with consumption habits becoming non-sustainable. Hopefully the scientific and technological innovation is going to help to defeat this challenge. Sustainable food production can only be achieved by reducing greenhouse gas emissions and reducing water usage. This growth must be achieved without further environmental damage. Sustainable intensification might be a way to ensure the necessary—and not overestimated—scale of production while mitigating environmental impacts. We must avoid further reducing our biodiversity for the easy profit of food production, not only because biodiversity provides numerous public goods that humankind relies on, but also because we have no right to deprive the future generation of the economic and cultural benefits. These challenges together represent the crucial problem that needs to be solved. To solve this crucial problem, we need a social revolution that breaks down the barriers between science and agriculture related to food production. The goal is not only to maximize productivity but also to optimize the results of production, environmental protection and social justice (fairness of food distribution) in a much more complex way.

According to the results, instead of the inclusion of additional agricultural area, further improved yields and food management will be necessary to provide sufficient amounts of additional food. This will require more efficient water and energy management as well as improvements in waste management. Due to the growing population and changing dietary habits, food supply (especially the animal protein-related consumption) is expected to increase the pressure on the environment. A higher share of plant-based consumption may help to reduce this pressure, but it is expected only in the developed areas with a relatively high GDP per capita. Climate change is the slowest changing component of the food supply, but its impact is felt globally. The right perception of climate change can have a serious impact on improving food security. Despite the overwhelming scientific evidence, there is often skepticism and emotional overtones in the debate surrounding climate change. However, effective solutions to problems require a united and cooperative approach. Coordinated restrictions on agricultural trade are essential in times of high and volatile food prices, which was often hampered by ad-hoc and unadvised trade restrictions in individual countries in the past. Higher food price volatility has become a feature of the liberalized agricultural market in the last decade. As price volatility cannot be reduced, the aim should be to spread and hedge the associated risks properly. Efficient future markets and different types of insurance could be useful tools to tackle these issues. Taking these factors into account is particularly important, since inadequate food supply is likely to lead to food-related riots and social unrest, which, in addition to their economic and social impact, have ethical and political implications as well.

Author Contributions: Conceptualization, D.F. and J.S.; Methodology, J.S and D.F.; Software, J.S.; Writing—original draft preparation, D.F.; J.S.; M.H.-R.; Writing—review and editing, D.F.; J.S.; M.H.-R.; Visualization, D.F; J.S..; Supervision, M.H.-R.

References

1. Knapp, S.; van der Heijden, M.G.A. A global meta-analysis of yield stability in organic and conservation agriculture. *Nat. Commun.* **2018**, *9*, 3632. [CrossRef] [PubMed]

2. Hofstra, N.; Vermeulen, L.C. Impacts of population growth, urbanisation and sanitation changes on global human Cryptosporidium emissions to surface water. *Int. J. Hyg. Environ. Health* **2016**, *219*, 599–605. [CrossRef] [PubMed]

3. Röös, E.; Bajželj, B.; Smith, P.; Patel, M.; Little, D.; Garnett, T. Greedy or needy? Land use and climate impacts of food in 2050 under different livestock futures. *Glob. Environ. Chang.* **2017**, *47*, 1–12. [CrossRef]

4. Kummu, M.; De Moel, H.; Salvucci, G.; Viviroli, D.; Ward, P.J.; Varis, O. Over the hills and further away from coast: Global geospatial patterns of human and environment over the 20th–21st centuries. *Environ. Res. Lett.* **2016**, *11*, 034010. [CrossRef]

5. Smith, P. Malthus is still wrong: We can feed a world of 9–10 billion, but only by reducing food demand. *Proc. Nutr. Soc.* **2015**, *74*, 187–190. [CrossRef]

6. d'Amour, C.B.; Reitsma, F.; Baiocchi, G.; Barthel, S.; Güneralp, B.; Erb, K.-H.; Haberl, H.; Creutzig, F.; Seto, K.C. Future urban land expansion and implications for global croplands. *Proc. Natl. Acad. Sci. USA* **2017**, *114*, 8939–8944. [CrossRef]

7. Popp, J.; Lakner, Z.; Harangi-Rakos, M.; Fari, M. The effect of bioenergy expansion: Food, energy, and environment. *Renew. Sustain. Energy Rev.* **2014**, *32*, 559–578. [CrossRef]

8. Wheeler, T.; Von Braun, J. Climate change impacts on global food security. *Science* **2013**, *341*, 508–513. [CrossRef]

9. Riggs, P.K.; Fields, M.J.; Cross, H.R. *Food and Nutrient Security for a Growing Population*; Oxford University Press US: Oxford, MS, USA, 2018.

10. Alexandratos, N.; Bruinsma, J. *World Agriculture towards 2030/2050: The 2012 Revision*; ESA Working Paper; FAO: Rome, Italy, 2012.

11. Westhoek, H.; Lesschen, J.P.; Rood, T.; Wagner, S.; De Marco, A.; Murphy-Bokern, D.; Leip, A.; van Grinsven, H.;

Sutton, M.A.; Oenema, O. Food choices, health and environment: Effects of cutting Europe's meat and dairy intake. *Glob. Environ. Chang.* **2014**, *26*, 196–205. [CrossRef]

12. The World Bank Homepage. Available online: https://databank.worldbank.org/home.aspx (accessed on 22 January 2019).

13. Database of Food and Agriculture Organization of the United Nations. Available online: http://www.fao.org/faostat/en/#data (accessed on 18 January 2019).

14. Boserup, E. *The Conditions of Agricultural Growth: The Economics of Agrarian Change under Population Pressure*; Routledge: London, UK, 2017. [CrossRef]

15. Davis, K.F.; Gephart, J.A.; Emery, K.A.; Leach, A.M.; Galloway, J.N.; D'Odorico, P. Meeting future food demand with current agricultural resources. *Glob. Environ. Chang.* **2016**, *39*, 125–132. [CrossRef]

16. Crist, E.; Mora, C.; Engelman, R. The interaction of human population, food production, and biodiversity protection. *Science* **2017**, *356*, 260–264. [CrossRef] [PubMed]

17. McLaughlin, D.; Kinzelbach, W. Food security and sustainable resource management. *Water Resour. Res.* **2015**, *51*, 4966–4985. [CrossRef]

18. Ramankutty, N.; Mehrabi, Z.; Waha, K.; Jarvis, L.; Kremen, C.; Herrero, M.; Rieseberg, L.H. Trends in global agricultural land use: Implications for environmental health and food security. *Annu. Rev. Plant Biol.* **2018**, *69*, 789–815. [CrossRef] [PubMed]

19. Blum, W.E. Functions of soil for society and the environment. *Rev. Environ. Sci. Bio/Technol.* **2005**, *4*, 75–79. [CrossRef]

20. Tilman, D.; Clark, M. Global diets link environmental sustainability and human health. *Nature* **2014**, *515*, 518. [CrossRef]

21. Godfray, H. The challenge of feeding 9–10 billion people equitably and sustainably. *J. Agric. Sci.* **2014**, *152*, 2–8. [CrossRef]

22. Arbuckle, J.G., Jr.; Morton, L.W.; Hobbs, J. Understanding farmer perspectives on climate change adaptation and mitigation: The roles of trust in sources of climate information, climate change beliefs, and perceived risk. *Environ. Behav.* **2015**, *47*, 205–234. [CrossRef]

23. WORLDOMETERS. Current World Population. 2019. Available online: https://www.worldometers.info/world-population/ (accessed on 7 January 2019).

24. Our World in Data. World Population over the Last 12,000 Years and UN Projection until 2100. 2018. Available online: https://ourworldindata.org/world-population-growth (accessed on 8 December 2018).

25. UN. World Population Prospects. 2019. Available online: https://population.un.org/wpp/Publications/Files/WPP2019_Highlights.pdf (accessed on 8 June 2019).

26. WORLDOMETERS. Current World Population. 2018. Available online: https://www.worldometers.info/world-population/world-population-by-year/ (accessed on 8 December 2018).

27. FAO. The Future of Food and Agriculture—Trends and Challenges. 2017. Available online: http://www.fao.org/3/a-i6583e.pdf (accessed on 8 December 2018).

28. Murchie, E.; Pinto, M.; Horton, P. Agriculture and the new challenges for photosynthesis research. *New Phytol.* **2009**, *181*, 532–552. [CrossRef]

29. Furbank, R.T.; Quick, W.P.; Sirault, X.R. Improving photosynthesis and yield potential in cereal crops by targeted genetic manipulation: Prospects, progress and challenges. *Field Crop. Res.* **2015**, *182*, 19–29. [CrossRef]

30. Foley, J.A.; Ramankutty, N.; Brauman, K.A.; Cassidy, E.S.; Gerber, J.S.; Johnston, M.; Mueller, N.D.; O'Connell, C.; Ray, D.K.; West, P.C.; et al. Solutions for a cultivated planet. *Nature* **2011**, *478*, 337. [CrossRef]

31. Lesk, C.; Rowhani, P.; Ramankutty, N. Influence of extreme weather disasters on global crop production. *Nature* **2016**, *529*, 84. [CrossRef] [PubMed]

32. Zhang, X.; Davidson, E.A.; Mauzerall, D.L.; Searchinger, T.D.; Dumas, P.; Shen, Y. Managing nitrogen for sustainable development. *Nature* **2015**, *528*, 51. [CrossRef] [PubMed]

33. Goucher, L.; Bruce, R.; Cameron, D.D.; Koh, S.L.; Horton, P. The environmental impact of fertilizer embodied in a wheat-to-bread supply chain. *Nat. Plants* **2017**, *3*, 17012. [CrossRef] [PubMed]

34. Dawson, C.J.; Hilton, J. Fertiliser availability in a resource-limited world: Production and recycling of nitrogen and phosphorus. *Food Policy* **2011**, *36*, S14–S22. [CrossRef]

35. Lamberth, C.; Jeanmart, S.; Luksch, T.; Plant, A. Current challenges and trends in the discovery of agrochemicals. *Science* **2013**, *341*, 742–746. [CrossRef]

36. Pittelkow, C.M.; Liang, X.; Linquist, B.A.; Van Groenigen, K.J.; Lee, J.; Lundy, M.E.; Van Gestel, N.; Six, J.; Venterea, R.T.; Van Kessel, C. Productivity limits and potentials of the principles of conservation agriculture. *Nature* **2015**, *517*, 365. [CrossRef]

37. Woolf, S.H. The meaning of translational research and why it matters. *JAMA* **2008**, *299*, 211–213. [CrossRef]

38. Jacobsen, S.-E.; Sørensen, M.; Pedersen, S.M.; Weiner, J. Feeding the world: Genetically modified crops versus agricultural biodiversity. *Agron. Sustain. Dev.* **2013**, *33*, 651–662. [CrossRef]

39. Hefferon, K.L.; Herring, R.J. The End of the GMO? Genome Editing, Gene Drives and New Frontiers of Plant Technology. *Journal* **2017**, *7*, 1–32.

40. Fairfield-Sonn, J.W. Political Economy of GMO Foods. *J. Manag. Policy Pract.* **2016**, *17*, 1.

41. FAO. *The State of Food Security & Nutrition around the World 2018*; FAO: Rome, Italy, 2018.

42. Jackson, P.; Ward, N.; Russell, P. Moral economies of food and geographies of responsibility. *Trans. Inst. Br. Geogr.* **2009**, *34*, 12–24. [CrossRef]

43. Warde, A. Consumption and theories of practice. *J. Consum. Cult.* **2005**, *5*, 131–153. [CrossRef]

44. Delormier, T.; Frohlich, K.L.; Potvin, L. Food and eating as social practice–understanding eating patterns as social phenomena and implications for public health. *Sociol. Health Illn.* **2009**, *31*, 215–228. [CrossRef] [PubMed]

45. Watson, M.; Meah, A. Food, waste and safety: Negotiating conflicting social anxieties into the practices of domestic provisioning. *Sociol. Rev.* **2012**, *60*, 102–120. [CrossRef]

46. West, P.C.; Gerber, J.S.; Engstrom, P.M.; Mueller, N.D.; Brauman, K.A.; Carlson, K.M.; Cassidy, E.S.; Johnston, M.; MacDonald, G.K.; Ray, D.K. Leverage points for improving global food security and the environment. *Science* **2014**, *345*, 325–328. [CrossRef]

47. Clark, M.; Tilman, D. Comparative analysis of environmental impacts of agricultural production systems, agricultural input efficiency, and food choice. *Environ. Res. Lett.* **2017**, *12*, 064016. [CrossRef]

48. OECD. *Education at a Glance 2018*; OECD: Paris, France, 2018. [CrossRef]

49. Blanchard, J.L.; Watson, R.A.; Fulton, E.A.; Cottrell, R.S.; Nash, K.L.; Bryndum-Buchholz, A.; Büchner, M.; Carozza, D.A.; Cheung, W.W.L.; Elliott, J.; et al. Linked sustainability challenges and trade-offs among fisheries, aquaculture and agriculture. *Nat. Ecol. Evol.* **2017**, *1*, 1240–1249. [CrossRef]

50. Cole, M.B.; Augustin, M.A.; Robertson, M.J.; Manners, J.M. The science of food security. *NPJ Sci. Food* **2018**, *2*, 14. [CrossRef]

51. Gustafsson, J.; Cederberg, C.; Sonesson, U.; Emanuelsson, A. *The Methodology of the FAO Study: Global Food Losses and Food Waste-Extent, Causes and Prevention—FAO, 2011*; SIK Institutet för livsmedel och bioteknik: Boras, Sweden, 2013.

52. Tirado, M.; Hunnes, D.; Cohen, M.; Lartey, A. Climate change and nutrition in Africa. *J. Hunger Environ. Nutr.* **2015**, *10*, 22–46. [CrossRef]

53. Holdsworth, M.; Kruger, A.; Nago, E.; Lachat, C.; Mamiro, P.; Smit, K.; Garimoi-Orach, C.; Kameli, Y.; Roberfroid, D.; Kolsteren, P. African stakeholders' views of research options to improve nutritional status in sub-Saharan Africa. *Health Policy Plan.* **2014**, *30*, 863–874. [CrossRef]

54. Godfray, H.C.J.; Beddington, J.R.; Crute, I.R.; Haddad, L.; Lawrence, D.; Muir, J.F.; Pretty, J.; Robinson, S.; Thomas, S.M.; Toulmin, C. Food security: The challenge of feeding 9 billion people. *Science* **2010**, *327*, 1185383. [CrossRef] [PubMed]

55. McGuire, S.; FAO; IFAD; WFP. *The State of Food Insecurity in the World 2015: Meeting the 2015 International Hunger Targets: Taking Stock of Uneven Progress*; FAO: Rome, Italy, 2015. [CrossRef]

56. Haddad, L.; Achadi, E.; Bendech, M.A.; Ahuja, A.; Bhatia, K.; Bhutta, Z.; Blössner, M.; Borghi, E.; Colecraft, E.; de Onis, M.; et al. The Global Nutrition Report 2014: Actions and Accountability to Accelerate the World's Progress on Nutrition. *J. Nutr.* **2015**, *145*, 663–671. [CrossRef] [PubMed]

57. Hengeveld, L.M.; Wijnhoven, H.A.; Olthof, M.R.; Brouwer, I.A.; Harris, T.B.; Kritchevsky, S.B.; Newman, A.B.; Visser, M.; Study, H.A. Prospective associations of poor diet quality with long-term incidence of protein-energy malnutrition in community-dwelling older adults: The Health, Aging, and Body Composition (Health ABC) Study. *Am. J. Clin. Nutr.* **2018**, *107*, 155–164. [CrossRef] [PubMed]

58. Lim, S.S.; Vos, T.; Flaxman, A.D.; Danaei, G.; Shibuya, K.; Adair-Rohani, H.; AlMazroa, M.A.; Amann, M.; Anderson, H.R.; Andrews, K.G. A comparative risk assessment of burden of disease and injury attributable to 67 risk factors and risk factor clusters in 21 regions, 1990–2010: A systematic analysis for the Global Burden of Disease Study 2010. *Lancet* **2012**, *380*, 2224–2260. [CrossRef]

59. Muthayya, S.; Rah, J.H.; Sugimoto, J.D.; Roos, F.F.; Kraemer, K.; Black, R.E. The global hidden hunger indices and maps: An advocacy tool for action. *PLoS ONE* **2013**, *8*, e67860. [CrossRef]

60. Eggersdorfer, M.; Akobundu, U.; Bailey, R.L.; Shlisky, J.; Beaudreault, A.R.; Bergeron, G.; Blancato, R.B.; Blumberg, J.B.; Bourassa, M.W.; Gomes, F. Hidden Hunger: Solutions for America's Aging Populations. *Nutrients* **2018**, *10*, 9. [CrossRef]

61. Trimmer, J.T.; Guest, J.S. Recirculation of human-derived nutrients from cities to agriculture across six continents. *Nat. Sustain.* **2018**, *1*, 427–435. [CrossRef]

62. Lobell, D.B.; Schlenker, W.; Costa-Roberts, J. Climate trends and global crop production since 1980. *Science* **2011**, *333*, 1204531. [CrossRef]

63. Diaz, D.; Moore, F. Quantifying the economic risks of climate change. *Nat. Clim. Chang.* **2017**, *7*, 774. [CrossRef]

64. Fróna, D. Globális kihívások a mezőgazdaságban. *Int. J. Eng. Manag. Sci.* **2018**, *3*, 195–205. [CrossRef]

65. Scialabba, N.E.-H.; Müller-Lindenlauf, M. Organic agriculture and climate change. *Renew. Agric. Food Syst.* **2010**, *25*, 158–169. [CrossRef]

66. Müller, C.; Robertson, R.D. Projecting future crop productivity for global economic modeling. *Agric. Econ.* **2014**, *45*, 37–50. [CrossRef]

67. Müller, C.; Bondeau, A.; Popp, A.; Waha, K.; Fader, M. Climate change impacts on agricultural yields. 2010. Available online: https://openknowledge.worldbank.org/handle/10986/9065?locale-attribute=en (accessed on 8 December 2018).

68. Challinor, A.J.; Watson, J.; Lobell, D.; Howden, S.; Smith, D.; Chhetri, N. A meta-analysis of crop yield under climate change and adaptation. *Nat. Clim. Chang.* **2014**, *4*, 287. [CrossRef]

69. Asseng, S.; Ewert, F.; Martre, P.; Rötter, R.P.; Lobell, D.; Cammarano, D.; Kimball, B.; Ottman, M.; Wall, G.; White, J.W. Rising temperatures reduce global wheat production. *Nat. Clim. Chang.* **2015**, *5*, 143. [CrossRef]

70. EASAC. *Opportunities and Challenges for Research on Food and Nutrition Security and Agriculture in Europe*; EASAC: Halle, Germany, 2017.

71. Lane, A.; Norton, M.; Ryan, S. *Water Resources: A New Water Architecture*; John Wiley & Sons: Hoboken, NJ, USA, 2017.

72. WHO. *Progress on Sanitation and Drinking Water: 2015 Update and MDG Assessment*; World Health Organization: Geneva, Switzerland, 2015.

73. Lu, C.; Tian, H. Global nitrogen and phosphorus fertilizer use for agriculture production in the past half century: Shifted hot spots and nutrient imbalance. *Earth Syst. Sci. Data* **2017**, *9*, 181–192. [CrossRef]

74. Cui, K.; Shoemaker, S.P. A look at food security in China. *NPJ Sci. Food* **2018**, *2*, 4. [CrossRef] [PubMed]

75. Qin, Y.; Zhang, X. The road to specialization in agricultural production: Evidence from rural China. *World Dev.* **2016**, *77*, 1–16. [CrossRef]

76. Kang, S.; Hao, X.; Du, T.; Tong, L.; Su, X.; Lu, H.; Li, X.; Huo, Z.; Li, S.; Ding, R. Improving agricultural water productivity to ensure food security in China under changing environment: From research to practice. *Agric. Water Manag.* **2017**, *179*, 5–17. [CrossRef]

77. Carter, C.A.; Zhong, F.; Zhu, J. Advances in Chinese agriculture and its global implications. *Appl. Econ. Perspect. Policy* **2012**, *34*, 1–36. [CrossRef]

78. Guan, X.; Wei, H.; Lu, S.; Dai, Q.; Su, H. Assessment on the urbanization strategy in China: Achievements, challenges and reflections. *Habitat Int.* **2018**, *71*, 97–109. [CrossRef]

79. Swinnen, J.; Squicciarini, P. Mixed messages on prices and food security. *Science* **2012**, *335*, 405–406. [CrossRef] [PubMed]

80. Calvo-Gonzalez, O.; Shankar, R.; Trezzi, R. *Are Commodity Prices More Volatile Now? A Long-Run Perspective*; The World Bank: Washington, DC, USA, 2010.

81. Baffes, J.; Haniotis, T. What explains agricultural price movements? *J. Agric. Econ.* **2016**, *67*, 706–721. [CrossRef]

82. Timmer, C.P. *Causes of High Food Prices*; ADB Economics Working Paper Series; ADB Economics: Manila, Philippines, 2008.

83. Imf, O.; Unctad, W. *Price Volatility in Food and Agricultural Markets: Policy Responses*; FAO: Roma, Italy, 2011.

84. Hochman, G.; Rajagopal, D.; Timilsina, G.; Zilberman, D. Quantifying the causes of the global food commodity price crisis. *Biomass Bioenergy* **2014**, *68*, 106–114. [CrossRef]

85. Serra, T.; Zilberman, D. Biofuel-related price transmission literature: A review. *Energy Econ.* **2013**, *37*, 141–151. [CrossRef]

86. Kristoufek, L.; Janda, K.; Zilberman, D. Correlations between biofuels and related commodities before and during the food crisis: A taxonomy perspective. *Energy Econ.* **2012**, *34*, 1380–1391. [CrossRef]

87. Kristoufek, L.; Janda, K.; Zilberman, D. Regime-dependent topological properties of biofuels networks. *Eur. Phys. J. B* **2013**, *86*, 40. [CrossRef]

88. Gilbert, C.L. How to understand high food prices. *J. Agric. Econ.* **2010**, *61*, 398–425. [CrossRef]

89. Opara, L.U. Traceability in agriculture and food supply chain: A review of basic concepts, technological implications, and future prospects. *J. Food Agric. Environ.* **2003**, *1*, 101–106.

90. Behzadi, G.; O'Sullivan, M.J.; Olsen, T.L.; Zhang, A. Agribusiness supply chain risk management: A review of quantitative decision models. *Omega* **2018**, *79*, 21–42. [CrossRef]

91. Horton, P.; Koh, L.; Guang, V.S. An integrated theoretical framework to enhance resource efficiency, sustainability and human health in agri-food systems. *J. Clean. Prod.* **2016**, *120*, 164–169. [CrossRef]

92. Farkasné Fekete, M.; Balyi, Z.; Szűcs, I. Az agrárgazdaság hatékonyságának néhány sajátos aspektusa. *Gazdálkodás Sci. J. Agric. Econ.* **2014**, *58*, 564–594.

93. Du, X.; Lu, L.; Reardon, T.; Zilberman, D. Economics of agricultural supply chain design: A portfolio selection approach. *Am. J. Agric. Econ.* **2016**, *98*, 1377–1388. [CrossRef]

94. Dunning, D.; Johnson, K.; Ehrlinger, J.; Kruger, J. Why people fail to recognize their own incompetence. *Curr. Dir. Psychol. Sci.* **2003**, *12*, 83–87. [CrossRef]

95. Cavicchi, C.; Vagnoni, E. Intellectual capital in support of farm businesses' strategic management: A case study. *J. Intellect. Cap.* **2018**, *19*, 692–711. [CrossRef]

12

Motives and Role of Religiosity towards Consumer Purchase Behavior in Western Imported Food Products

Faheem Bukhari [1], Saima Hussain [2,*], Rizwan Raheem Ahmed [3], Dalia Streimikiene [4,*], Riaz Hussain Soomro and Zahid Ali Channar [6]

[1]   Faculty of Business Administration, IQRA University, Defense view Shaheed-e-Millet Road (Ext), Karachi 75500, Pakistan; faheembukhari@iqra.edu.pk

[2]   Faculty of Management Sciences, Shaheed Zulfiqar Ali Bhutto Institute of Science and Technology, SZABIST 90, Block 5, Clifton Karachi 75600, Pakistan

[3]   Faculty of Management Sciences, Indus University, Karachi 75300, Pakistan; rizwanraheemahmed@gmail.com

[4]   Institute of Sport Science and Innovations, Lithuanian Sports University, Kaunas 44221, Lithuania

[5]   Institute of Health Management, Dow University of Health Sciences, Karachi 74200, Pakistan; riaz.soomro@duhs.edu.pk

[6]   Department of Business Administration, Sindh Madressatul Islam University, Karachi 74000, Pakistan; drzahidalic@gmail.com

*   Correspondence: saima.hussain@szabist.edu.pk (S.H.); dalia@mail.lei.lt (D.S.)

Abstract: The undertaken study examines the influence of the marketing mix, consumer attributes, and the role of religiosity towards consumer purchase behavior regarding western imported food products in Pakistan. The study has used the theory of planned behaviors as underpinning foundations for testing factors. In total, 1080 respondents from eight cities in Pakistan—Karachi, Lahore, Islamabad, Quetta, Peshawar, Hyderabad, Larkana, and Faisalabad—were part of this study. Path analysis performed through SEM (structural equation modeling). The result unveiled that product attributes, price, self-concept, brand trust, personality, and religiosity positively correlated with consumer's purchase intention in a Muslim country. The result of this study will also help potential future candidates for the food industry, especially those aimed at using the Asian consumer market. The penetration of western imported food may also bring convergence where the nation can feel upgraded and privileged. The study also adds to the academic literature on Muslim consumer behavior by combining numerous factors on a single model, grounded in the theory of planned behavior. Limited study has analyzed religiosity and other factors in context with a Muslim majority population. This study is a preliminary effort to understand the Muslim consumer food purchase behavior inadequately investigated by the consumer researcher.

Keywords: religiosity; western imported food; theory of planned behavior; Muslim consumers; consumer buying motive

1. Introduction

Religion is documented as an essential factor that profoundly influences consumer buying decisions [1]. Religion may serve to link consumers to a style of life that determines the pattern of consumption. Studies examining the effect or impact of religion on consumer behavior are based on two aspects: religious affiliation and religiosity [2,3]. The religious association mainly explored in comparison with the denominational association or the religious identification of a person (e.g., Catholic, Protestant, and Jewish). Although religiosity (in other words, religious commitment) is a significant construct to identify the effect of ethical behavior on a consumer's consumption and purchase behavior [4,5].

Islam, as a religion, presents a comprehensive way of life and controls the behavior of Muslim buyers, to achieve satisfaction with this life and hereafter [6]. Religious beliefs (e.g., concerning halal food) are the best guiding principle to identify food consumption choices for Muslims who actively follow religious guidelines as these rules address the Islamic tenets of food consumption [7]. Over the next 40 years, Islam will grow more rapidly than any other dominant faith. If current trends persist, by 2050, there will be nearly as many Muslims as Christians in the world [8]. This rapid increase of the global Muslim population indicates an opportunity for researchers to investigate more about Islam and Muslim consumers' behavior in various contexts such as food consumption. Investigation of Islamic consumption patterns may add value to the academic literature on consumer behavior [9].

Among Muslim consumers, Islamic rules administer the culture, which assists as a direction in their daily lives. Muslims must spend their money for explicit purposes only such as for general living, education, health, and aiding the poor and those in need. Hence, the concept of moderation is encouraged and Muslims are told to base their usage on strict observance to this practice [10]. The Pakistani population is 97 percent Muslim with different religious beliefs as compared to the western part of the world [11]. In spite of these Islamic guidelines on appropriate food consumption and moderate spending, money spent on western imported food has been increasing in Pakistan [12]. Therefore, it is advantageous to know the consumer perception or motives behind the purchase of western imported food. Especially, as this is an increasing trend as the population grows and the general economy has developed so that there is increased discretional personal spending for the middle and upper classes in Pakistan [12,13].

Pakistani Muslim consumers may have different perceptions about western imported food products (concerning the marketing mix, personal, social, and cultural elements). Their religious commitment indeed expresses the intensity of their faith and is indicated in part through their consumption choices about western imported food products [14]. Hence, it is particularly interesting to investigate the Pakistani Muslim consumer's motives behind the purchase of western imported food products and the role of religiosity in determining their purchase behavior. The reasons for this behavior have scarcely been studied in prior research and yet there is an opportunity to explore the factors mentioned above for the western imported food products in a Muslim dominated country like Pakistan [15–19]. This research proposes an essential contribution to the field of Muslim consumer's behavior, and it also adds value to the literature on consumer behavior by employing a model wherein the single conceptual framework tests the elements, based on the theory of planned behavior. Religion is a critical element of Pakistani culture that directly affects the behavior of Pakistani consumers [20].

Furthermore, religion expressively governs cultural and social behaviors in Asian and Middle Eastern societies as compared to western nations. For these reasons, the level of religiosity needs research as an essential factor in shaping Muslim's purchasing behavior regarding western food items in Pakistan. This area is still under-researched and many studies have suggested exploring the influences of religiosity in defining Muslim consumer's purchase behavior in these regions in particular [21–24].

The national religion in Pakistan is Islam, in which 97% of the population is Muslim—numbering 207,774,520 inhabitants [11]. Interestingly, the people of Pakistan spend 42% of their income on food-related items; the total trade and wholesale of 17% consists of food items [11]. Pakistan's middle and upper class spend money on imported western food items [25,26]. Throughout the first six months of the 2018 fiscal year, Pakistan spent US$312.5 million on the importing of coffee, tea, beverage whiteners, and spices, the second-largest spending category in the food products sector [27]. Pakistan has spent US$908.9 million on the import of animal or vegetable fats and oil products. In this period, the country also spent US$500.9 million on imported oilseeds and 'oleaginous fruit', which is the part of a plant used to produce vegetable oil [27]. It can be a fruits (e.g., olive), seeds (e.g., sesame), or nuts (e.g., walnut) [27]. Pakistan is an emerging market for the consumption of imported food items [28]. The market context for the study discussed next with an explanation of its research objectives.

- To investigate factors affecting Muslim consumer's purchase behavior in western imported food products in Pakistan.
- To identify the relationships that exist between each of these factors and western imported food products purchase behavior.
- To determine the moderation of demographic profile on a relationship between contributory factors and consumer's purchase behavior.

Integration of Theory of Planned Behavior

The undertaken study employed the theory of planned behavior (TPB) to know the behavior intentions of Muslim consumers of western imported food products. This study is an attempt to comprehend the purchasing intention of food items consumers using the TPB model from a Pakistani perspective. According to Donald et al. [29] and Armitage and Conner [30], the TPB model mainly dominated by attitude, and several psychological new dimensions have studied with the TPB model. Such as in organic food products' research by Robinson and Smith [31] investigated self-congruity concerning environmental consumerism. Moral obligations are tested for organic food by Arvola et al. [32], Reviews by Armitage and Conner [30] explains that TPB framework accounts for 39–50% variations in intention leading to 27–36% of the difference in a consumer's behavior. However, Ajzen's [33] original model assumptions verified by several studies that antecedents potentially correlate with each other [34–36]. Hence, the present research has included constructs such as product attributes, price, promotion, brand loyalty, brand trust, customer satisfaction, religiosity, subjective norms, self-concept, personality, lifestyle, and social class besides the TPB factors for evaluating Muslims' customers' purchasing aim in the context of western imported food products. This study is an attempt to comprehend the consumers' purchase behavior employing the TPB framework in the Pakistani Muslim consumer's context. The element of religiosity is a combination of religious dimensions, which was described in Glock and Stark's [37] study. Since this study is based on Muslim consumer behavior in the context of western imported food, the combine notion of beliefs, practices, and knowledge about Islam and its linkage with food buying behavior is presented to explain how Muslim consumers take their food-related purchase decisions. New constructs added to TPB in prolific recent literature [34,35,37] specific to various domains. Extensions of the Ajzen [33] model over a few years have proved an improvement in the explanatory power of the framework.

The rest of the paper organized as follows: Section 2 demonstrates the review of literature and hypotheses formulation, Section 3 deals with materials and methods, Section 4 comprises of results, Section 5 deals with discussions, and Section 6 presents the conclusion. Next, the study presents practical applications and followed by theoretical contributions. Lastly, limitations and future research directions have discussed.

2. Literature Review and Hypotheses Development

2.1. Product Attributes

Product attributes are the features of products through which brands recognized and distinguished. In other words, product attributes denoted to be the descriptive aspect of a marketing plan that characterizes the consumer's evaluative standards when selecting specific goods or services [38]. Product characteristics are discussed in terms of being either intrinsic or extrinsic. Intrinsic product attributes are specific to a product, unchangeable, and comprise features such as form, ingredients, flavor, color, and smell. Extrinsic characteristics are not a crucial part of the physical product such as value, brand name, and country of origin [39]. A study from Norway Torjusen et al. [40] reported that the traditional food quality aspects such as appearance, freshness, and taste, which they named 'observation traits', were important to all respondents. Most of the respondents were concerned about aspects related to food production and processing, they chose food with no harmful substances and the least possible additives. According to Dahm, Samonte, and Shows [41], the taste is as equally as an important attribute as quality, followed by price, appearance, and availability [42]. Knight [43] highlights the importance of factors like 'country of the producer' and 'product quality' and its impact on buying decision making in globally available product classes. The researchers reported that, when the imported goods are of a higher value, customers are willing to pay a higher price. Product attributes play a vital role in this research, which investigates these key attributes' influence on consumer buying behavior concerning western imported food. These have identified during the qualitative focused interviews as flavor, taste, nutritious value, and healthiness. Understanding these attributes from a consumer's perspective may assist the manufacturers in developing a refined marketing strategy. Thus, the above information leads to the development of the following hypothesis:

Hypothesis 1 (H1): *Product attributes are positively associated with consumer's buying intention towards western food products.*

2.2. Price

Price has always remained a cornerstone for any food item in every society; therefore, pricing strategy always considers segmentation, market condition, trade margins, competitors' price, and marketing and internal cost [43]. It is directed at distinct consumers alongside competitors [44]. Price is a major factor in determining consumers' choices. Even though many other factors unrelated to price are important in the literature, the price is the main determinant of the purchase decision for large segments of consumers across many countries [45,46]. Thus, the above information leads to the development of the following hypothesis:

Hypothesis 2 (H2): *Price is positively associated with consumer's buying intention towards western food products.*

2.3. Promotion

Marketing communication has a positive and vital impact on consumers' purchase intentions and companies' sales volumes. In particular, the advertising plan has an influence on the attitude and the purchase intention towards a brand [46]. Another study by Song, Safari and Mansori [47] reveals the effects of five marketing stimuli, which include marketing communication and promotion of the food items on the perceived value of consumers. Afterward, the effect of this perceived value on the actual purchase decision examined. The results showed a relationship between marketing communication and perceived customer value among organic food consumers of Malaysia. In contrast to the existing literature, findings for the same study revealed no relationship between sales promotion and product perceived value. [43,45]. Hence, the above information leads to the development of the following hypothesis:

Hypothesis 3 (H3): *Promotion is positively associated with consumer's buying intention towards western food products.*

2.4. Personality

Marketers accept the buyers experienced brands as a means to comprehend their personalities. We have taken a modified questionnaire in which we used the modified items for the brand personality from the previous study [48]. According to Banerjee [49], both individual and brand personalities have an important influence on brand preference in the consumer's mind. Consumer Preference implies that at the time of brand choice, consumers give prominence to individual personality and the personality of the chosen brand. In food products, such a relationship has also been found. According to Chang,

Tseng, and Chu [50], few consumer traits lead to a positive consumer perception about food traceability, which means food items processing, production, and delivery to the consumers. The researchers used the Big Five Factor model to assess various traits of consumers and a 3M framework of motivation and personality (market, means, and motivation) for analyzing consumer's perceptions regarding food traceability [47,50]. Among elemental traits, it was found that openness, conscientious and extroverted personalities, combined with actual material and bodily needs, tend to be linked with compound traits such as health consciousness and the need for learning. These compound traits influence situational traits (consumer perceptions of food traceability and the concern for food value) and initiate the intention to purchase. Hence, the above information leads to the development of the following hypothesis:

Hypothesis 4 (H4): *Personality is positively associated with consumer's buying intention towards western food products.*

2.5. Lifestyle

The lifestyle to a certain extent defines patterns or trends of consumption. It is observed by looking at individuals' organization, space and time, leisure activities, working hours, housing, appearance, and other daily activities. In other words, lifestyle is one important variable, which expresses consumer's choices [51,52]. Ahaiwe et al. [53] reported several factors, which may influence consumers' buying behavior and their brand preferences for goods and services. Among these are cultural factors, social class, values and beliefs, interests, lifestyle, and personality. These factors jointly referred to as psychographic variables, which play a considerable role in consumers' preferences for products [54]. In a Chinese study comparing lifestyles and their impact on purchase intentions for domestic and imported food products, three groups identified which each had different behaviors: risk-takers, traditionalists, and experiencers. It was found that risk-takers and traditionalists were associated with purchasing imported fruits [55]. Another study suggests a food-related lifestyle model comprised of five components to explain consumption behavior: quality, methods to shopping, food consumption situation, manner of cooking, and purchasing motives [51,56]. Hence, the above information leads to the development of the following hypothesis:

Hypothesis 5 (H5): *Lifestyle is positively associated with consumer's buying intention towards western food products.*

2.6. Family (Subjective Norms)

A subjective norm generally explained as a person's awareness about what essential others consider the individual should comply with [57]. The association between subjective norms and attitudes towards behavior has been verified and tested. For example, researchers have established the pathway from subjective norms to attitudes towards behavior and found it significant [58]. Within the framework of subjective norms, reference groups somehow affect the values and behavior of others. Reference groups, particularly buyer groups of references—for example, institution and trade, professional institutions, social organizations, friends, family members—influence product selection and the choice of a specific brand. Most purchases influenced by the opinions of the groups of references, which include friends and professional institutions [59]. Parents have an impact on a person's purchase decision. Moreover, the dominance of the preference of the husband or the wife differs from the product category. Thus, in food items, the wife is predominantly the key decision-maker. However, children also influence at the time of purchasing [60]. Hence, the above information leads to the development of the following hypothesis:

Hypothesis 6 (H6): *Family is positively associated with consumer's buying intention towards western food products.*

2.7. Social Class

Social class is used as a basis for market segmentation because members of different classes reflect different consumption patterns [61]. Especially in those countries where class differences exist, social class significantly impacts on consumer decisions [62]. When it comes to consumers' response to a new product, research indicates differences between the low and high socioeconomic classes. The low socioeconomic class is less likely to purchase a new product or one with new technologies. According to Majabadi et al. [63], there is a class difference in consumer preferences towards food products, prices, and concepts of value. This social class research has been investigated as an important factor to explore if it influences consumer's purchase behavior. Social class also is shown to connect with patterns of media usage, language patterns, source credibility, and spending behavior [56,62]. Social class is yet another important factor that needs to be studied in the context of food buying behavior. In a country such as Pakistan, with its complex history and deeply rooted perceptions of social class, social class may impact the purchase of western imported food products. Since this study based on the purchase behavior of western imported food in Pakistan, understanding the social classes in this context is a prerequisite. Hence, the above information leads to the development of the following hypothesis:

Hypothesis 7 (H7): *Social class is positively associated with consumer's buying intention towards western food products.*

2.8. Self-Concept

The congruence of brand personality and the consumer's self is a way to create an emotional attachment to the brand and other various brand-related outcomes [64]. Additionally, various research studies have posited a significant relationship between this congruity and a positive brand attitude, positive brand perception, and the intention to buy [65]. Another study also supports the view that consumers prefer those products, which match (somehow) with their self-concept [66]. Self-concept has been identified as playing a mediating role concerning the underdog brand effect and the intention to buy [67]. The research of Hoonsopon [68] also found and used the self-concept as a moderating variable in the relationship between consumer innovativeness and new product purchase intentions. Hence, it concluded that even in new product adoption, consumers have a unique self-concept, and there is a chance the new product will be adopted when it fits with the self-concept of the consumer. Regardless of the rising research regarding self-concept and consumer behavior, there are still areas in the literature, which need further exploration. Earlier researchers have not taken into account the self-concept and its influence on consumer's buying behavior for the food segment [65,68]. To fill this gap, this study expands on earlier research examining the role of the self-concept in consumer behavior by ascertaining its influence on buying behavior concerning western imported food. In the case of western imported food products, this self-concept is an important factor in the buyer's purchase decision. This includes the consumer's attitude and perception, and if both of these are positive towards the product, the consumer may end up finally deciding to purchase the product. Hence, the above information leads to the development of the following hypothesis:

Hypothesis 8 (H8): *Self-concept is positively associated with consumer's buying intentions towards western food products.*

2.9. Brand Trust

In the case of food purchase behavior, self-assurance in credence features, and could lead to brand trust. A food-related research study conducted by Chen and Lee [69] reported that brand trust plays an important role. There is an important relationship amid various kinds of trust and consumers' perceptions of safety concerning food items. The researchers specified two main types of trust: general and specific, specific trust further classified into supplier level, and industry-level trusts. Using the survey method in Beijing, results indicated an affirmative link amid consumer's brand trust in food producers and retailers with their perceptions of food safety. When marketers aim to increase specific trust, they should address the integrity and ability of producers [70]. Drescher et al. [71] studied Canadian households' perceptions of processed meat and their levels of trust. The research suggested customers with the highest level of brand trust spend more on processed meat than those with a low level. In the past literature, the importance of brand trust has highlighted along with its impact on creating highly valued relationships between consumers and firms. In this research, the element of brand trust has been added to determine the extent of consumers' brand trust in western imported food and if this is influencing their purchase behavior. Brand trust is essential for making sure that the consumer makes the purchase. In the case of western imported food products, brand trust is essential for purchase behavior to take place. Once the consumer trusts the brand, he or she is willing to pay higher prices to procure the western imported food product. Henceforth, the above information leads to the development of the following hypothesis:

Hypothesis 9 (H9): *Brand trust is positively associated with consumer's buying intentions towards western food products.*

2.10. Religiosity

The religious practices influence the imminence sited on attitudes and factual life towards possessing and consuming goods and services [72]. Religiosity, a central point of any religion, has a close association with consumer behavior. Thus, an exploration of religiosity allows in-depth investigation of consumer behavior [73]. Abundant literature is available regarding the association of religiosity and consumers' behavior. For example, in a practical study regarding consumers' intentions and religiosity amongst 602 typically Protestant customers. Rakrachakarn et al. [74] reported the noteworthy inference that religiosity influences numerous aspects of customers' lifestyles that ultimately reshape the selection behavior. In Vitell's [75] evaluation of religiosity and consumer behavior, one observation was prominent: that the number of academic studies has been inadequate in clarifying customers' norms and religious views. This was associated with the arguments of Hannah, Avolio, and May [76] who specified that norms and capabilities of views accounted for 20% of the difference in the behavior explained. Henceforth, the above information leads to the development of the following hypothesis:

Hypothesis 10 (H10): *Religiosity is positively associated with consumer's buying intentions towards western food products.*

2.11. Purchase Intention

Numerous factors influence purchase intention such as subjective norms, attitudes, and perceived behavioral control [77]. The consumers of imported food perceive these items as better quality compared to locally produced food brands and this positive attitude has affected their purchase intention [78]. Many previous studies further endorsed that factors such as subjective norms, health consciousness, and brand familiarity somehow influenced the purchase intention [76,78]. The results are consistent with several developed countries' literature in which perceived value was quite important and had an impact on the food buying behavior and consumers were willing to pay extra to avail the maximum benefits [79]. A perception of better quality was also one of the key aspects of shaping consumers' purchase intention. Those with a positive mindset concerning western food brands were likely to possess a positive intention to purchase it [80]. A positive attitude thus found to serve as an important stimulus and possibly influenced the consumers' purchase intention. However, in most of the cases and especially in food buying behavior, it has been noted that purchase intention is a primary indicator and a leading factor towards a final purchase behavior [78,81]. Henceforth, the above information leads to the development of the following hypothesis:

Hypothesis 11 (H11): *Purchase intention is positively associated with consumer's buying behavior towards western food products.*

Thus, the conceptual frame of the undertaken study is shown in Figure 1.

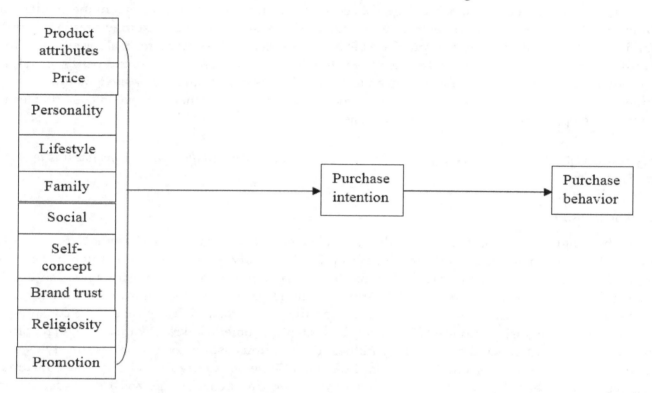

Figure 1. Conceptual framework.

3. Materials and Methods

A structured questionnaire has tested the model established in this study. The primary survey conducted in the eight metropolitan cities of Pakistan, Karachi, Lahore, Islamabad, Peshawar, Hyderabad, Faisalabad, Quetta, and Larkana which together cover four regions (Sindh, Punjab, Baluchistan, and Khyber Pakhtunkhwa). The main reason for selecting eight metropolitan cities of Pakistan was to ensure the sample represented the key Pakistani urban cities with diversified socio-economic classes and having the awareness and usage of western imported food, guaranteeing the generalizability of the research outcome [82]. The responses were collected online and in-person for the pilot study conducted in Karachi with a sample of 375 respondents. A pilot study has enhanced the precision of the survey instrument. A sample size of 375 deemed appropriate since there were 75 scale items and exploratory factor analysis requires a minimum of 5:1 ratio of respondents to items [83]. To run the factor analysis, a sample size smaller than 100 considered to be "dangerous", and sample sizes larger than 200 are considered safe for adequate conclusions [84,85]. The final survey administered to 1080 respondents through in-person and online media such as email and social media. The sample size was calculated based on the ratio of 20 respondents per item, which is the prescribed ideal sample size needed to conduct SEM [85,86]. Before inputting the data, the questionnaires checked for any missing data. This initial screening revealed that out of 1080 surveys, 927 filled in. Hence, the response rate was 86 percent. Data normality observed by reviewing residual plots, which appeared to be reasonable, and the skewness and kurtosis values were near zero. Thus, the notion of normality undisrupted.

Measurement Scaling for Constructs and Items

The data is collected from 927 respondents for the undertaken study by using a structured and modified questionnaire. We used adapted measurement scales for constructs and items from the previous literature such as modified items of product attributes are taken from previous studies [38–43], modified items for marketing mix such as price and promotion have been taken from the previous literature [43–46]. However, the adapted items of lifestyle and personality are considered from the previous studies [51–56]. The subjective norms of the family items have been modified from previous studies such as Al-Swidi et al. [57] and Kautonen et al. [59]. The adapted items for the social class have been taken from previous studies such as Rani [61], Durmaz and Taşdemir [62], and Majabadi et al. [63]. The modified items of self-concept are taken from previous literature [64–68]. The items for brand trust have taken from previous studies such as Chen and Lee [69] and Drescher et al. [71]. Religiosity is the most important variable in this research, we have taken adapted items from the previous literature [72–76]. We used modified items for purchase intention and purchase behavior constructs from previous studies [77–81].

4. Results and Discussions

The survey forms distributed to 1080 respondents from eight cities of Pakistan, Karachi, Hyderabad, Larkana, Lahore, Islamabad, Faisalabad, Quetta, and Peshawar. A sample of 927 individuals consisting of professionals, university students, and homemakers responded to the survey, which presented a response rate of 86 percent. The main demographic variables included in this study are gender, religious identity, and age, level of education, monthly income, marital status, and employment status. Table 1 exhibits the frequencies and corresponding percentages of each demographic variable.

Table 1. Demographics.

Item	Characteristic	No. of Respondents	Percentage Response
Gender	Male	542	58
	Female	385	41
Age	18–24	541	58
	25–35	313	34
	36–45	55	05
	46–55	17	18
	55+	03	01
Level of education	Primary	02	01
	Secondary or high school	54	05
	Diploma/vocational education and training	11	11
	Undergraduate (Bachelor)	308	33
	Postgraduate (Masters)	530	57
	PhD	19	02
	Others	03	01
Per month income in Pak rupees	Less than 40,000	10	01
	40,001–64,000 PKR	448	48
	64,001–150,000 PKR	327	35
	150,001–250,000 PKR	81	08
	250,001+ PKR	61	07
Marital status	Single	659	71
	Married	256	27
	Divorced	06	01
	Others	06	01

Table 1. *Cont.*

Item	Characteristic	No. of Respondents	Percentage Response
Employment status	Full time	395	42
	Part-time	81	08
	Self employed	55	06
	Unemployed	396	42
City	Karachi	393	42
	Hyderabad	41	04
	Larkana	37	04
	Lahore	282	30
	Islamabad	30	03
	Quetta	23	02
	Faisalabad	73	08
	Peshawar	48	05

As Table 1 indicates, of the 927 respondents, 542 were male, and 385 were females. The age group of 18 to 24 was the largest, with over half of the respondents 58 percent. About 48 percent had a monthly income between PKR 40,001–64,000 (equivalent to US$310–497). Concerning academic qualifications, the majority had a postgraduate degree (57 percent). Most were single (71 percent), 42 percent were employed full time, and 42 percent resided in Karachi, followed by Lahore (with 30 percent).

4.1. Factor Loading and Path Analysis

The framework of this research tested in SPSS and AMOS 23.0 version. Analysis of a multi-stage SEM (structural equation modeling) process comprising of measurement and structural model conducted by path analysis [85,86]. Moreover, an exploratory factor analysis (EFA) test has explored several factors for this study. Almost forty factors recognized with the commonalities greater than 0.5. Table 2 illustrates the results obtained through factor analysis. Furthermore, the sphericity and KMO tests recognized a significant relationship between the variables to authenticate the operationalization of factor analysis [86,87] as shown in Tables 3 and 4.

Table 2. Component matrix with rotations.

	Component											
	1	**2**	**3**	**4**	**5**	**6**	**7**	**8**	**9**	**10**	**11**	**12**
Product attribute	0.620											
Product attribute	0.712											
Product attribute	0.656											
Price		0.504										
Price		0.595										
Price		0.500										
Promotion			0.862									
Promotion			0.912									
Promotion			0.833									
Promotion			0.616									
Lifestyle				0.608								
Lifestyle				0.889								
Lifestyle				0.866								
Personality					0.806							
Personality					0.910							
Personality					0.922							
Personality					0.884							
Family						0.828						
Family						0.889						
Family						0.866						

Table 2. *Cont.*

	1	2	3	4	5	6	7	8	9	10	11	12
							Component					
Social-class							0.535					
Social-class							0.706					
Social-class							0.630					
Social-class							0.529					
Self-concept								0.747				
Self-concept								0.891				
Self-concept								0.760				
Brand trust									0.705			
Brand trust									0.782			
Brand trust									0.701			
Religiosity										0.845		
Religiosity										0.798		
Religiosity										0.648		
Religiosity										0.592		
PI											0.550	
PI											0.774	
PI											0.801	
PB												0.738
PB												0.846
PB												0.773

Table 3. KMO analysis Bartlett's sphericity test.

KMO Measure of Sampling Adequacy		0.905
Bartlett's test of sphericity	Approx. Chi-square	20295.022
	Df	1035
	Sig.	0.000

Table 4. Total variances.

Component	Eigenvalues		
	Total	% of Variance	Cumulative %
1	11.594	21.47	21.47
2	5.573	10.319	31.789
3	3.389	6.275	38.064
4	2.11	3.908	41.972
5	1.849	3.424	45.396
6	1.792	3.319	48.716
7	1.682	3.115	51.83
8	1.491	2.761	54.592
9	1.368	2.534	57.126
10	1.168	2.163	59.289
11	1.089	2.017	61.306
12	1.021	1.89	63.196

4.2. Structural Model

The goodness of fit indices was fit for the model Chatfield [87] in Table 5.

Table 5. Overall model fit indices.

Indices	Reported Value	Recommended Value
Chi-square/DF ratio (CMIN/DF)	1.984	<3
GFI	0.932	0.90
AGFI	0.920	0.90
NFI	0.921	0.90
IFI	0.959	0.90
TLI	0.954	0.90
CFI	0.959	0.90
RMSEA	0.033	<0.08
ECVI	1.759 (Default model) 1.771 (Saturated model) 18.964 (Independence model)	Default model should report the smallest value
HOELTER	527 at 0.01 level	>200 at 0.01 level

The reported CMIN/DF (Chi-square (χ2)/df = 1.984 displays a good model fit since the value is less than 3, which is a statistical benchmark. The remaining fit indices also report that the model fits the research data. CFI, IFI, and TLI values are above the minimum threshold level 0.9 indicating a good model fit. AGFI and GFI values are as per the minimum threshold, which is 0.90. The RMSEA value is 0.033 which is <0.08 [88]. Besides, the ECVI indicates the smaller value, which presents that the model crosschecked by utilizing similar cases from the same target group. Lastly, the HOETLER figure of 527 at 0.01 level directed suitable sample suitability for the model; As per Hoelter [89], >200 would specify good sample appropriateness for the model. Hence, the figures deliver sufficient indication to support for a good model fit. Consequently, all the above-stated fit indices were acceptable well the above-suggested criteria.

4.3. Validity and Reliability

Reliabilities, factor loadings, and Average variance extracted (AVE) displayed in Table 6. Besides, the constructs' validity comprises 'convergent validity' and 'discriminant validity', and Table 6 has all values equal and above the prescribed threshold of 0.72 to 0.9 [90]. The criterion for convergent and discriminant validity is as per the benchmark suggested by [91] and henceforth established convergent validity, as mentioned in Table 6.

Table 6. Validity and reliability measures.

Scales	Variable	Factor Loading	AVE	Composite Reliability	Cronbach's α
Product attribute		0.681			
Product attribute	Product attribute	0.807	0.555	0.788	0.711
Product attribute		0.741			
Price		0.704			
Price	Price	0.664	0.500	0.749	0.715
Price		0.751			
Promotion		0.794			
Promotion	Promotion	0.802	0.689	0.898	0.834
Promotion		0.844			
Promotion		0.879			

Table 6. *Cont.*

Scales	Variable	Factor Loading	AVE	Composite Reliability	Cronbach's α
Lifestyle	Lifestyle	0.927	0.750	0.900	0.784
Lifestyle		0.839			
Lifestyle		0.828			
Personality	Personality	0.794	0.679	0.894	0.897
Personality		0.841			
Personality		0.843			
Personality		0.817			
Family	Family	0.802	0.500	0.855	0.786
Family		0.793			
Family		0.847			
Social class	Social class	0.702	0.584	0.849	0.840
Social class		0.693			
Social class		0.744			
Social class		0.692			
Self-concept	Self-concept	0.711	0.515	0.761	0.737
Self-concept		0.722			
Self-concept		0.720			
Brand trust	Brand trust	0.772	0.620	0.830	0.870
Brand trust		0.752			
Brand trust		0.836			
Religiosity	Religiosity	0.591	0.552	0.737	0.731
Religiosity		0.618			
Religiosity		0.738			
Religiosity		0.616			
PI	Purchase intention	0.705	0.500	0.723	0.766
PI		0.687			
PI		0.655			
PB	Purchase behavior	0.831	0.631	0.837	0.825
PB		0.720			
PB		0.831			

The constructs stated in Table 7 fulfill the criterion of discriminant validity that is the square root of AVE > all possible squared correlations, as suggested by Hair et al. [85].

Table 7. Measurement model discriminant validity measures.

Component	PA	PRC	PROMO	PER	LIFE	FAM	SC	SELF	BT	REL	PI	PB
Product attribute	0.745											
Price	0.442	0.707										
Promotion	0.218	0.220	0.830									
Personality	0.076	0.236	0.303	0.866								
Lifestyle	0.240	0.134	0.171	0.054	0.824							
Family	0.182	0.200	0.348	0.422	0.098	0.707						
Social class	0.094	0.206	0.348	0.661	0.058	0.501	0.764					
Self-concept	0.395	0.292	0.125	0.095	0.282	0.099	0.142	0.718				
Brand trust	0.500	0.403	0.129	0.199	0.206	0.173	0.213	0.417	0.787			
Religiosity	0.135	−0.029	0.163	−0.055	0.270	0.125	−0.038	0.099	−0.007	0.743		
Purchase intention	0.496	0.453	0.192	0.205	0.310	0.229	0.232	0.424	0.564	0.084	0.707	
Purchase behavior	0.471	0.427	0.164	0.179	0.291	0.190	0.195	0.387	0.568	0.102	0.704	0.794

4.4. Outcomes of Hypothesized Relationships

Table 8 displays all relationships within independent, intervening, and dependent variables. The findings from the quantitative data analysis reveal that product attributes are positively associated with consumer purchase intentions for the sustenance of western imported food products, as exhibited in Table 8. It means that consumers give importance to product features at the time of purchase. The outcome of this hypothesis is also in line with the previous literature, for instance, Lian and Yoong [92] reported that product features such as taste, freshness, and packaging influenced consumers' attitudes to purchasing food products. It was further endorsed by Wee et al. [79] who found that product safety related to food brands had a significant influence on consumers' purchasing behavior of imported food products. Further research by Nasution and Rossanty [93] confirmed the long-established view that product labeling had an essential association with consumers' attitudes to the purchasing behavior of food products.

Table 8. Standardized regression weights and p-values.

Hypothesis			Estimates	p-Value
Ha1	Supported	Product attribute -> PI	0.217	<0.001 ***
Ha2	Supported	Price -> PI	0.204	<0.001 ***
Ha3	Not Supported	Promotion -> PI	0.012	0.704
Ha4	Supported	Personality -> PI	0.108	0.001 ***
Ha5	Not Supported	Lifestyle -> PI	−0.018	0.745
Ha6	Not Supported	Family -> PI	−0.015	0.699
Ha7	Not Supported	Social class -> PI	0.097	0.127
Ha8	Supported	Self-concept -> PI	0.066	0.091 *
Ha9	Supported	Brand trust -> PI	0.466	<0.001 ***
Ha10	Supported	Religiosity -> PI	0.135	<0.001 ***
Ha11	Supported	PI-> purchase behavior	0.91	<0.001 ***

Note: PI = Purchase intention; * Significant at 10% ($p < 0.1$); *** Significant at 1% ($p < 0.01$).

The second association investigated was between price and consumer purchase intention for western imported food. The quantitative data results illustrate a positive association between western imported food prices and consumer purchase intention. The previous literature has suggested that imported food products had a higher rate compared to local food products [46]. The main reasons offered were that these products perceived to have better overall quality, with quality ingredients, a country of origin, and an established product name [94,95]. Hence, consumers are eager to pay more price for such food brands due to the perception of their superior value [96,97].

The next relationship investigated was between promotion and consumer purchase intention. Results from the quantitative findings indicate that promotion is not associated with consumer purchase intention, which differs from previous literature. Research conducted by Kazmi et al. [19] reported that product communication plays an imperative role in persuading consumers' attitudes to a product and thus adds brand recall and awareness about the food brands. Effective communication through various channels such as magazines, newspapers, social media, and television thus enhances food brand awareness, sustenance in the market, and purchase intention, particularly among young consumers [97].

The results showed that the impact of lifestyle does not have any effect on consumer's purchase intention, and showing no association between them. It suggests that Muslim consumer's lifestyles indirectly connected with the purchase intention in the context of western imported food products in Pakistan. On the contrary, the findings from the past literature revealed that consumers' health consciousness to food ingredients strongly motivate and influence consumers to engage in purchasing food brands manufactured outside the country. However, the overall quality and taste are also significant contributors to the purchase decision. Hence, elevating western imported food products [95].

The relationship between family and consumer purchase intention explored. Results indicated neither causal relationship nor any positive association between them. It contrasted with Tsang et al. [96] study which found a correlation between subjective norms and consumer food-buying behavior. Moreover, earlier research by Montano and Kasprzyk [36] endorsed the view that subjective criteria impact consumer food purchase behavior. The quantitative results of the current study of an absence of association can be due to different cultural and societal settings.

The construct of social class was also investigated, with no association found with the consumer buying intention for western imported food in Pakistan. This result differs from those of the literature. Durmaz and Taşdemir [62], for instance, reported that this factor had a substantial influence on customer buying behavior as this influence started throughout childhood, and the family shaped it. The researchers further stated that young people from the upper social class were more brands conscious and likely to seek information about the brand before making the final purchase decision as compared to their lower-class counterparts.

Brand trust was also investigated during the quantitative data analysis, indicating an affirmative corroboration between brand trust of organic food and consumer's intention to purchase. In past research, the construct of brand trust widely discussed, and similar findings revealed. The study conducted by Flavián and Guinalíu [98] reported that the trust element linked with the brand image which reduces the consumers' risk at the time of purchase. Therefore, a positive brand image leads to brand trust. Hence, the improved brand image of a company leads to positive and confident consumer behavior about that brand's products related to food. Other research findings have reinforced a positive association between the brand image of a company and a consumer's brand trust.

The construct of religiosity also tested and it discovered to have a positive association with the intentions of consumers' buying regarding organic food, which aligned with the past literature. Mathras et al. [14] reported that halal consciousness and product elements have considerably influenced Muslims' intention to purchase packaged food that is halal produced by Muslim or non-Muslim producers. Quantitative results confirm earlier research that in Muslim consumers, the religiosity behaves like an intervening factor within the different exogenous and endogenous variable [99]. Therefore, religious ethics perform integral in deciding consumer behavior. Religion guides Muslim consumers to take actions as per religious principles [15,22,23]. Thus, the quantitative analysis aligned with the previous literature wherein the connection of religiosity with consumer buying behavior well established.

Self-concept investigated, and results revealed it associated with the consumer purchase intention of western imported food, which aligns with the previous literature. Sirgy [100] reported that consumers preferred those products, which matched (somehow) with their self-concept. Self-concept identified as playing a mediating role concerning the brand effect and the intention to buy. Hoonsopon and Puriwat [101] also confirmed self-concept as a controlling variable for the innovative product purchase intentions.

4.5. Moderation Results

The moderation analysis aimed to establish if gender, income, and city moderate the relationship and impact among independent and mediation, leading to dependent variables. Firstly, gender, income, and city coded as dummy variables. A correlation analysis performed between the dependent variable and dummy coded variables from gender, income, and city to establish which variables might be potential moderators. The only significant correlation obtained between PB and the dummy variable of Income Group 3, i.e., with an income between PKR 64,001-150,000 ($r = 0.065$, $p = 0.047$). Therefore, the dummy variable of Income Group 3 shortlisted as a potential moderator variable. First, a regression model (Block 1) fitted to predict PB using PI (purchase intention) and the dummy variable of Income Group 3. The regression model found to be significant (r-square = 0.586). The model coefficients are shown in Table 9; both the main effects of PI and the dummy variable of Income Group 3 were found to be significant.

Table 9. Block 1 regression model.

Model	B	Std. Error	T	Sig.
(Constant)	0.807	0.075	10.775	<0.001
PI	0.772	0.021	36.072	<0.001
Dummy variable indicating that INCOME = 3.0 ("64,001–150,000").	0.079	0.035	2.257	0.024

Next, the interaction between PI and the dummy variable of Income Group 3 added to the Block 1 model, this constituted the Block 2 model. The regression model was found to be significant (r-square = 0.586). The model coefficients shown in Table 10, after adding the interaction term, the main effect of PI was still substantial but the main impact of the dummy variable of Income Group 3 was not significant anymore. It indicates that the dummy variable of Income Group 3 moderates the relationship between PI and PB, but the main effect of PI is also substantial.

Table 10. Block 2 regression model.

Model		t	Sig.
(Constant)	0.812	9.615	<0.001
PI	0.771	31.772	<0.001
Dummy variable indicating that INCOME = 3.0 ("64,001–150,000")	0.058	0.318	0.751
PI_×_Income_3	0.006	0.121	0.903

5. Conclusions

Pakistan, as a developing country, has shown growth in imported food consumption while experiencing economic growth and stability. Comprehending the Pakistani Muslim consumer means understanding that Pakistan is a Muslim dominated society with 97 percent of its population. This study has examined the consumer motives behind the purchase of western imported food products. Even though prices are on the higher side compared to local food products, western imported food is making inroads into Pakistani food purchasing behavior. The findings concluded that product attributes are positively associated with consumer purchase intentions for the sustenance of western imported food products. The second association demonstrated a positive association between western imported food prices and consumer purchase intention. The study also revealed that promotion is not associated with consumer purchase intention. The results showed that the impact of lifestyle does not have any impact on consumer's purchase intention, and no association was found between them. It suggests that Muslim consumers' lifestyles are indirectly connected with the purchase intention in the context of western imported food products in Pakistan. However, the overall quality and taste are also important contributors to the purchase decision. The relationship between family and consumer purchase intention indicated neither causal relationship nor any positive association between them. The construct of social class also investigated, with no association found with the consumer buying intention for western imported food in Pakistan. The researchers further concluded that young people from the upper social class were more brands conscious and likely to seek information about the brand before making the final purchase decision as compared to their lower-class counterparts. The brand trust concluded an affirmative corroboration between brand trust of organic food and consumer's intention to purchase. Hence, the improved brand image of a company leads to positive and confident consumer behavior about that brand's products related to food. The construct of religiosity concluded a positive association with the intentions of consumers' buying regarding organic food. The study provides a holistic picture of cultural understanding, wherein religiosity was one of the central points of discussion to understand Pakistani Muslim consumers' purchase intentions about western imported food. Thus, it provides a strong empirical contribution to research in the context of a Muslim population in the sub-continent by identifying the factors stated above. From a demographic perspective, this study developed a profile of consumers of western imported food. This overall information adds value to the literature on consumer behavior. This profile can serve as a learning paradigm for future researchers interested in working on consumers' food purchasing patterns.

6. Practical Implications

The western world always seeks opportunities to export food products to emerging markets. With this strategic export vision, Asian consumer markets present great opportunities for many western food growers and businesses to explore [12]. Hence, this study may facilitate western food producers and exporters to understand Asian consumer behavior in particular western imported food products in Muslim Markets. They may then adjust their current marketing strategy, enabling them to export to such consumer markets. The outcome of this study brings in numerous opportunities for those western marketing practitioners interested in exploring and developing the Asian Muslim majority consumer market. Western food producers may differentiate their product offerings by emphasizing key attributes extracted from this study such as taste, quality, attractive packaging, and ingredients. Having a differentiated product may add a meaningful benefit which may enhance consumers' quality perception or may decrease the perceived risk associated with the use of western imported food products. The enhancement in product attributes may develop brand trust and justification of premium price charged by these western food producers. Since this research is based on a Muslim majority population country, western producers must consider the element of religiosity at the time of developing their marketing strategy. Incorporating the halal logo or stamp, which displays halal product authentication, is highly important for western imported food producers as this builds trust among Muslim consumers to purchase and consume such food. The findings witnessed a positive association with purchase intention towards western food items. Thus, this validates the importance of religiosity and, in particular, the halal authentication is the most important factor for the prospective consumer.

7. Theoretical Contribution

The theory of planned behavior once again verified as a powerful model for testing consumer purchase intention by combining several factors on a single model [30]. New constructs added in TPB to enhance the explanatory power [32,34,35] in various domains. Thus, the present study has included marketing mix (product, price, and promotion); followed by consumer-related factors such as personality, social class, brand trust, and lifestyle; under the antecedents of attitude for family and friends; the antecedents of subjective norms such as self-concept and religiosity; and the antecedents of behavioral control. These tested concepts, for purchase intentions, are a strong contribution in this area of consumer behavior literature. The researchers have considered the above-stated factors as it plays an imperative role in food purchase intention. Therefore, understanding Muslim consumers' perception of western imported food products is significant. The undertaken study also provided the existence of religiosity as a dimension of behavioral control. Adding in this way to the stated theory in the Muslim context where religiosity impacts purchasing and, in particular, food buying choices.

8. Limitations and Areas of Further Research

This research mainly restricted to the eight urban metropolitan cities of Pakistan, and this results in a lack of reliability and credibility of the study outcome. In future studies, researchers can work on the number of cities and make a comparison between cities by utilizing the factors addressed in this study. A cross-country comparison among the cities would be a new area of investigation for researchers and an opportunity for marketers to alter their food product categories accordingly. There is a lack of concentration on a specific food category. This study-addressed consumers' motives behind the purchase of western imported food in general hence generalization to a specific food category is weak. Moving ahead, researchers can choose a specific category, for example, within fast-moving consumer goods (FMCG), a category of milk or any specific food item such as chocolates or biscuits explored. Within this category, a comparison study developed between the local and imported product categories. This research was limited to several consumer participants' groups such as homemakers, university students, and professionals. Pakistan represents many ethnic cultures by adding more participant ethnic groups in the future with specific race/ethnicity, which may provide a better insightful representation of consumer perception in the respective buying behavior as well as facilitate more reliable comparisons.

Author Contributions: Conceptualization, F.B. and S.H.; Methodology, R.R.A.; Software, R.H.S.; Validation, D.S., R.R.A., and Z.A.C.; Formal analysis, S.H.; Investigation, F.B.; Resources, D.S.; Data curation, R.H.S.; Writing—original draft preparation, F.B.; Writing—review and editing, Z.A.C.; Visualization, D.S.; Supervision, R.R.A.; Project administration, D.S. All authors have read and agreed to the published version of the manuscript.

Acknowledgments: We acknowledge the support of IQRA University that provided during this project including administrative and technical support.

References

1. Forghani, M.H.; Kazemi, A.; Ranjbarian, B. Religion, peculiar beliefs and luxury cars' consumer behavior in Iran. *J. Islam. Mark.* **2019**, *10*, 673–688. [CrossRef]
2. Agarwala, R.; Mishra, P.; Singh, R. Religiosity and consumer behavior: A summarizing review. *J. Manag. Spiritual. Relig.* **2019**, *16*, 32–54. [CrossRef]
3. Choi, Y.; Paulraj, A.; Shin, J. Religion or religiosity: Which is the culprit for consumer switching behavior? *J. Int. Consum. Mark.* **2013**, *25*, 262–280.
4. Shukor, S.A.; Jamal, A. Developing scales for measuring religiosity in the context of consumer research. *Middle-East. J. Sci. Res.* **2013**, *13*, 69–74.
5. Quoquab, F.; Pahlevan, J.; Ramayah, T.M. Factors affecting consumers' intention to purchase a counterfeit product: An empirical study in the Malaysian market. *Asia Pac. J. Mark. Logist.* **2017**, *29*, 837–853. [CrossRef]
6. Swimberghe, K.; Flurry, L.A.; Parker, J.M. Consumer religiosity: Consequences for consumer activism in the United States. *J. Bus. Ethics* **2011**, *103*, 453–467. [CrossRef]
7. Asri, N.M.; Aziz, A.A. Halal Dietary Supplement Products in Malaysia. In *Management of Shari'ah Compliant Businesses*; Springer: Berlin/Heidelberg, Germany, 2019; pp. 133–138.
8. Hackett, C.; Lipka, M. The demographic factors that make Islam the world's fastest-growing major religious group. (The Religious and Ethnic Future of Europe). *Scr. Inst. Donneriani Abo.* **2018**, *28*, 11–14. [CrossRef]
9. Hackett, C.; Connor, P.; Stonawski, M.; Skirbekk, V. *The Future of World Religions: Population Growth Projections, 2010–2050*; Pew Research Center: Washington, DC, USA, 2015.
10. Al Harethi, A.R.S. The Role of The Islamic Economy in Rationalizing Consumer Behavior. *J. Islam. Bus. Econ. Rev.* **2019**, *2*, 13–17.

11. Zahid, M.M.; Ali, B.; Ahmad, M.S.; Thurasamy, R.; Amin, N. Factors Affecting Purchase Intention and Social Media Publicity of Green Products: The Mediating Role of Concern for Consequences. *Corp. Soc. Responsib. Environ. Manag.* **2018**, *25*, 225–236. [CrossRef]

12. Hayat, N.; Hussain, A.; Yousaf, H. Food demand in Pakistan: Analysis and projections. *South Asia Econ. J.* **2016**, *17*, 94–113. [CrossRef]

13. Ramli, A.M.; Mirza, A.A.I. The Theory of Consumer Behavior: Conventional vs Islamic. In *Proceedings of the 2nd Islamic Conference (iECONS2007)*; Faculty of Economics and Muamalat, Islamic Science University of Malaysia: Nilai, Malaysia, 2010.

14. Mathras, D.; Cohen, A.; Mandel, N.; Mick, D.G. The effects of religion on consumer behavior: A conceptual framework and research agenda. *J. Consum. Psychol.* **2016**, *26*, 298–311. [CrossRef]

15. Khalek, A.A.; Sharifah, H.S.; Hairunnisa, M.I. A study on the factors influencing young Muslims' behavioral intention in consuming halal food in Malaysia. *J. Syariah* **2015**, *23*, 79–102.

16. Bornemann, T.; Schöler, L.; Homburg, C. In the eye of the beholder? The effect of product appearance on shareholder value. *J. Prod. Innov. Manag.* **2015**, *32*, 704–715. [CrossRef]

17. Chamhuri, N.; Batt, P.J. Consumer perceptions of food quality in Malaysia. *Br. Food J.* **2015**, *117*, 1168–1187. [CrossRef]

18. Kazmi, S.A.Z.; Naarananoja, M.; Kytola, J. Harnessing new product development processes through strategic thinking initiatives. *Int. J. Strateg. Decis. Sci.* **2015**, *6*, 28–48. [CrossRef]

19. Kazmi, S.A.Z.; Naaranoja, M.; Kytola, J.; Kantola, J. Effective Corporate Communication: A Solution to Foster New Product Idea Generation Dynamics. In *Advances in Human Factors, Business Management, Training, and Education*; Springer: Cham, Gremany, 2017; pp. 1033–1045.

20. Shah, S.A.M.; Amjad, S. Investigating moral ideology, ethical beliefs, and moral intensity among consumers of Pakistan. *Asian J. Bus. Ethics* **2017**, *6*, 153–187. [CrossRef]

21. Yousaf, S.; Malik, M.S. Evaluating the influences of religiosity and product involvement level on the consumers. *J. Islam. Mark.* **2013**, *4*, 163–186. [CrossRef]

22. Wan Ismail, W.R.; Othman, M.; Rahman, R.A.; Kamarulzaman, N.H.; Ab Rahman, S.B. "Is sharing really caring?" The impact of eWoM on halal tolerance among Malay Muslim consumers. *J. Islam. Mark.* **2019**, *10*, 394–409. [CrossRef]

23. Hosseini, S.M.P.; Mirzaei, M.; Iranmanesh, M. Determinants of Muslims' willingness to pay for halal-certified food. *J. Islam. Mark.* **2019**. [CrossRef]

24. Souiden, N.; Jabeur, Y. The impact of Islamic beliefs on consumers' attitudes and purchase intentions of life insurance. *Int. J. Bank Mark.* **2015**, *33*, 423–441. [CrossRef]

25. Ali, S.; Ullah, H.; Akbar, M.; Akhtar, W.; Zahid, H. Determinants of Consumer Intentions to Purchase Energy-Saving Household Products in Pakistan. *Sustainability* **2019**, *11*, 1462. [CrossRef]

26. Euro Monitor. Packaged Food in Australia 2018. 2018. Available online: https://www.euromonitor.com/packaged-food-in-australia/report (accessed on 20 September 2019).

27. Khan, S.N.; Mohsin, M. The power of emotional value: Exploring the effects of values on green product consumer choice behavior. *J. Clean. Prod.* **2017**, *150*, 65–74. [CrossRef]

28. Ali, S.; Danish, M.; Khuwaja, F.M.; Sajjad, M.S.; Zahid, H. The Intention to Adopt Green IT Products in Pakistan: Driven by the Modified Theory of Consumption Values. *Environments* **2019**, *6*, 53. [CrossRef]

29. Donald, B.; Glasmeier, A.; Gray, M.; Lobao, L. Austerity in the City: Economic Crisis and Urban Service Decline? *Camb. J. Reg. Econ. Soc.* **2014**, *7*, 3–15. [CrossRef]

30. Armitage, C.J.; Conner, M. Efficacy of the theory of planned behaviour: A meta-analytic review. *Br. J. Soc. Psychol.* **2001**, *40*, 471–499. [CrossRef]

31. Robinson, R.; Smith, C. Psychosocial and demographic variables associated with consumer intention to purchase sustainably produced foods as defined by the Midwest Food Alliance. *J. Nutr. Educ. Behav.* **2002**, *34*, 316–325. [CrossRef]

32. Arvola, A.; Vassallo, M.; Dean, M.; Lampila, P.; Saba, A.; Lahteenmaki, L.; Shepherd, R. Predicting intentions to purchase organic food: The role of affective and moral attitudes in the Theory of Planned Behaviour. *Appetite* **2008**, *50*, 443–454. [CrossRef]

33. Ajzen, I. The theory of planned behavior. *Organ. Behav. Hum. Decis. Process.* **1991**, *50*, 179–211. [CrossRef]

34. Yadav, R.; Pathak, G.S. Young consumers' intention towards buying green products in a developing nation: Extending the theory of planned behavior. *J. Clean. Prod.* **2016**, *135*, 732–739. [CrossRef]

35. Yadav, R.; Pathak, G.S. Determinants of consumers' green purchase behavior in a developing nation: Applying and extending the theory of planned behavior. *Ecol. Econ.* **2017**, *134*, 114–122. [CrossRef]

36. Montano, D.E.; Kasprzyk, D. Theory of Reasoned Action, Theory of Planned Behavior, and the Integrated Behavioral Model. Chapter 6 Health Behavior. In *Theory, Research, and Practice Book*, 5th ed.; Karen, G., Barbara, R., Viswanath, K., Eds.; Jossey-Bass: San Francisco, CA, USA, 2015; pp. 95–124.

37. Glock, C.; Stark, R. The New Denominationalism. *Rev. Relig. Res.* **1965**, *7*, 8–17.

38. Shamsher, R. Relationship between Store Characteristics and Store Loyalty: An Explorative Study. *Int. J. Econ. Empir. Res.* **2014**, *2*, 431–442.

39. Jamal, A.; Goode, M. Consumers, and brands: A study of the impact of self-image congruence on brand preference and satisfaction. *Mark. Intell. Plan.* **2001**, *19*, 482–492. [CrossRef]

40. Torjusen, H.; Lieblein, G.; Wandel, M.; Francis, C.A. Food system orientation and quality perception among consumers and producers of organic food in Hedmark County, Norway. *Food Qual. Prefer.* **2001**, *12*, 207–216. [CrossRef]

41. Dahm, M.J.; Samonte, A.V.; Shows, A.R. Organic foods: Do eco-friendly attitudes predict eco-friendly behaviors? *J. Am. Coll. Health* **2009**, *58*, 195–202. [CrossRef]

42. Chen, Y.S.; Chang, C.H. Enhance green purchase intentions: The roles of green perceived value, green perceived risk, and green trust. *Manag. Decis.* **2012**, *50*, 502–520. [CrossRef]

43. Knight, B.A. Towards Inclusion of Students with Special Educational Needs in the Regular Classroom. *SfL-NASEN* **1999**, *14*, 3–7. [CrossRef]

44. Kotler, P.; Keller, K.L. *Marketing Management Global*, 14th ed.; Pearson Education Limited: London, UK, 2012.

45. Kotler, P. *Marketing Management*; Asia and North America Edition; Pearson Prentice Hall: New York, NY, USA, 2009.

46. Sharma, Y.; Nasreen, R.; Mishra, V. Impact of Consumer-Centric Marketing-Mix on Purchase Behavior of Non-Core Food Items: An Empirical Study of Urban Subsistence Marketplace. *Asian J. Manag. Sci.* **2017**, *6*, 28–41.

47. Song, B.L.; Safari, M.; Mansori, S. The marketing stimuli factors influencing consumers' attitudes to purchase organic food. *Int. J. Bus. Manag.* **2016**, *11*, 109–119.

48. Chryssochoidis, G.; Krystallis, A.; Perreas, P. Ethnocentric-beliefs and country-of-origin (COO) effect—Impact of country, product and product attributes on Greek consumers' evaluation of food products. *Eur. J. Mark.* **2007**, *41*, 1518–1544. [CrossRef]

49. Banerjee, S. Influence of consumer personality, brand personality, and corporate personality on brand preference: An empirical investigation of the interaction effect. *Asia Pac. J. Mark. Logist.* **2016**, *28*. [CrossRef]

50. Chang, A.; Tseng, C.H.; Chu, M.Y. Value creation from a food traceability system based on a hierarchical model of consumer personality traits. *Br. Food J.* **2013**, *115*, 1361–1380. [CrossRef]

51. Bolton, R.N.; Parasuraman, A.; Hoefnagels, A.; Kabadayi, S.; Gruber, T.; Loureiro, Y.K.; Solent, D. Understanding Generation Y and their use of Social Media: A Review and Research Agenda. *J. Serv. Manag.* **2013**, *24*, 245–267. [CrossRef]

52. Goddard, A.; Morrow, D. Assessing the impact of motivational interviewing via co-active life coaching on engagement in physical activity. *Int. J. Evid.-Based Coach. Mentor.* **2015**, *13*, 101–121.

53. Ahaiwe, E.; Onwumere, J.; Agodi, J. Analysis of determinants of Brand Preference for Cosmetics in Abia State, Nigeria. *Int. J. Bus. Manag.* **2015**, *3*, 244–250.

54. Qing, P.; Lobo, A.; Chongguang, L. The impact of lifestyle and ethnocentrism on consumers' purchase intentions of fresh fruit in China. *J. Consum. Mark.* **2012**, *29*, 43–52. [CrossRef]

55. Lassoued, R.; Hobbs, J. Consumer confidence in credence attributes: The role of brand trust. *Food Policy* **2015**, *52*, 99–107. [CrossRef]

56. Szakály, Z.; Szente, V.; Kövér, G.; Polereczki, Z.; Szigeti, O. The influence of lifestyle on health behavior and preference for functional foods. *Appetite* **2012**, *58*, 406–413. [CrossRef]

57. Al-Swidi, A.K.; Behjati, S.; Shahzad, A. Antecedents of Online Purchasing Intention among MBA Students: The Case of University Utara Malaysia Using the Partial Least Squares Approach. *Int. J. Bus. Manag.* **2012**, *7*, 35–49. [CrossRef]

58. Yazdanpanah, M.; Forouzani, M. Application of the Theory of Planned Behaviour to Predict Iranian Students' Intention to Purchase Organic Food. *J. Clean. Prod.* **2015**, *107*, 342–352. [CrossRef]

59. Kautonen, T.; van Gelderen, M.W.; Fink, M. Behavior Robustness of the Theory of Planned in predicting entrepreneurial intentions and actions. *Entrep. Theory Pract.* **2015**, *39*, 655–674. [CrossRef]

60. Armstrong, E.A.; Hamilton, L.T.; Armstrong, E.M.; Seeley, J.L. "Good Girls": Gender, Social Class, and Slut Discourse on Campus. *Soc. Psychol. Q.* **2014**. [CrossRef]

61. Rani, P. Factors influencing consumer behaviour. *Int. J. Curr. Res. Acad. Rev.* **2014**, *2*, 52–61.

62. Durmaz, Y.; Taşdemir, A.A. Theoretical Approach to the Methods Introduction to International Markets. *Int. J. Bus. Soc. Sci.* **2014**, *5*, 47–54.

63. Majabadi, H.A.; Solhi, M.; Montazeri, A.; Shojaeizadeh, D.; Nejat, S.; Farahani, F.K.; Djazayeri, A. Factors Influencing Fast-Food Consumption Among Adolescents in Tehran: A Qualitative Study. *Iran. Red Crescent Med. J.* **2016**, *18*, e23890. [CrossRef]

64. Gonzalez-Jimenez, H. The self-concept life cycle and brand perceptions: An interdisciplinary perspective. *AMS Rev.* **2017**, *7*, 67. [CrossRef]

65. Sung, Y.; Choi, S.M. The influence of self-construal on self-brand congruity in the United States and Korea. *J. Cross-Cult. Psychol.* **2012**, *43*, 151–166. [CrossRef]

66. Lee, J. Universals and specifics of math self-concept, math self-efficacy, and math anxiety across 41 PISA 2003 participating countries. *Learn. Individ. Differ.* **2009**, *19*, 355–365. [CrossRef]

67. Tuškej, U.; Golob, U.; Podnar, K. The role of consumer-brand identification in building brand relationships. *J. Bus. Res.* **2013**, *66*, 53–59. [CrossRef]

68. Hoonsopon, D. The Moderating Effects of Self-Brand Concept and Reference Group on Consumer Innovativeness Toward Purchase Intention. In *Marketing Challenges in a Turbulent Business Environment. Developments in Marketing Science: Proceedings of the Academy of Marketing Science*; Groza, M., Ragland, C., Eds.; Springer: Cham, Gremany, 2016.

69. Chen, W.; Lee, K. Sharing, liking, commenting, and distressed? The pathway between Facebook interaction and psychological distress. *Cyberpsychol. Behav. Soc. Netw.* **2013**, *16*, 728–734. [CrossRef]

70. Su, J.; Tong, X. Brand personality, consumer satisfaction, and loyalty: A perspective from denim jeans brands. *Fam. Consum. Sci. Res. J.* **2016**, *44*, 427–446. [CrossRef]

71. Drescher, J.; Cohen-Kettenis, P.; Winter, S. Minding the body: Situating gender identity diagnoses in the ICD-11. *Int. Rev. Psychiatry* **2012**, *24*, 568–577. [CrossRef]

72. Shaharudin, M.R.; Pani, J.J.; Mansor, S.W.; Elias, S.J. Purchase Intention of Organic Food in Kedah, Malaysia; A Religious Overview. *Int. J. Mark. Stud.* **2010**, *2*, 96–103.

73. Chaudhry, S.; Razzaque, M.A. Religious Commitment and Muslim Consumers: A Model. to Study the Consumer Decision Making Process. In *Proceedings of the 2010 Academy of Marketing Science (AMS) Annual Conference. Developments in Marketing Science: Proceedings of the Academy of Marketing Science*; Deeter-Schmelz, D., Ed.; Springer: Cham, Gremany, 2015.

74. Rakrachakarn, V.; Moschis, G.P.; Ong, F.S.; Shannon, R. Materialism and life satisfaction: The role of religion. *J. Relig. Health* **2015**, *54*, 413–426. [CrossRef] [PubMed]

75. Vitell, S.J. The role of religiosity in business and consumer ethics: A review of the literature. *J. Bus. Ethics* **2009**, *90*, 155–167. [CrossRef]

76. Hannah, S.T.; Avolio, B.J.; May, D.R. Moral Maturation and Moral Conation: A Capacity Approach to Explaining Moral Thoughts and Action. *Acad. Manag. Rev.* **2011**, *36*. [CrossRef]

77. Vassallo, M.; Scalvedi, M.L.; Saba, A. Investigating psychosocial determinants in influencing sustainable food consumption in Italy. *Int. J. Consum. Stud.* **2016**, *40*, 422–434. [CrossRef]

78. Smith, S.; Paladino, A. Eating clean and green? Investigating consumer motivations towards the purchase of organic food. *Australas. Mark. J.* **2010**, *18*, 93–104. [CrossRef]

79. Wee, C.S.; Ariff, M.S.B.M.; Zakuan, N.; Tajudin, M.N.M. Consumers perception, purchase intention and actual purchase behavior of organic food products. *Rev. Integr. Bus. Econ. Res.* **2014**, *3*, 378–397.

80. Peter, J.P.; Olson, J.C. *Consumer Behavior and Marketing Strategy*, 8th ed.; McGraw-Hill, International Edition: Singapore, 2008; p. 400.

81. Krystallis, A.; Fotopoulos, C.; Zotos, Y. Organic Consumers' Profile and Their Willingness to Pay (WTP) for Selected Organic Food Products in Greece. *J. Int. Consum. Mark.* **2006**, *19*, 81–106. [CrossRef]

82. Gobo, G. Upside down. Reinventing research design. Chapter 5. In *Handbook of Qualitative Data Collection*; Flick, U., Ed.; Sage: London, UK, 2018; pp. 65–83.

83.   Hair, J.; Risher, J.; Sarstedt, M.; Ringle, C. When to use and how to report the results of PLS-SEM. *Eur. Bus. Rev.* **2019**, *31*, 2–24. [CrossRef]

84.   Ahmed, R.R.; Streimikiene, D.; Berchtold, G.; Vveinhardt, J.; Channar, Z.A.; Soomro, R.H. Effectiveness of Online Digital Media Advertising as A Strategic Tool for Building Brand Sustainability: Evidence from FMCGs and Services Sectors of Pakistan. *Sustainability* **2019**, *11*, 3436. [CrossRef]

85.   Hair, J.F., Jr.; Black, W.C.; Babin, B.J.; Anderson, R.E.; Tatham, R.L. *Multivariate Data Analysis*, 6th ed.; Prentice-Hall: Upper Saddle River, NJ, USA, 2006.

86.   Ahmed, R.R.; Vveinhardt, J.; Štreimikiene, D. The direct and indirect impact of Pharmaceutical industry in Economic expansion and Job creation: Evidence from Bootstrapping and Normal theory methods. *Amfiteatru Econ.* **2018**, *20*, 454–469. [CrossRef]

87.   Chatfield, C. *Introduction to Multivariate Analysis*; Routledge: New York, NY, USA, 2018.

88.   Hair, J.F., Jr.; Hult, G.T.M.; Ringle, C.M.; Sarstedt, M. *A Primer on Partial Least Squares Structural Equation Modeling (PLS-SEM)*, 2nd ed.; Sage Publications: Thousand Oaks, CA, USA, 2016.

89.   Hoelter, J.W. The analysis of covariance structures: Goodness-of-fit indices. *Sociol. Methods Res.* **1983**, *11*, 325–344. [CrossRef]

90.   Kline, T.J.B.; Sulsky, L.M.; Rever-Moriyama, S.D. Common method variance and specification errors: A practical approach to detection. *J. Psychol.* **2000**, *134*, 401–421. [CrossRef]

91.   Fornell, C.; Larcker, D.F. Evaluating structural equation models with unobservable variables and measurement error. *J. Mark. Res.* **1981**, *39*, 39–50. [CrossRef]

92.   Lian, S.B.; Yoong, L.C. Assessing the Young Consumers' Motives and Purchase Behavior for Organic Food: An Empirical Evidence from a Developing Nation. *Int. J. Acad. Res. Bus. Soc. Sci.* **2019**, *9*, 69–87. [CrossRef]

93.   Nasution, M.D.T.P.; Rossanty, Y. Country of origin as a moderator of a halal label and purchase behaviour. *J. Bus. Retail Manag. Res.* **2018**, *12*, 194–201. [CrossRef]

94.   Tomić, M.; Alfnes, F. Effect of Normative and Affective Aspects on Willingness to Pay for Domestic Food Products—A Multiple Price List Experiment. *J. Food Prod. Mark.* **2018**, *24*, 681–696. [CrossRef]

95.   Nguyen, H.V.; Nguyen, N.; Nguyen, B.K.; Lobo, A.; Vu, P.A. Organic food purchases in an emerging market: The influence of consumers' personal factors and green marketing practices of food stores. *Int. J. Environ. Res. Public Health* **2019**, *16*, 1037. [CrossRef]

96.   Tsang, M.M.; Ho, S.C.; Lian, T.P. Consumer attitudes toward mobile advertising: an empirical study. *Int. J. Elec. Com.* **2004**, *8*, 65–78. [CrossRef]

97.   Cui, L.; Jiang, H.; Deng, H.; Zhang, T. The influence of the diffusion of food safety information through social media on consumers' purchase intentions: An empirical study in China. *Data Technol. Appl.* **2019**, *53*, 230–248. [CrossRef]

98.   Flavián, C.; Guinalíu, M.; Torres, E. The influence of corporate image on consumer trust: A comparative analysis in traditional versus internet banking. *Internet Res.* **2005**, *15*, 447–470. [CrossRef]

99.   Garg, P.; Joshi, R. Purchase intention of "Halal" brands in India: The mediating effect of attitude. *J. Islam. Mark.* **2018**, *9*, 683–694. [CrossRef]

100.   Sirgy, M.J. Self-congruity theory in consumer behavior: A little history. *J. Glob. Sch. Mark. Sci.* **2018**, *28*, 197–207. [CrossRef]

101.   Hoonsopon, D.; Puriwat, W. The effect of reference groups on purchase intention: Evidence in distinct types of shoppers and product involvement. *Australas. Mark. J.* **2016**, *24*, 157–164. [CrossRef]

13

Perception of Older Adults about Health-Related Functionality of Foods Compared with Other Age Groups

Dávid Szakos *, László Ózsvári and Gyula Kasza

Department of Veterinary Forensics an Economics, University of Veterinary Medicine Budapest, 1078 Budapest, Hungary; ozsvari.laszlo@univet.hu (L.Ó.); kasza.gyula@univet.hu (G.K.)
* Correspondence: szakos.david@univet.hu

Abstract: The proportion of older adults in the population is significantly growing in the EU, therefore, wellbeing of the older population has become a social challenge. Functional foodstuffs are food products with nutritional composition that may reduce the risk of diet-related diseases or enhance physiological functions. Therefore, they could play an important role in prevention and mitigation of health-related problems, and in promotion of healthy ageing. The aim of this study is to present the impact of age on consumer preferences about functionality of foods, covering attitude aspects, nutrition claims, possible carriers, some particular health problems and expectations about sustainable production. The results are based on a representative quantitative survey. Findings highlight statistically significant ($p < 0.05$) differences in preferences of older adults compared to other age segments. They generally accept functional foods, especially when functionality is attached to increased vitamin, protein, and fiber content. Older adults also prefer products with lower salt and sugar content, which were less relevant for other age groups. Products of fruit and vegetable origin are distinguished as carriers of functional traits. Compared to other segments, older adults accept products of animal origin (especially milk products) and even breakfast products on a higher level. The paper provides details about particular health issues that could be addressed by functional foods based on actual consumer concerns.

Keywords: functional food; consumer survey; nutrition claims; health claims; older adults; healthy diet; healthy ageing

1. Introduction

The relationship between health and nutrition has come to the forefront of scientific research due to global health trends and lifestyle changes. According to WHO data, chronic non-communicable diseases (CNDs) are the leading cause of death worldwide [1]. In 2016, they were responsible for 71% (41 million) of the 57 million deaths which occurred globally, and 94% of the number of deaths in Hungary. Major CNDs are cardiovascular diseases (44% of all CND deaths), cancers (22%), chronic respiratory diseases (9%), and diabetes (4%), all of which are strongly connected to dietary factors, among others [2]. Therefore, WHO formed a guideline for healthy diet to prevent chronic diseases worldwide, and national level health prevention programs also emerged [3–5]. Demographic statistics related to ageing shows that life expectancy (LE) and proportion of older adults in the population are increasingly growing both at global and EU level. Between 2000 and 2016, global LE at birth increased by 5.5 years, from 66.5 to 72.0 years [6]. In the EU, almost one fifth of the population (19.7%) was over the age of 65 in 2018, and the relative share of the population is projected to reach 28.5% until 2050 [7]. According to the latest country reports of the European Health & Life Expectancy Information System (EHLEIS) based on 2015 data, Hungarian LE was 18.2 years (21.2 for women and 17.9 for men) at the age 65 [8]. This index, compared to LE at birth, give better estimation to older adults, but do not give information about the quality of those years. The same report presents another indicator: the healthy life years (HLY, also called healthy life expectancy or disability-free life expectancy), which was 5.9 years in Hungary at the age 65, so 68% of elderly years (approximately 12 years) are usually spent with disabilities. LE at birth in Hungary was 75.7 years in 2015, which was nearly 5 years below the EU average, mainly due to higher death rates from cardiovascular diseases and cancer [9]. The same study highlights that only slightly more than half (56%) of Hungarians consider themselves to be in good health, which is one of the lowest rates in the EU. Besides new ways in the investigation of health-related issues [10–12], sustainability of food consumption (including food security) is also becoming an increasingly prominent topic for the scientific community [13–16]. Furthermore, the harmonization of a balanced and sustainable diet opened a new research regime [17].

Nowadays, a rising number of consumers follow a special or consciously composed diet because of health issues or lifestyle decisions, which have opened new opportunities for food business operators. During the last decades, a special focus was given to the health-related functionality of foodstuffs [18]. Functional foods with high added value have become the fastest growing area of the food industry, although the market share varies greatly from country to country, and there is not one generally accepted definition of functional foods in the industry, so different market data are available due to different interpretations of the category [18,19]. A study reviewed over one hundred different definitions to determine the boundaries of functional food better [20]. One even argued that functional foods might not be handled as a well-definable separate product category [21]. Although the definitions help scientific and professional dialogues, they do not have a particularly significant role from the perspective of consumers. Instead of legal definitions, consumers receive information about the functional properties of food through advertisements and labels. Regarding labels, nutrition and health claims in the EU may appear on products by following the indications of Regulation (EC) No. 1924/2006 [22] and Regulation (EU) 432/2012 [23] based on the scientific advice of the European Food Safety Authority (EFSA) [24,25].

The increasing importance of the functionality of foodstuffs was recognized even before the turn of the millennium by the food industry, which has accelerated the development of new products. However, new products had a high failure rate on the market in the 1990s, because most of them were not preceded by a deeper exploration of consumer needs [19,26]. Developing functional food is often a far more complex issue than introducing a new variation of generalized food products, which was realized by researchers and company experts in the 2000s. Many consumer-related studies emerged about functional foods from that time. The first consumer studies related to functional foods tried to explore the effect of socio-demographics factors [27–29], attitudinal profiles and motivations [21,30–34], and reactions connected to health and nutrition claims [35–38]. Based on the results of the studies that focused on the concept of functional foods in general, later studies targeted more specific product categories and novel concepts [39–44]. It also means that the focus shifted to market-related surveys and product development aspects.

The relatively few consumer-related articles, which focus on older European consumers, also follow a marketing approach, and they are connected to protein-enriched functional foods majorly [45–47]. In Eastern Europe, a Polish study, based on a nationwide representative consumer survey in 2009, found significant differences between age groups in functional food consumption, awareness, and perceived barriers to health improvement. Qualitative consumer studies that support the food product development for older adults have also been published [48,49]. In Hungary, however, a few consumer studies have recently emerged on the relationship between health and food consumption [50–52], and the perspective of older adults on functional foods have not been analyzed yet.

Based on the previous findings, physical and psychological wellbeing of the older population has become a globally significant social challenge. The aim of this study is to give an overall picture about the impact of age on consumer expectations about the functionality of foods, which covers attitude, health-related lifestyle factors, nutrition claims, carriers, health problems, and known diseases as well.

It was also an important goal of our study to investigate the most common sustainability markers used on food products.

2. Materials and Methods

The results of this study are based on a quantitative consumer survey conducted between 11 July and 14 August in 2018 with 1002 respondents. For data collection, personal sampling method was used with a questionnaire designed to be suitable for self-administered completion. Research was conducted at crowded traffic junctions in different Hungarian cities: Budapest, Dombóvár, Eger, Füzesabony, Győr, Kiskunfélegyháza, Miskolc, Siófok, Szeged, Székesfehérvár, Szolnok, and Veszprém. In terms of sex, age, and geographical distribution (NUTS-2) of the respondents, the sample is representative of the total adult population of Hungary, based on the latest census [53] at the time of the data collection (Table 1). To ensure representativeness, we employed quota sampling. During the research design, besides general socio-demographic characteristics, we aimed to collect data on some further particular conditions that may affect food consumption directly according to literature [29] (Table 2).

In the beginning of the interview, the respondents were informed about the aim of the research and the management of anonymous data. If the respondents were willing to participate, before the research questions were asked, the quota parameters (age, sex, geographical location) had been recorded, that allowed the quota numbers to be tracked by the interviewers to ensure an appropriate level of representability. Although the questionnaire was designed to be self-administered, interviewers provided help to fill the questionnaire, which was important in the case of older respondents.

The questionnaire contained 288 variables, from attitude-related questions through to nutritional claims and carrier foods to questions focusing on diseases. The questionnaire employed closed-form questions predominantly. Many questions were measured on five-point Likert scale, where grade 1 meant "strongly disagree" and grade 5 meant "strongly agree." Table A1 in Appendix A shows the content of the questionnaire in terms of all variables used in this study.

Statistical analysis of the data was carried out by IBM SPSS Statistics 22.0 software package. Beyond descriptive analysis, Kruskal-Wallis test and Pearson's chi-square test (CI: 95%) were used to analyze data on ordinal scale when the distribution of data did not meet the criteria for normal distribution [54]. Factor analysis (principal component analysis—PCA) was used to explore overlaps and to combine correlated variables [55].

Table 1. Representative socio-demographic characteristics of the sample (% of respondents, n = 1002).

Socio-demographic Categories		Sample	Population *
Sex	Female	53.19	53.07
	Male	46.81	46.93
Age	18–29	17.96	17.59
	30–39	16.97	17.04
	40–59	34.53	33.83
	>60	30.54	31.54
Geographical distribution (NUTS-2)	Central Hungary	31.04	30.75
	Central Transdanubia	10.78	10.80
	Western Transdanubia	10.18	10.03
	Southern Transdanubia	8.68	9.13
	Northern Hungary	11.48	11.62
	Northern Great Plain	15.07	14.90
	Southern Great Plain	12.77	12.78

* Latest census data of Hungarian Central Statistical Office to adult Hungarian population [53].

Table 2. Further socio-demographic characteristics of the sample (valid % of respondents).

Socio-demographic Categories		%
Place of living	Village	15.49
	Another city	61.54
	Capital city	22.98
Highest accomplished qualification	Primary and vocational school	11.46
	High school (graduated)	33.37
	Higher education	55.17
Income level (subjective estimation)	Below average	13.11
	Average	68.16
	Above average	17.17
Economic status	Active worker	54.64
	Entrepreneur	6.25
	Retiree	27.12
	Job seeker	1.51
	Homemaker	1.41
	Student	9.07
Children under 15 years of age in the household	Yes	20.04
	No	79.96
Number of persons living in the household	1	16.48
	2	40.88
	3	17.83
	4	14.18
	5 or more	10.63
Special dietary needs	Respondent	49.50
	Another person in the family	10.68
	No	34.53
	Did not respond	5.29

3. Results

3.1. Attitudes and Lifestyle Factors toward Nutrition

At the beginning of the survey, attitude-related 1–5 Likert questions were listed in order to characterize different age groups based on their opinion about health and age-related aspects of nutrition (Table 3).

Table 3. Attitudes toward nutrition in different age groups (level of agreement, 1–5 Likert scale).

Variables	Total Sample	18–29	30–39	40–59	>60	Sig.
	M (SD)					
Nutrition has a direct impact on health	4.60 (0.783)	4.51 (0.897)	4.60 (0.751)	4.61 (0.744)	4.64 (0.773)	0.362
Healthy diet has great impact on the prevention of diseases in older adults	4.50 (0.822)	4.57 (0.703)	4.43 (0.868)	4.50 (0.830)	4.49 (0.852)	0.534
For older adults, diet has a more important role in health	4.31 (0.976)	4.34 (0.871)	4.16 (1.046)	4.28 (0.971)	4.42 (0.976)	0.003
Healthy diet is important for me	4.28 (0.885)	4.05 (0.878)	4.16 (0.879)	4.31 (0.827)	4.45 (0.920)	<0.0001

According to the results, each age group perceived strong connection between nutrition and health. While all age groups agreed that "healthy diet has a great impact on the prevention of diseases in older adults," the oldest group attached significantly more importance to "for older adults, diet has a more important role in health" compared to other age groups. The importance of a healthy diet increases with age according to the responses.

The questionnaire also contained lifestyle-related multiple choice questions, which allowed further differentiation of age groups (Table 4).

Table 4. Perception of health-related lifestyle factors in different age groups.

Variables	Total Sample	18–29	30–39	40–59	>60	Sig.
			%			
I feel healthy in general	59.96	67.05	57.23	57.85	59.72	0.184
I want to lose weight, and I do something about this	41.63	38.07	39.76	44.77	41.13	0.461
I exercise regularly	37.64	48.30	43.03	35.17	30.85	0.001
I have a stressful lifestyle	36.02	36.93	49.40	43.60	18.37	<0.0001
I do not sleep enough	33.75	31.82	40.96	39.53	23.67	<0.0001
I do not exercise enough	32.57	32.39	34.34	38.19	24.82	0.005
I pay more attention to my diet than average	30.03	22.73	34.34	28.78	33.57	0.049
I use dietary supplements	25.70	22.16	28.92	27.33	24.03	0.400
I can spend only a little time on eating and cooking	19.09	33.52	25.90	19.19	6.01	<0.0001
I smoke every day	15.17	20.45	21.08	16.57	6.71	<0.0001
I try to consume less alcohol	17.44	21.59	17.47	16.57	15.90	0.431

Significant difference between age groups was not found in terms of self-estimation of health, the need for weight loss, use of dietary supplements, and alcohol consumption habits. Younger respondents exercise more often, although they still tend to think it is below the required level. Stressful lifestyle and not enough sleep are the most common problems reported by the middle-aged groups. Older adults rarely smoke and this group has significantly more time to eat and cook than the younger respondents. Consumers between 30 and 39 years and over 60 years state that they pay more attention to diet compared to the average.

3.2. Nutrition Claims

The questionnaire contained 39 nutrition claims in total, covering all options listed by the Regulation (EC) No. 1924/2006. In some cases, claims were presented through an example, such as "source of calcium," while others used a generalized form, for instance, "source of vitamins." EU and national level food law allow the use of the terms salt and sodium as synonyms in labelling, so both terms were included in the questionnaire. Besides the claims listed in the regulation, some other elements were also included (for example, prebiotic, contains antioxidants, etc.). Respondents expressed their opinion on 1–5 Likert scale about their preference of the listed nutrition claims shown in Figure 1. For better interpretation, PCA was used to reduce 39 items to 8 well-distinguishable nutrition claim categories (Table 5). As expected on the basis of previous relevant studies, the created categories highlighted that consumers did not perceive significant differences between multiple level claims [40,56]. Accordingly, the categories give a robust representation for multiple level claim groups (for instance, with no added sugar, low in sugar, sugar-free, within the factor named sugar).

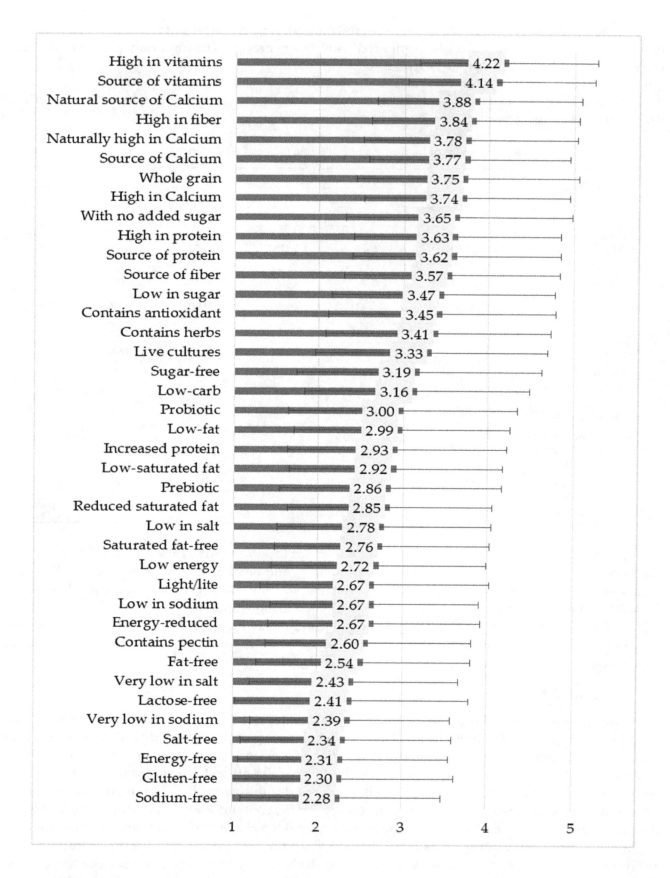

Figure 1. Preference of nutrition claims on food (1–5 Likert scale, where 5 means the highest level of preference).

Table 5. Nutrition claims factor categories based on PCA (Rotation method: Varimax; KMO: 0.944; Bartlett: < 0.0001).

Factor Names	Included Variables	Total Variance Explained	
		% of Variance	Cumulative %
Vitamins and minerals	High in vitamins Source of vitamins Natural source of Calcium Naturally high in Calcium Source of Calcium High in Calcium	13.469	13.469
Salt	Low in salt Low in sodium Very low in salt Very low in sodium Salt-free Sodium-free	13.353	26.822
Not listed claims	Whole grain Contains antioxidant Contains herbs Live cultures Probiotic Prebiotic Contains pectin	11.714	38.536
Light	Low energy Light/lite Energy-reduced Energy-free	9.601	48.137
Sugar	With no added sugar Low in sugar Sugar-free Low-carb	8.451	56.588
Fat	Low-fat Low-saturated fat Reduced saturated fat Saturated fat-free Fat-free	8.108	64.697
Protein and fiber	High in fiber High in protein Source of protein Source of fiber Increased protein	7.369	72.065
Free from	Lactose-free Gluten-free	4.693	76.758

In general, nutrition claims related to vitamins and minerals received the highest preference scores. Protein, fiber and sugar content also seem to be important for the respondents. Claims related to fat content, energy, and salt can typically be found in the middle section of the list. PCA analysis clearly indicated a group constituted by those claims that are not listed in the Regulation (EC) No. 1924/2006 (Table 5). Lactose-free and gluten-free are at the bottom of the list. In terms of salt and sodium, the former one is more preferred by the consumers, although both terms indicate the same nutritional element (Figure 1).

Figure 2 shows the differences between age groups in regard to the nutrition claims categories composed with PCA.

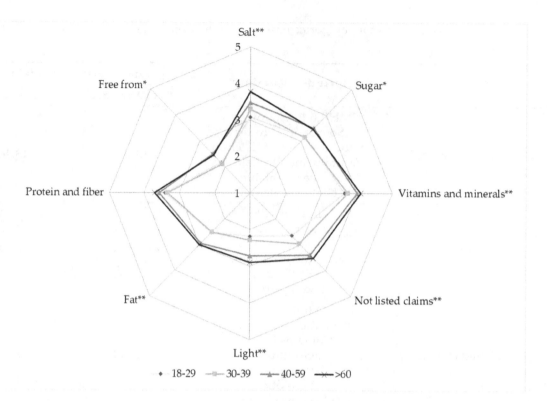

Figure 2. Preference of nutrition claims categories composed with PCA, between different age groups (* $p < 0.05$; ** $p < 0.01$).

Differences between preferences of age groups are significant in all cases, except nutrition claims related to protein and fiber. Respondents over 60 years typically have stronger preference of the listed claims than the younger age groups. The most significant difference was found in the case of salt-related claims.

3.3. Carrier Foods

Previous studies highlighted the importance of the type of carrier food products regarding acceptance of health benefits by consumers [30,40,57,58]. During data collection, respondents could express their health-related preference about 25 types of foods on 1-5 Likert scale (Figure 3). Table 6 shows the 5 carrier categories composed by PCA.

Eating fruits and vegetables is the best way for keeping a healthy diet according to the respondents. "Fruits and vegetables" form an independent group by PCA, which contains processed products and mushrooms, too. The following categories are "meat, fish, and egg," "natural products," and "dairy products." "Natural products" is a heterogeneous group compared to the others. It contains juice; honey; tea; nuts and other oily seeds, muesli; and herbal products. "Breakfast products," namely fruit jam, bakery products, and margarine are at the end of the preference list.

Figure 4 shows the differences between age groups connected to carrier categories composed with PCA.

Differences between preferences of age groups are significant in cases of "dairy products," "breakfast products," and "meat, fish, and eggs." "Fruits and vegetables" and "natural products" are fairly important for all age groups. Older adults preferred "dairy products" and "breakfast products" to a greater extent than others.

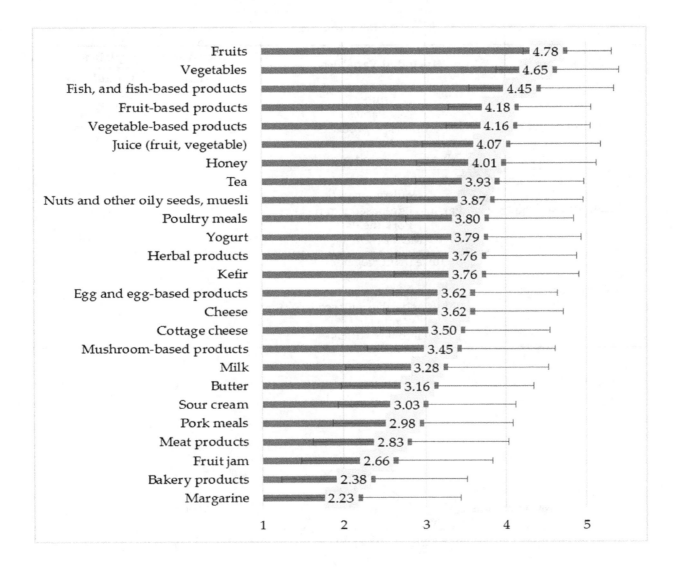

Figure 3. Preference of different food carriers (1-5 Likert scale, where 5 means the highest level of preference).

Table 6. Carrier factor categories based on PCA (Rotation method: Varimax; KMO: 0.848; Bartlett: < 0.0001).

Factor Names	Included Variables	Total Variance Explained	
		% of Variance	Cumulative %
Dairy products	Yogurt Kefir Cheese Cottage cheese Milk Butter Sour cream	14.541	14.541
Breakfast products	Margarine Bakery products Fruit jam	10.256	24.797

Table 6. *Cont.*

Factor Names	Included Variables	Total Variance Explained	
		% of Variance	Cumulative %
Fruits and vegetables	Fruits Vegetables Fruit-based products Vegetable-based products Mushroom-based products	10.038	34.835
Meat, fish, and eggs	Fish and fish-based products Poultry meals Egg and egg-based products Pork meals Meat products	9.783	44.619
Natural products	Juice (fruit, vegetable) Honey Tea Nuts and other oily seeds, muesli Herbal products	9.349	53.967

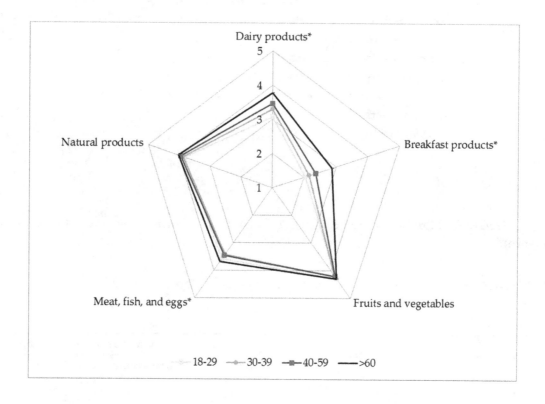

Figure 4. Preference of food carrier categories composed with PCA, between different age groups (* $p < 0.05$).

3.4. Health problems and acceptance of functional foods

The main health problems people are most affected by and worried about compared to the acceptance of mitigation and prevention with functional foods are shown in Figure 5.

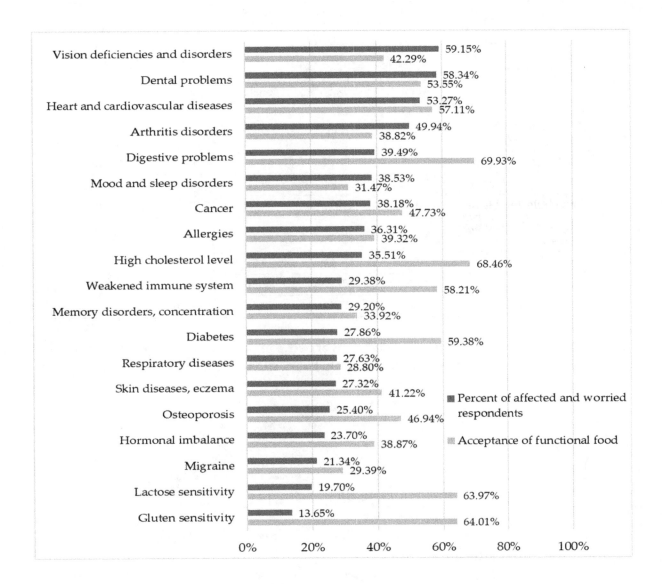

Figure 5. Consumer concerns in regard to certain health problems and the perceived suitability of functional foods to prevent or mitigate these problems.

According to the results, the Hungarian population is mainly concerned/worried about the following health problems: vision deficiencies and disorders, dental problems, and heart and cardiovascular diseases. Results also highlight that diets containing functional foods for the mitigation and prevention of health problems are mainly related to digestive problems, high cholesterol level, lactose sensitivity, and gluten sensitivity according to the opinion of the respondents.

In the case of several health problems, the age of respondents has been a significant factor, shown in Table 7.

In the vast majority of cases where significant differences were detected, older adults are more affected or worried about the certain health problems. The only exception is migraine, which worries and affects younger people more.

Table 7. Presence of particular health concerns in different age groups.

Health Problem	18–29	30–39	40–59	>60	Sig.
			%		
Vision deficiencies and disorders	46.06	43.04	64.57	72.97	<0.0001
Dental problems	40.85	57.86	65.91	61.21	<0.0001
Heart and cardiovascular diseases	27.53	45.24	48.86	54.93	<0.0001
Arthritis disorders	27.11	37.34	53.97	69.70	<0.0001
Digestive problems	29.45	35.67	45.00	42.36	0.0063
High cholesterol level	19.88	31.45	34.56	51.60	<0.0001
Memory disorders, concentration	23.49	26.42	28.19	38.02	0.0142
Diabetes	17.58	26.58	29.77	34.50	0.0033
Osteoporosis	13.25	20.13	27.65	36.79	<0.0001
Migraine	22.42	25.16	24.91	10.86	0.0017

Figure 6 shows the proportion of affected/worried consumers in the age groups, who would accept food as a solution to prevent and/or mitigate the particular health problem.

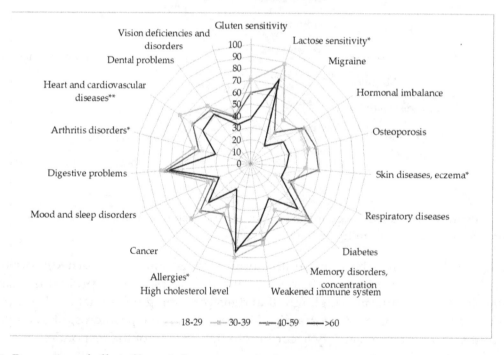

Figure 6. Proportion of affected/worried consumers in the age groups, who accepted food as a solution to prevent and/or mitigate a particular health problem (* $p < 0.05$; ** $p < 0.01$).

Among the affected/worried consumers, significant differences were detected between age groups in the case of heart and cardiovascular diseases, arthritis disorders, allergies, skin diseases and eczema, and lactose sensitivity. In the majority of these cases, older adults are characterized by a lower level of acceptance than the younger ones. In general, affected young adults and middle-aged adults show the highest level of acceptance of functional food products: younger adults particularly accept these in case of heart and cardiovascular diseases, and lactose sensitivity, while middle-aged adults would prefer functional food to prevent/mitigate the effect of skin diseases and eczema and allergies.

3.5. Sustainability Factors

In the context of functional food preferences, the possible value-added characteristics of certain sustainability factors were also analyzed. It gives an opportunity to identify further consumer expectations about "healthy" food products. Results about consumer perceptions are presented in Table 8.

Table 8. Preference of certain sustainability factors in regard to functional food products in different age groups.

Variables	Total Sample	18–29	30–39	40–59	>60	Sig.
			M (SD)			
Domestic product	3.85 (1.213)	3.24 (1.370)	3.86 (1.086)	3.85 (1.170)	4.21 (1.085)	<0.0001
Small-scale production	3.77 (1.215)	3.06 (1.289)	3.71 (1.224)	3.92 (1.105)	4.05 (1.125)	<0.0001
Animal welfare considerations	3.73 (1.228)	3.51 (1.403)	3.77 (1.179)	3.73 (1.150)	3.83 (1.220)	0.143
Traditional product	3.53 (1.159)	2.87 (1.257)	3.43 (1.257)	3.64 (1.067)	3.87 (0.981)	<0.0001
Organic product	3.47 (1.281)	3.08 (1.379)	3.29 (1.291)	3.61 (1.203)	3.65 (1.246)	<0.0001
Produced with modern technology	2.93 (1.212)	2.52 (1.178)	2.69 (1.231)	2.95 (1.225)	3.29 (1.217)	<0.0001

According to the preference of the consumers, domestic origin is the most prominent aspect, followed by small-scale production and animal welfare considerations. Except in the case of animal welfare, differences between age groups were proven to be significant by using a confidence interval of 95%. All listed sustainability factors were more preferred by consumers over the age of 60, which indicates that the combination of sustainability labels (especially in regard to domestic origin) and health and nutrition claims on the package could bear a recognized value for senior conscious consumers.

4. Discussion

This paper aims to give an overall picture on the impact of age on consumer expectations about the functionality of foods based on a nationwide representative consumer survey. Besides a description of significant differences between age groups about health-related attitudes and lifestyle factors, our study analyzes the most important aspects of functionality of foods through quantification of consumer expectations and preferences.

A special focus was placed on older consumers in the analysis. The importance of the wellbeing of older adults is increasing, as their social representation grows. While there is a big variety of functional foodstuffs on the market already, their contribution to disability-free life years also depends on consumer choice. The combination of functional foods with scientifically proven health and nutrition claims and appropriate consumer perception would deliver significant social benefits.

Thirty-eight nutrition claims and 25 carrier food types were categorized with PCA to identify the most prominent decision points of older adults compared to other age groups. Respondents expressed their involvement and concerns about 19 health problems, and also gave their opinion about the appropriateness of food as a solution to prevent and/or mitigate the particular health problem. Our study also investigated the preference of the most common sustainability markers for functional foods.

Significant differences were found between age groups in consumer preferences about the functionality of foods. Results are harmonizing with previous studies, which pointed out that older adults have a more positive attitude toward functionality of foods in general [33,59–61].

"Vitamins and minerals," "protein and fiber," and claims related to sugar content were found to be the most preferred nutrition claims. Differences between preferences of age groups were significant in all cases, except nutrition claims related to protein and fiber. Respondents over 60 years typically prefer the listed claims more than younger age groups. The most important significant difference was found about salt-related claims. A recent Italian study with similar methodology also identified significant differences between age groups in regard to the preference of nutrition claims, and found similarly that claims related to vitamins are the most preferred ones by the consumers [62]. The same study found a lower level of preference about salt-related claims, which can be explained by country differences described in previous cross-country research [36]. Moreover, previous studies indicated a connection between health status and the importance people attach to nutrition and health claims [33,35]. In this case, it is important to mention that the proportion of death caused by cardiovascular diseases—where the reduction of salt intake would be one of the most important dietetic factors—is four times higher in Hungary than in Italy [63].

Based on the opinion of respondents, the following food categories are the most suitable for a healthy diet: "fruits and vegetables," "meat, fish, and egg," "natural products," (e.g., juices, tea, honey), and "dairy products." Older adults preferred the "dairy products" and "breakfast products" significantly more than the younger respondents. Related studies mostly analyzed carrier food products combined with some particular claims that contributed for product development in a more direct manner [30,40,57,58]. These studies concluded that the type of the carrier had a greater effect on the acceptance of health benefits by consumers than the type of claim. Furthermore, consumers showed higher acceptance toward the functionality of foods, where the functional ingredient was inherently contained to some extent [30,40].

A Dutch study, which analyzed health claims, pointed out that health claims related to physiological health problems (e.g., heart and cardiovascular diseases, osteoporosis, cancer) are of greater importance among such claims stated on the labels of foodstuffs than those related to psychological problems (e.g., stress, fatigue), which are in line with our results [35]. A recent Hungarian study which examined the effect of socio-demographic factors in the case of functional foods also highlighted the importance of age regarding to health problems [50]. Our results indicate that the Hungarian population is mainly concerned about the following health problems: vision deficiencies and disorders, dental problems, and heart and cardiovascular diseases. According to the respondents, functional foods are most suitable for the mitigation and prevention of digestive problems, high cholesterol level, lactose sensitivity, and gluten sensitivity. In the vast majority of cases where significant differences are detected, older adults are more affected or worried about health problems. After filtering the sample only for the affected/worried consumers, less significant differences between age groups were detected. Where significant differences were found, younger and middle-aged adults are more likely to accept food as a solution to prevent and/or mitigate the particular health problem.

In terms of key health-related sustainability factors of food products, domestic origin played the most important role, followed by small-scale production and animal welfare, which are in line with previous studies [64]. Animal welfare was the only factor found to be universally appreciated, while other factors are preferred by the older adults to a higher extent.

The results of this quantitative study highlighted the importance of considering the wellbeing of older adults during product development. The investigation proven that significant differences in attitudes and preferences do exist and might be used for product differentiation. The paper contains a detailed data set about possible carrier food and functionality combinations that might be used for subsequent academic studies and for field experts as well. However, it is important to consider country-level differences that might be reflected in the preference of carrier food and functionality combinations. In this survey, we have collected a representative sample of the Hungarian population that served the purpose of demographical comparison well, and was also able to deliver some health status specific results according to the most frequent problems. This investigation was limited to respondents in relatively good health conditions, due to the methodology of data collection, which was performed at crowded traffic junctions. To reach older adults who are affected by serious health problems, investigations must be expanded to retirement homes and hospitals. However, during our research, the main focus was on disability-free life years, which required the answers of persons with seemingly normal health conditions. The research was conducted in summer, and seasonality may effect consumer preference on foodstuffs—especially in terms of locally produced fruits and vegetables—according to certain studies [65]. However, questions were aimed to measure general attitudes. Additionally, seasonality tends to be less important in the last decades, especially in the urbanized population of economically developed countries.

This explorative study can be used as a basis for a subsequent research focusing on the ageing consumers to provide an in-depth insight into their food consumption behavior and perception of the link between nutrition and health. It must be considered for further research that a larger sample size of older adults would allow the use of sophisticated multivariate statistics methods, such as cluster analysis and structural equation modelling. By this investigation, further segmentation could be conducted to identify possible gaps in education, availability of expected health-promoting products and innovation areas. This research can also reveal behavioral reasons behind health-related food consumption habits of older adults, which, besides fostering product development, could lay the foundations of social and health-related policy actions as well.

Author Contributions: D.S. and G.K. conceived and designed the experiments; D.S. analyzed the data; D.S., L.Ó. and G.K. contributed to conceptualization and writing the paper. All authors have read and agreed to the published version of the manuscript.

Appendix A

Table A1. Summary of the questionnaire for all variables used in this study.

Questions	Set of Values	Listed Variables
To what extent do you agree with the following statements?	1-5 Likert scale	Attitude factors presented in Table 3
Which of the following lifestyle statements are relevant for you?	Multiple choice	Lifestyle factors presented in Table 4
To which extent do you prefer the following nutrition claims while shopping?	1-5 Likert scale	Nutrition claims presented in Figure 1
How much do you think the consumption of the following foods contribute to your health?	1-5 Likert scale	Carrier foods presented in Figure 3
How important is it for you that a "healthy food" has the following properties?	1-5 Likert scale	Sustainability factors presented in Table 8
Which health problems do you worry about?	Not concerned/ Concerned	Health-related problems presented in Figure 5
Would you choose "healthier foods" to prevent or mitigate the following health problems?	Yes/No	
Socio-demographic parameters presented in Tables 1 and 2.		

References

1. World Health Organization. Noncommunicable Diseases Country Profiles 2018. Available online: https://www.who.int/nmh/publications/ncd-profiles-2018/en/ (accessed on 18 February 2020).
2. World Health Organization Global Health Estimates 2016: Death by Cause, Age, Sex, by Country and by Region, 2000–2016. Geneva. Available online: https://www.who.int/healthinfo/global_burden_disease/estimates/en/index1.html (accessed on 18 February 2020).
3. World Health Organization Diet, Nutrition, and the Prevention of Chronic Diseases: Report of a Joint WHO/FAO Expert Consultation. Available online: https://www.who.int/dietphysicalactivity/publications/trs916/en/ (accessed on 18 February 2020).
4. Jankovic, N.; Geelen, A.; Streppel, M.T.; de Groot, L.C.P.G.M.; Orfanos, P.; van den Hooven, E.H.; Pikhart, H.; Boffetta, P.; Trichopoulou, A.; Bobak, M.; et al. Adherence to a Healthy Diet According to the World Health

Organization Guidelines and All-Cause Mortality in Elderly Adults From Europe and the United States. *Am. J. Epidemiol.* **2014**, *180*, 978–988. [CrossRef] [PubMed]

5. Kiss, A.; Popp, J.; Oláh, J.; Lakner, Z. The Reform of School Catering in Hungary: Anatomy of a Health-Education Attempt. *Nutrients* **2019**, *11*, 716. [CrossRef] [PubMed]

6. World Health Organization. World health statistics 2019: Monitoring Health for the SDGs, Sustainable Development Goals. Available online: https://www.who.int/gho/publications/world_health_statistics/2019/en/ (accessed on 18 February 2020).

7. OECD/European Observatory on Health Systems and Policies (2017), Hungary: Country Health Profile 2017, State of Health in the EU, OECD Publishing, Paris/European Observatory on Health Systems and Policies, Brussels. Available online: http://dx.doi.org/10.1787/9789264283411-en (accessed on 18 February 2020).

8. Eurostat Healthy Life Years statistics—Statistics Explained. Available online: https://ec.europa.eu/eurostat/statistics-explained/index.php?title=Healthy_life_years_statistics#Healthy_life_years_at_birth (accessed on 18 February 2020).

9. Eurostat. *Ageing Europe—Looking at the Lives of Older People in the EU*, Statistical books, 2019th ed.; Publications Office of the European Union: Luxembourg, 2019; ISBN 978-92-76-09815-7.

10. Barabási, A.-L.; Menichetti, G.; Loscalzo, J. The unmapped chemical complexity of our diet. *Nat. Food* **2019**, *1*, 33–37. [CrossRef]

11. Jacobs, D.R.; Tapsell, L.C. Food synergy: The key to a healthy diet. *Proc. Nutr. Soc.* **2013**, *72*, 200–206. [CrossRef] [PubMed]

12. Giacalone, D.; Wendin, K.; Kremer, S.; Frøst, M.B.; Bredie, W.L.P.; Olsson, V.; Otto, M.H.; Skjoldborg, S.; Lindberg, U.; Risvik, E. Health and quality of life in an aging population – Food and beyond. *Food Qual. Prefer.* **2016**, *47*, 166–170. [CrossRef]

13. Vermeir, I.; Verbeke, W. Sustainable Food Consumption: Exploring the Consumer "Attitude – Behavioral Intention" Gap. *J. Agric. Environ. Ethics* **2006**, *19*, 169–194. [CrossRef]

14. Reisch, L.; Eberle, U.; Lorek, S. Sustainable food consumption: An overview of contemporary issues and policies. *Sustain. Sci. Pract. Policy* **2013**, *9*, 7–25. [CrossRef]

15. Popp, J.; Oláh, J.; Kiss, A.; Lakner, Z. Food Security Perspectives in Sub-Saharan Africa. *Amfiteatru Econ. J.* **2019**, *21*, 361–376. [CrossRef]

16. Benedetti, I.; Laureti, T.; Secondi, L. Choosing a healthy and sustainable diet: A three-level approach for understanding the drivers of the Italians' dietary regime over time. *Appetite* **2018**, *123*, 357–366. [CrossRef]

17. EAT-Lancet Commission. Food Planet Health - Healthy Diets from Sustainable Food Systems. Available online: https://eatforum.org/eat-lancet-commission/eat-lancet-commission-summary-report/ (accessed on 18 February 2020).

18. Siró, I.; Kápolna, E.; Kápolna, B.; Lugasi, A. Functional food. Product development, marketing and consumer acceptance—A review. *Appetite* **2008**, *51*, 456–467.

19. Menrad, K. Market and marketing of functional food in Europe. *J. Food Eng.* **2003**, *56*, 181–188. [CrossRef]

20. Bigliardi, B.; Galati, F. Innovation trends in the food industry: The case of functional foods. *Trends Food Sci. Technol.* **2013**, *31*, 118–129. [CrossRef]

21. Urala, N.; Lähteenmäki, L. Reasons behind consumers' functional food choices. *Nutr. Food Sci.* **2003**, *33*, 148–158. [CrossRef]

22. European Commission. Regulation (EC) No 1924/2006 of the European Parliament and of the Council of 20 December 2006 on Nutrition and Health Claims Made on Foods. Available online: https://eur-lex.europa.eu/legal-content/EN/TXT/PDF/?uri=CELEX:02006R1924-20141213 (accessed on 18 February 2020).

23. European Union. Commission Regulation (EU) No 432/2012 of 16 May 2012 Establishing a List of Permitted Health Claims Made on Foods, other than Those Referring to the Reduction of Disease Risk and to Children's Development and Health. Available online: https://eur-lex.europa.eu/legal-content/EN/TXT/HTML/?uri=CELEX:32012R0432&from=EN (accessed on 18 February 2020).

24. Verhagen, H.; Vos, E.; Francl, S.; Heinonen, M.; van Loveren, H. Status of nutrition and health claims in Europe. *Arch. Biochem. Biophys.* **2010**, *501*, 6–15. [CrossRef]

25. Verhagen, H.; van Loveren, H. Status of nutrition and health claims in Europe by mid 2015. *Trends Food Sci. Technol.* **2016**, *56*, 39–45. [CrossRef]

26. Hilliam, M. The Market for Functional Foods. *Int. Dairy J.* **1998**, *8*, 349–353. [CrossRef]

27. Gilbert, L.C. The functional food trend: What's next and what Americans think about eggs. *J. Am. Coll. Nutr.* **2000**, *19*, 507S–512S. [CrossRef]

28. Childs, N.M. Functional foods and the food industry: Consumer, economic and product development issues. *J. Nutraceuticals Funct. Med. Foods* **1997**, *1*, 25–43. [CrossRef]

29. Verbeke, W. Consumer acceptance of functional foods: Socio-demographic, cognitive and attitudinal determinants. *Food Qual. Prefer.* **2005**, *16*, 45–57. [CrossRef]

30. Bech-Larsen, T.; Grunert, K.G. The perceived healthiness of functional foods: A conjoint study of Danish, Finnish and American consumers' perception of functional foods. *Appetite* **2003**, *40*, 9–14. [CrossRef]

31. Landström, E.; Hursti, U.-K.K.; Becker, W.; Magnusson, M. Use of functional foods among Swedish consumers is related to health-consciousness and perceived effect. *Br. J. Nutr.* **2007**, *98*, 1058–1069. [CrossRef] [PubMed]

32. Niva, M. 'All foods affect health': Understandings of functional foods and healthy eating among health-oriented Finns. *Appetite* **2007**, *48*, 384–393. [CrossRef] [PubMed]

33. Urala, N.; Lähteenmäki, L. Consumers' changing attitudes towards functional foods. *Food Qual. Prefer.* **2007**, *18*, 1–12. [CrossRef]

34. Szakály, Z.; Szente, V.; Kövér, G.; Polereczki, Z.; Szigeti, O. The influence of lifestyle on health behavior and preference for functional foods. *Appetite* **2012**, *58*, 406–413. [CrossRef]

35. Van Kleef, E.; van Trijp, H.C.; Luning, P. Functional foods: Health claim-food product compatibility and the impact of health claim framing on consumer evaluation. *Appetite* **2005**, *44*, 299–308. [CrossRef]

36. Van Trijp, H.C.; Van der Lans, I.A. Consumer perceptions of nutrition and health claims. *Appetite* **2007**, *48*, 305–324. [CrossRef]

37. Behrens, J.H.; Villanueva, N.D.; Da Silva, M.A. Effect of nutrition and health claims on the acceptability of soyamilk beverages. *Int. J. Food Sci. Technol.* **2007**, *42*, 50–56. [CrossRef]

38. Urala, N.; Arvola, A.; Lähteenmäki, L. Strength of health-related claims and their perceived advantage. *Int. J. Food Sci. Technol.* **2003**, *38*, 815–826. [CrossRef]

39. Banovic, M.; Arvola, A.; Pennanen, K.; Duta, D.E.; Brückner-Gühmann, M.; Lähteenmäki, L.; Grunert, K.G. Foods with increased protein content: A qualitative study on European consumer preferences and perceptions. *Appetite* **2018**, *125*, 233–243. [CrossRef]

40. Verbeke, W.; Scholderer, J.; Lähteenmäki, L. Consumer appeal of nutrition and health claims in three existing product concepts. *Appetite* **2009**, *52*, 684–692. [CrossRef]

41. Wortmann, L.; Enneking, U.; Daum, D. German Consumers' Attitude towards Selenium-Biofortified Apples and Acceptance of Related Nutrition and Health Claims. *Nutrients* **2018**, *10*, 190. [CrossRef] [PubMed]

42. Micale, R.; Giallanza, A.; Russo, G.; La Scalia, G. Selection of a Sustainable Functional Pasta Enriched with Opuntia Using ELECTRE III Methodology. *Sustainability* **2017**, *9*, 885. [CrossRef]

43. Sagan, A.; Blicharz-Kania, A.; Szmigielski, M.; Andrejko, D.; Sobczak, P.; Zawiślak, K.; Starek, A. Assessment of the Properties of Rapeseed Oil Enriched with Oils Characterized by High Content of α-linolenic Acid. *Sustainability* **2019**, *11*, 5638. [CrossRef]

44. Annunziata, A.; Vecchio, R. Consumer perception of functional foods: A conjoint analysis with probiotics. *Food Qual. Prefer.* **2013**, *28*, 348–355. [CrossRef]

45. van der Zanden, L.D.; van Kleef, E.; de Wijk, R.A.; van Trijp, H.C. Knowledge, perceptions and preferences of elderly regarding protein-enriched functional food. *Appetite* **2014**, *80*, 16–22. [CrossRef]

46. van der Zanden, L.D.; van Kleef, E.; de Wijk, R.A.; van Trijp, H.C. Understanding heterogeneity among elderly consumers: An evaluation of segmentation approaches in the functional food market. *Nutr. Res. Rev.* **2014**, *27*, 159–171. [CrossRef]

47. van der Zanden, L.D.; van Kleef, E.; de Wijk, R.A.; van Trijp, H.C. Examining heterogeneity in elderly consumers' acceptance of carriers for protein-enriched food: A segmentation study. *Food Qual. Prefer.* **2015**, *42*, 130–138. [CrossRef]

48. Doma, K.M.; Farrell, E.L.; Leith-Bailey, E.R.; Soucier, V.D.; Duncan, A.M. Older Adults' Awareness and Knowledge of Beans in Relation to Their Nutrient Content and Role in Chronic Disease Risk. *Nutrients* **2019**, *11*, 2680. [CrossRef]

49. Collins Orla; Bogue Joe Designing health promoting foods for the ageing population: A qualitative approach. *Br. Food J.* **2015**, *117*, 3003–3023. [CrossRef]

50. Plasek, B.; Lakner, Z.; Kasza, G.; Temesi, Á. Consumer Evaluation of the Role of Functional Food Products in Disease Prevention and the Characteristics of Target Groups. *Nutrients* **2020**, *12*, 69. [CrossRef]

51. Temesi, Á.; Bacsó, Á.; Grunert, K.G.; Lakner, Z. Perceived correspondence of health effects as a new determinant influencing purchase intention for functional food. *Nutrients* **2019**, *11*, 740. [CrossRef] [PubMed]

52. Dávid Szakos; Ózsvári László; Kasza Gyula Consumer demand analysis in the Hungarian functional food market focused on the main health problems. *Gradus* **2020**, *7*, 62–66.

53. Hungarian Central Statistical Office (HCSO). Hungarian Census Data 2016. Available online: https://www.ksh.hu/mikrocenzus2016/kotet_3_demografiai_adatok (accessed on 18 February 2020).

54. Clason, D.L.; Dormody, T.J. Analyzing data measured by individual Likert-type items. *J. Agric. Educ.* **1994**, *35*, 4. [CrossRef]

55. Grafen, A.; Hails, R. *Modern Statistics for the Life Sciences*, 2002nd ed.; Oxford University Press: Oxford, UK, 2002; Volume 351.

56. Hooker, N.H.; Teratanavat, R. Dissecting Qualified Health Claims: Evidence from Experimental Studies. *Crit. Rev. Food Sci. Nutr.* **2008**, *48*, 160–176. [CrossRef] [PubMed]

57. Ares, G.; Gámbaro, A. Influence of gender, age and motives underlying food choice on perceived healthiness and willingness to try functional foods. *Appetite* **2007**, *49*, 148–158. [CrossRef]

58. Williams, P.; Ridges, L.; Batterham, M.; Ripper, B.; Hung, M.C. Australian consumer attitudes to health claim—food product compatibility for functional foods. *Food Policy* **2008**, *33*, 640–643. [CrossRef]

59. Urala, N.; Lähteenmäki, L. Attitudes behind consumers' willingness to use functional foods. *Food Qual. Prefer.* **2004**, *15*, 793–803. [CrossRef]

60. Bimbo, F.; Bonanno, A.; Nocella, G.; Viscecchia, R.; Nardone, G.; De Devitiis, B.; Carlucci, D. Consumers' acceptance and preferences for nutrition-modified and functional dairy products: A systematic review. *Appetite* **2017**, *113*, 141–154. [CrossRef]

61. Messina, F.; Saba, A.; Turrini, A.; Raats, M.; Lumbers, M. Older people's perceptions towards conventional and functional yoghurts through the repertory grid method: A cross-country study. *Br. Food J.* **2008**, *110*, 790–804. [CrossRef]

62. Cavaliere, A.; Ricci, E.C.; Banterle, A. Nutrition and health claims: Who is interested? An empirical analysis of consumer preferences in Italy. *Food Qual. Prefer.* **2015**, *41*, 44–51. [CrossRef]

63. European Hearth Network European Cardiovascular Disease Statistics 2017. Available online: http://www.ehnheart.org/cvd-statistics.html (accessed on 18 February 2020).

64. Grunert, K.G.; Hieke, S.; Wills, J. Sustainability labels on food products: Consumer motivation, understanding and use. *Food Policy* **2014**, *44*, 177–189. [CrossRef]

65. Wilkins, J.L. Seasonality, food origin, and food preference: A comparison between food cooperative members and nonmembers. *J. Nutr. Educ.* **1996**, *28*, 329–337. [CrossRef]

14

Environmental Determinants of a Country's Food Security in Short-Term and Long-Term Perspectives

Alina Vysochyna [1], Natalia Stoyanets [2], Grzegorz Mentel [3,*] and Tadeusz Olejarz [4]

[1] Department of Accounting and Taxation, Sumy State University, 40000 Sumy, Ukraine; a.vysochyna@uabs.sumdu.edu.ua

[2] Faculty of Economics and Management, Sumy National Agrarian University, 40000 Sumy, Ukraine; natalystoyanez@gmail.com

[3] Department of Quantitative Methods, The Faculty of Management, Rzeszow University of Technology, 35-959 Rzeszow, Poland

[4] Department of Management Systems and Logistics, The Faculty of Management, Rzeszow University of Technology, 35-959 Rzeszów, Poland; olejarz@prz.edu.pl

* Correspondence: gmentel@prz.edu.pl

Abstract: About 10% of the world population suffered from hunger in 2018. Thereby, the main objective of this research is the identification of environmental drivers and inhibitors of a country's food security in the short and long run. The Food Security Index (FSI) was constructed from 19 indicators using Principal Component Analysis. Identification of the short- and long-run relationships between the FSI and environmental factors was realized with the pooled mean-group estimator for 28 post-socialistic countries for 2000–2016. Empirical research results showed that a country's food security in the short run is affected by greenhouse gas emissions but boosted by the increase of renewable energy production. Reduction of carbon dioxide emissions, electrification of rural populations, access to clean fuels, renewable energy production, arable land, and forest area growth might be essential tasks in order to ensure countries' food security in the long-run.

Keywords: food security; food availability; food access; food stability; food utilization; environmental determinants; sustainable development

1. Introduction

Global economic development at the end of the XX century led to the boosting of industrial and technological development. However, these processes also triggered numerous destructive trends, especially for the environment. In turn, the scale of the environmental problems needed cooperation of the global community to solve them, so the Agenda 21 and, recently, the Millennium Development Goals were developed in order to coordinate the efforts of different countries on the way to elimination of global damages and implementation of sustainable development. At the Millennium Summit in 2000, eight Millennium Development Goals were developed, aimed at poverty, hunger and child mortality reduction, decrease of different diseases, expansion of education, banning of gender inequality, triggering of cooperation of local community, and promotion of sustainable environment development. All of these goals have quantitative measures that needed to be achieved in order to fulfill the goals. Global community cooperation during the last decades allowed the partial fulfilment of these goals. Nevertheless, considering such achievements and newly appeared damages in 2015 at the United Nations General Assembly, the 17 Sustainable Development Goals by 2030 were introduced [1]. It is worth noting that most of the Sustainable Development Goals focus on food security or environmental issues that clarify their urgency and importance both at national (local) and supranational levels.

Elimination of hunger and different forms of malnutrition in order to overcome food insecurity continues to be an urgent global task because of the insufficient economic growth dynamics in different countries, climate change, existence of war conflicts and political instability zones, etc. Namely, according to the Food and Agriculture Organization of the United Nations (FAO) report [2] in 2000, there were 792 million people in 98 countries who met food insecurity problems, while, in 2018 [3], more than 820 million people were still suffering from hunger. Such a situation proves the extreme urgency of the need for the global community's cooperation in order to fulfill the Zero Hunger goal by 2030. Moreover, it is also essential to continue scientific research aiming at clarification of factors strengthening or worsening country food security. That might help to develop a more well thought out and scientifically grounded economic policy at both national and supranational levels.

Particularly, according to the FAO [4], nowadays, "food security exists when all people, at all times, have physical and economic access to sufficient, safe, and nutritious food that meets their dietary needs and food preferences for an active and healthy life". Moreover, in terms of the FAO approach, food security has four dimensions, namely food availability, food access, food stability, and food utilization. Food availability is about physical existence of foodstuffs of appropriate quality that might be supplied to the population. Food access characterizes the possibility of getting food considering legal, political, economic, and social conditions. Food utilization illustrates rationality and effectiveness of consumption, sanitation, and water access conditions. Food stability is about ensuring foodstuff provision at any time range, even in cases of insufficient economic situations or realization of some other risks [4]. As it becomes evident from the essence of the food security perspectives, some of them mostly dependent on economic conditions, but the majority of pillars are reliant on environmental preconditions. Consequently, environmental determinants play a crucial role in foodstuff production, distribution, and the quality of its consumption. However, the functioning of food-producing enterprises is quite often (especially in developing countries) accompanied by numerous adverse environmental effects (air pollution, soil degradation, elimination of certain species of flora and fauna, reduction of forest area, greenhouse gas emissions increase, etc.). On the contrary, spurring environmental problems would likely lead to an increase in food insecurity and disruption of sustainability of the national economy.

It should be noted that there is plenty of research that specifies the influence of social, economic, and environmental factors on a country's food security as a whole and on its perspectives separately, but they sometimes contradict each other. In addition, different groups of scientists focused on various environmental aspects and food security pillars, so it might be hard to see the situation comprehensively. Therefore, from both theoretical and empirical points of view, it is crucial to identify the impact of environmental (ecological) factors on a country's food security in the short-run and long-run perspectives using up-to-date data and scientific approaches. Specifically, this research aimed at clarifying several important issues:

(1)   Identification of the relevant environmental factors that influence a country's food security (we used to think that some environmental problems might damage foodstuff production, distribution, and consumption value, but the existence of contradictory empirical research results about such an impact reveals the necessity of further theoretical and empirical findings in this direction);

(2)   Comprehensiveness: As a rule, empirical research is narrow and focused on the clarification of influence of some certain environmental determinant on a country's food security or its pillars; however, we try to consider the vast majority of potential environmental factors mentioned in previous empirical research; this approach might be useful from the regulatory perspective because it could help to identify the priorities of environmental, economic, and social government policy (to some extent);

(3)   Clarification of the short- and long-run impacts of environmental factors on a country's food security (basically, most of the empirical research is based on classical regression analysis and aimed at confirmation or rejection of some empirical hypothesis, but it should be taking into

consideration that environmental factors likely have no immediate influence on a country's food security; thus, it is by far more valuable to clarify this impact in different time perspectives).

Moreover, the food security concept originated in 1974 during the World Food Conference, but gained its modern features in 1996 at the World Food Summit [4]. Despite the conceptual clarification from the mid-1990s of the XX century, the possibility of tracking countries' progress in terms of food security appeared only in 2012 with the launching of the Global Food Security Index. Thus, there is no considerable amount of similar research results aimed at testing the influence of the environmental factors on a proxy of countries' food security, especially in different time perspectives.

Consequently, this research might have significant theoretical and empirical value both in terms of development of countries' environmental and food security policies and tracking of changes the environmental determinants of countries' food security.

2. Literature Review

In order to fulfill the task of comprehensiveness of the research, it is necessary to generalize potential environmental determinants influencing a country's food security (alternatively, foodstuff production, agribusiness performance, etc.) that were previously mentioned by scientists. Basically, some theoretical and empirical findings confirm the hypothesis about social, economic, or environmental factors' impacts on countries' food security as a whole or its particular perspective.

It should be noted that there is a set of scientific research that, in general terms, supports the hypothesis about the influence of environmental factors on a country's food security or its proxies. Namely, Musová, Musa, and Ludhova [5], Dwikuncoro and Ratajczak [6], and Vasa [7] researched factors influencing food purchasing (food utilization) in the Slovak Republic, Poland, and Hungary. They found out that consumer behavior is mostly driven by economic factors (quality and prices of products, household income). However, environmental factors also matter—69% of respondents mentioned that they prefer environmentally friendly goods. Moreover, Jakubowska and Radzymińska [8] found out that Czech students, who participated in the research, declare environmental motives as dominant in their consumer choices. Dabija, Bejan, and Dinu [9] also identified that consumers of Generation Z prefer green suppliers. In turn, Gadeikienė, Dovalienė, Grase, and Banytė [10], Arslan [11], Olasiuk and Bhardwaj [12], and Ahmad [13] reveal that environmental preconditions and comprehensive nutrition knowledge play an important role in ensuring sustainable consumption. Thus, this group of scientists supports the idea that environmental image and responsibility are impactful for food consumption (food utilization proxy of a country's food security).

In terms of discussing the impact of environmental determinants on the performance of food producers and foodstuff trade, i.e., food availability and partial food access, Morkūnas, Volkov, and Pazienza [14], Morkūnas et al. [15], and Tomchuk et al. [16] mentioned that economic and environmental factors have an impact for resilience of agricultural enterprises. Similarly, Handayani, Wahyudi, and Suharnomo [17], Mikhaylova et al. [18], Akhtar [19], Kheyfets and Chernova [20], Stjepanović, Tomić, and Škare [21], Cismas et al. [22], Jayasundera [23], and Harold [24] proved that green innovations positively influence business performance, sustainability of agriculture, and food security. Haninun, Lindrianasari, and Denziana [25] mentioned that environmental performance has an effect on financial performance. Ortikov, Smutka, and Benešován [26] reveal that increase of innovativeness and eco-friendliness might be among essential preconditions of an increase of competitiveness of Uzbekistan's agrarian foreign trade. However, Shuquan [27] empirically proved the existence of the relationship between international trade and countries' environmental performance (case of China). In turn, Smutka, Maitah, and Svatoš [28], Falkowski [29], and Kadochnikov and Fedyunina [30] pointed out that, in the case of Russian foodstuff imports, not environmental, but economic and political factors matter. However, in the case of Russia's exports to EU countries, political and environmental determinants play a more significant role. This block of research supports the idea that eco-friendliness and environmental responsibility are not just influencing consumers' motives, but also argue that agricultural enterprises are also driven by environmental motives. Nevertheless, these researches also allow us to conclude that environmental factors play a prior role in foodstuff trade in developed countries, but a secondary role in developing countries.

The third set of researches is mainly focused on clarification of state regulations' influence on a country's food security. In turn, Krajnakova, Navickas, and Kontautiene [31] mentioned that environmental regulation might be a trigger of a country's competitiveness and sustainability. Similarly, Grenčíková et al. [32], Bilan et al. [33–36], Lyulyov et al. [37], Akhmadeev et al. [38], Bhandari [39], Bello, Galadima, and Jibrin [40], Sokolenko, Tiutiunyk, and Leus [41], Lizińska, Marks-Bielska, and Babuchowska [42], Vacca and Onishi [43], Kostyuchenko et al. [44], and Popp et al. [45] found out that different environment-related institutional factors significantly influence countries' sustainability and food security.

Previous parts of the literature review proved the hypothesis that environmental (ecological) factors, in general terms, do influence a country's food security and its perspectives. Moreover, this allows the revelation that environmental responsibility is triggered by regulatory and institutional preconditions and is an essential determinant of consumer choice and agricultural business performance. Thus, it creates a background for more in-depth analysis regarding the identification of specific environmental factors that have impacts on a country's sustainable development and food security. In this perspective, it should be mentioned that Vasylyeva and Pryymenko [46], Mekhum [47], Lu et al. [48], Androniceanu and Popescu [49], Lyeonov et al. [50], Abdimomynova et al. [51], and Mentel et al. [52] clarify renewable energy production and consumption as among key environmental determinants. Additionally, Aitkazina et al. [53] pointed out that an increase in greenhouse gas emissions by agrarian enterprises and expansion of use of chemical fertilizers create threats for sustainable development and, consequently, a country's food security. Similarly, Sibanda and Ndlela [54], Dkhili and Dhiab [55], Mačaitytė and Virbašiūtė [56], and Odermatt [57] also argue that increase of carbon emissions negatively influences company performance, countries' food security, and sustainability. In turn, Vasylieva [58] mentioned that a country's food security is dependent on yields, rational land use, development of innovations, and infrastructure. However, Aliyas, Ismail, and Alhadeedy [59] supposed that a country's food security and agricultural sustainability are based on environmental friendliness, decrease of chemical fertilizers, and effective ecological state policy.

Consequently, a comprehensive analysis of the theoretical and empirical research results aimed at clarifying factors affecting countries' food security leads to the conclusion that economic factors are still among key determinants of foodstuff consumption (it mostly depends on prices of goods and household income) and agribusiness performance (as a key sphere of food production and distribution). At the same time, there is a considerable block of research proving that the influence of ethical, institutional, and specific environmental factors on a country's food security become more significant. In turn, among major environmental determinants affecting a country's food security, scientists mention water and soil usage, energetic issues (expansion of renewable and traditional energy production and consumption), greenhouse gas emission, fertilizer usage, etc. Nevertheless, the influence of these factors on a country's food security is revealed, but scientists have no unified position about the scale and character of such an impact, so it might be valuable, from both theoretical and practical perspectives, to identify which factors are more influential in the long run and which in the short run.

3. Materials and Methods

Previous studies [60] were mainly related to primary empirical research. Specifically, they allowed the identification of the potential blocks of environmental determinants affecting a country's food security, such as: (1) Measures concerning natural resource availability and usage; (2) energy production and consumption items; (3) fertilizer usage; (4) greenhouse gas emissions by agricultural enterprises; (5) parameters of agribusiness yield. In turn, as a result of this literature review, a set of 37 environmental determinants was collected from the World Bank DataBank [61] and the United Nations Environment Program Data Explorer [62]. Correlation analysis helped to select the most influential factors and eliminate multicollinearity problems. It allowed the choosing of 14 out of 37 environmental factors. Additionally, two of these 14 variables were eliminated because they had negative influences on regression model quality parameters. Therefore, previous research [60] helped to clarify a set of environmental factors that do have an impact on a country's food security.

The realization of this research task implied the need for several stages: (1) Construction of the comprehensive food security indicator; (2) identification of certain ecological factors influencing food security in the short and long run.

In general terms, the research was based on data collected from public sources (the World Bank DataBank [61], the United Nations Environment Programme Data Explorer [62], and the Food and Agriculture Organization of the United Nations database (FAOSTAT) [63]) for 28 post-socialistic countries (Albania, Armenia, Azerbaijan, Belarus, Bosnia and Herzegovina, Bulgaria, Croatia, Czech Republic, Estonia, Georgia, Hungary, Kazakhstan, Kyrgyz Republic, Latvia, Lithuania, Macedonia, Moldova, Montenegro, Poland, Romania, Russia, Serbia, Slovak Republic, Slovenia, Tajikistan, Turkmenistan, Ukraine, and Uzbekistan) from 2000 to 2016.

As for the first stage, it might be noted that The Economist in cooperation with the FAO have developed the Global Food Security Index, which consists of 28 measurement indicators of affordability, availability, quality, and safety of food. Nevertheless, this index has been calculated from 2012, which is too small a period for gaining reliable modeling results. That is why the Food Security Index (FSI) was constructed. The FSI consists of 19 indicators of food availability, food access, food stability, and food utilization. The FAO officially identifies these parameters as measures of food security. The descriptions of the indicators used for the FSI's construction are in Table 1.

Table 1. Measurement indicators of the Food Security Index (FSI).

Perspective of Food Security	Indicators of Food Security Measurement
Food availability	Average dietary energy supply adequacy, % (ADESA); Average value of food production, USD per capita (FoodProd); Share of dietary energy supply derived from cereals, roots, and tubers, % (CRT); Average protein supply, gr/capita/day (Protein); Average supply of proteins of animal origin, gr/capita/day (AnProt).
Food access	Rail line density, total route in km per 100 square km of land area (Railway); GDP per capita, USD (GDPpc); Prevalence of undernourishment, % (Under); Depth of the food deficit, kcal/capita/day (FoodDef).
Food stability	Cereal import dependency ratio, % (Cereals); Percentage of arable land equipped for irrigation, % (Irrig); Value of food imports over total merchandise exports, % (ImEx); Political stability and absence of violence, index (PolStab); Per capita food production variability, thousand USD (FPV); Per capita food supply variability, kcal/capita/day (FoodSup).
Food utilization	Percentage of population with access to improved drinking water sources, % (ImWater); Percentage of population with access to sanitation facilities, % (Sanit); Prevalence of obesity in the adult population (18 years and older), % (Obesity); Prevalence of anemia among women of reproductive age (15-49 years), % (Anemia).

The FAO does not clarify a certain algorithm for aggregation of food availability, food access, food stability, and food utilization indicators. Therefore, the Principal Component Analysis (PCA) in Stata software was used to realize this particular task. Namely, the eigenvalues of the first principal component were used as weighted coefficients for the FSI's construction. It is worth noting that we use the PCA method rather than the Analytic Hierarchy Process (AHP) because it is a rather complicated task for realizing pairwise judgments to prioritize measures of food security on a scale of 1 to 9. Thus, we decided to apply not a subjective, but a more objective method (PCA), which aimed at clarification of data trends and identification of weight coefficients based on them [64]. In addition, before applying the PCA, all of the above-mentioned indicators were primarily normalized considering their stimulating or unstimulating influence on the state of countries' food security. The normalization process allows us to arrange them from 0 to 1.

In turn, the second stage of the research is focused on the identification of environmental determinants influencing a country's food security in short- and long-run perspectives. As the research sample includes rather huge number of observations, both in terms of periods, countries, and independent variables (panel data sample), a pooled mean-group (PMG) estimator, developed by Pesaran, Shin, and Smith [65], was used. Traditionally, in research based on panel data with a large number of cross-sections but a small number of time observations, fixed effects are applied, as well as random effects estimators or generalized method of moments. However, an increase in the number of time observations might result in non-stationarity. As this research covers a rather large number of cross-sectional observations and time observations, it is better to apply the PMG estimator. Moreover, this research method allows us to manage the problem of non-stationarity and better fits heterogeneous panels. In addition, the PMG estimator considers both pooling and averaging approaches (it allows short-run coefficients to differ across countries, but long-run coefficients might be equal for the whole panel). Thus, it helps to mix some technical aspects from the mean group estimator and fixed effects estimator [66].

The PMG estimator allows testing of the hypothesis about the existence of influence on food security (specifically, the FSI) in the long-term and short-term perspectives of the following environmental indicators: X1—access to clean fuels and technologies for cooking (% of population); X2—access to electricity in rural areas (% of rural population); X3—agricultural methane emissions (% of total); X4—agricultural nitrous oxide emissions (% of total); X5—arable land (% of land area); X6—cereal yield (kg per hectare); X7—CO_2 emissions (metric tons per capita); X8—electric power transmission and distribution losses (% of output); X9—electricity production from renewable sources, excluding hydroelectric (% of total); X10—fertilizer consumption (kilograms per hectare of arable land); X11—forest area (% of land area); X12—renewable electricity output (% of total electricity output). The summative statistics for the set of dependent and independent variables are in Table 2.

Table 2. Summative statistics for the set of variables.

Variable	Obs	Mean	Std. Dev.	Min	Max
FSI	448	1.295	0.565	0.16	2.25
X1	476	82.774	16.484	38.07	100.00
X2	476	99.724	0.519	95.68	100.00
X3	237	36.122	16.434	0.00	75.29
X4	237	63.263	15.924	0.00	87.68
X5	464	23.172	15.259	0.58	56.23
X6	464	3310.283	1250.055	804.10	6742.3
X7	465	5.642	3.669	0.29	17.31
X8	471	13.835	7.341	1.82	72.90
X9	471	1.553	3.677	0.00	29.99
X10	364	95.155	85.911	0.84	495.23
X11	476	29.539	17.141	1.23	62.12
X12	476	29.985	30.526	0.00	100.00

Notes: X1—access to clean fuels and technologies for cooking (% of population); X2—access to electricity in rural areas (% of rural population); X3—agricultural methane emissions (% of total); X4—agricultural nitrous oxide emissions (% of total); X5—arable land (% of land area); X6—cereal yield (kg per hectare); X7—CO_2 emissions (metric tons per capita); X8—electric power transmission and distribution losses (% of output); X9—electricity production from renewable sources, excluding hydroelectric (% of total); X10—fertilizer consumption (kilograms per hectare of arable land); X11—forest area (% of land area); X12—renewable electricity output (% of total electricity output); Obs—amount of observations; Std. Dev.—Standard deviation.

Based on the results presented in Table 2, it should be noted that the number of observations differs for some variables. Nevertheless, the panel is strongly balanced, which allows us to get reliable and significant empirical research results.

4. Results

Taking into account weight coefficients (Table 3), the FSI was constructed with the PCA approach. It is also worth noting that the calculated FSI is quite representative. Its comparison with the Global Food Security Index for those 13 countries, which are matched in both samples (Belarus, Kazakhstan, Poland, Hungary, Poland, Hungary, Poland, Hungary, Russia, Serbia, Slovakia, Tajikistan, Ukraine, and Uzbekistan) for the years 2012–2016, revealed a correlation of 90.20%. Consequently, the FSI allows the characterization of the same trends as those displayed by the Global Food Security Index.

Table 3. Weight coefficients of indicators of Food Security Index.

Indicator	Eigenvalue	Indicator	Eigenvalue	Indicator	Eigenvalue
ADESA	0.2571	Under	0.2779	FoodSup	−0.0198
FoodProd	0.2452	FoodDef	0.2266	ImWater	0.2247
CRT	−0.2904	Cereals	0.1765	Sanit	0.0920
Protein	0.2994	Irrig	−0.2673	Obesity	−0.2855
AnProt	0.2761	ImEx	0.1203	Anemia	0.1908
Railway	0.2126	PolStab	0.2416		
GDPpc	0.2834	FPV	0.1295		

Analysis of the FSI level in 2016 shows that the highest level of food security is in the Czech Republic (2.25 from 2.39), and the lowest is in Tajikistan (0.16). It is also worth noting that such countries as Albania, Armenia, Azerbaijan, Bosnia and Herzegovina, Georgia, Kyrgyz Republic, Macedonia, Moldova, Serbia, Tajikistan, Turkmenistan, Ukraine, and Uzbekistan have less-than-average levels of national food security. The rest of the countries have higher-than-average levels of national food security. In terms of the characteristics of the dynamics of the FSI level, it might be highlighted that Azerbaijan (566.19%), Tajikistan (520.97%), Uzbekistan (182.79%), Armenia (178.68%), Turkmenistan (97.80%), Georgia (83.93%), and Albania (74.63%) have the best growth dynamics in comparison with 2001, while for the other countries, the growth rate fluctuates in almost the same range (about 31.66%).

The next step is the identification of the relationship between the relevant environmental determinants and the FSI. It is based on the panel data regression analysis (PMG estimator). Practically, it was implemented with the help of the "xtpmg" add-on of the Stata software. The results of the regression analysis are given in Table 4.

Therefore, the following conclusions can be made. The vast majority of the environmental factors have a statistically significant long-term impact on countries' food security (significant at the 10%, 5%, or 1% level). Environmental determinants that have no statistically significant impact on the FSI level in the long-term perspective are as follows: Agricultural methane emissions (% of total emissions) (X3); agricultural nitrous oxide emissions (% of total emissions) (X4); cereal yield (kg per hectare) (X6); electric power transmission and distribution losses (% of output) (X8). Thus, the absence of a statistically significant impact of the growth of greenhouse gas emissions by agricultural enterprises on the level of countries' food security is mostly explained by the intensified efforts of the world community on the reduction of such emissions (according to the Kyoto Protocol, countries are obliged to reduce greenhouse gas emissions by 2100). Additionally, agro-industrial enterprises provide only 10%–12% of the total emissions, while transport, industrial, construction, and energy enterprises have a greater impact on the ecosystem. The reduction in the net carbon dioxide emissions of the agro-industrial sector was largely explained by the decline of deforestation and the increase in forest plantations.

However, the increase of carbon dioxide emissions per capita from all sources of pollution (X7) remains a strong factor of the negative impact on countries' food security in the long-run perspective. Namely, an increase of this independent variable by a point results in a decrease of country food security level by 0.0886 (or 3.71% of the maximum possible FSI value).

Table 4. Results of identifying short- and long-run coefficients of environmental factors' influence on the FSI.

Variable	Coefficient	Std. Deviation	Z	P > |z|
	Long-Run Perspective			
X1	0.0105	0.0076	1.38	0.168
X2	0.0811	0.0513	1.58	0.114
X3	−0.0055	0.0127	−0.43	0.667
X4	0.0041	0.0050	0.82	0.410
X5	0.0285	0.0175	1.62	0.104 *
X6	0.0000	0.0000	0.87	0.384
X7	−0.0886	0.0248	3.57	0.000 ***
X8	0.0028	0.0019	1.40	0.161
X9	−0.0262	0.0168	−1.56	0.218
X10	0.0020	0.0009	2.23	0.026 **
X11	0.0948	0.0561	1.69	0.091 *
X12	0.0154	0.0054	2.86	0.004 ***
	Short-Run Perspective			
X1	−0.0134	0.0180	−0.74	0.457
X2	−0.0101	0.0074	−1.36	0.173
X3	−0.0109	0.0041	−2.86	0.007 ***
X4	0.0001	0.0007	0.12	0.906
X5	−0.0024	0.0049	−0.48	0.629
X6	-4.86×10^{-6}	5.20×10^{-6}	−0.93	0.350
X7	−0.0190	0.0082	−2.31	0.021 **
X8	−0.0001	0.0004	−0.31	0.755
X9	0.0181	0.0043	4.26	0.000 ***
X10	−0.0003	0.0003	−0.91	0.364
X11	0.0514	0.0619	0.83	0.406
X12	−0.0021	0.0010	−2.08	0.038 **
Cons	−3.6307	1.2189	−2.98	0.003 ***
ec	−0.2920	0.0863	−3.38	0.001 ***
Countries	28	28	28	28
Observations	476	476	476	476

Notes: X1—access to clean fuels and technologies for cooking (% of population); X2—access to electricity in rural areas (% of rural population); X3—agricultural methane emissions (% of total); X4—agricultural nitrous oxide emissions (% of total); X5—arable land (% of land area); X6—cereal yield (kg per hectare); X7—CO_2 emissions (metric tons per capita); X8—electric power transmission and distribution losses (% of output); X9—electricity production from renewable sources, excluding hydroelectric (% of total); X10—fertilizer consumption (kilograms per hectare of arable land); X11—forest area (% of land area); X12—renewable electricity output (% of total electricity output); *—significance at 10% level; **—significance at 5% level; ***—significance at 1% level.

In turn, some factors have a positive impact on the countries' food security, such as:

- Access to clean fuels and technologies for cooking (% of population) (X1)—an increase of the environmental factor by a point results in strengthening of a country's food security by 0.0105 (or 0.43% of the maximum possible FSI value);

- Access to electricity in rural areas (% of rural population) (X2)—an increase of the environmental factor by a point results in strengthening of a country's food security by 0.0811 (or 3.39% of the maximum possible FSI value), which means that further electrification of rural areas using environmentally friendly technologies should be a priority direction of public policy.

This statement is also confirmed by a positive and statistically significant impact of expanding renewable electricity output (% of total electricity output) (X12) on the country's food security in the long-run perspective. Namely, its increase by a point leads to strengthening of a country's food security by 0.0154 (or 0.64% of the maximum possible FSI value). Experts note [67–69] that the expansion of land for growing biofuel plants might have some negative consequences. It leads to the elimination of the land from the process of food production and may harm a country's food security. Consequently, this damage might not be offset by the positive environmental impact of using biofuels instead of traditional fuels.

In addition, the statistical significance of the long-term effects of arable land growth (X5) and forest area growth (X11) was confirmed at the 10% level. Particularly, an increase by a point of one of these particular environmental factors (X5 and X11) results in an increase in a country's food security by 0.0285 and 0.0948, respectively. Such a trend is quite natural, since the expansion of arable land will increase the volume of food products. However, such a scenario can have negative consequences and requires a well-thought-out and scientifically grounded approach. In particular, an intensive approach to the agricultural sector's development is preferable. It helps to ensure an increase of agricultural production without large-scale use of additional land resources. It is also equally important to use the most environmentally friendly tools for increasing agribusiness productivity and yields. While there is no widespread expansion of an intensive model of agricultural management, extensive technologies still do not lose their relevance. This is also confirmed by the statistically significant impact of the indicator "fertilizer consumption (kilograms per hectare of arable land)" (X10) on a country's food security (at the 5% level). Its increase by a point results in the FSI increase by 0.0020 (0.08% of maximum FSI value).

It is worth noting that most of the short-run coefficients are not statistically significant. However, the variables "agricultural methane emissions (% of total)" (X3) and "CO_2 emissions (metric tons per capita)" (X7) have a statistically significant negative impact on the food security index at the 1% and 5% levels, respectively. In addition, the positive impact of growth in electricity production from renewable sources is confirmed (both without hydroelectric power—variable X9 and with hydroelectric power—variable X12).

However, in most cases, the particular environmental factors are statistically significant only in short or long run. Consequently, we cannot compare statistically significant results with insignificant ones. Hence, we mainly focused only on the analysis and practical implications of only statistically significant research results. Nonetheless, it is worth noting that the increase in renewable electricity output (% of total electricity output) (X12) has a positive long-term but negative short-term influence on a country's food security. These findings might be partially explained by the specificity of the sample of countries. Namely, most of 28 post-socialistic countries have triggered more intensive economic, environmental, and technological development only for the last three decades. That is the main reason for the absence of a highly productive network of renewable energy stations. Consequently, the expansion of renewable electricity output leads to an immediate negative impact on a country's food security because of the partial elimination of land and water resources from foodstuff production and the worsening of its quality. Otherwise, in the long run, renewable energy outcompetes traditional energy production, which is more harmful to the environment and countries' food security. Familiar trends were also mentioned in the FAO report "Impacts of Bioenergy on Food Security" [70].

In turn, the increase of CO_2 emissions negatively influences a country's food security both in the short and long run. However, the scale and significance of this factor's effect become more influential in the long-term perspective.

5. Discussion

Aggregation of these empirical research results aimed at the identification of the influence of environmental (ecological) factors on a country's food security in short- and long-run perspectives allows the confirmation of trends and cohesions identified by other scientists. Specifically, Sola et al. [71] analyzed 132 articles about the influence of access to clean fuels and technologies for cooking on food security measures. Researchers mentioned that, in general, most of the scientists argued that this factor has a positive impact on food security and nutrition. However, there are no numerous empirical pieces of evidence of it. However, our research results allow us to quantitatively clarify such an impact: An increase of the factor by a point results in the strengthening of a country's food security by 0.0105 in the long run. Moreover, the FAO [72] also actively supports the idea that access to clean fuels leads to better nutrition and less environmental damage.

In addition, our empirical results about the impact of access to electricity in rural areas on food security also correlate with the FAO's findings. Namely, in publication [72], it is mentioned that access to electricity is crucial for a country's food security because electricity is necessary at each stage of foodstuff production. Moreover, access to electricity in rural areas might become a driver of agricultural productivity, efficiency, and food security.

In turn, Wambua, Omoke, and Telesia [73] found empirical pieces of evidence that lack of arable lands and other familiar resources are preconditions of food insecurity in Kenya. Mbuthia, Kioli, and Wanjala [74] highlighted the importance of the other resource factors. Namely, they revealed that the prohibition of cutting trees (forest areas) has a positive influence on household food security. Thereby, our research results form empirical evidence of the relationships that were previously identified at a theoretical level.

Moreover, Wambua, Omoke, and Telesia [73] also revealed that using animal manure or industrial fertilizers allows an increase in agricultural crops. Hence, the authors pointed out that households using fertilizers for agricultural issues did not face the problem of food insecurity even in periods of unfavorable weather and climate conditions (based on 66 households' self-assessment). In our research, the hypothesis about the long-run positive impact of fertilizer consumption on a country's food security measures was also confirmed.

In the research, it was revealed that CO_2 emissions have a negative influence on a country's food security, as was also highlighted in other research by Sibanda and Ndlela [54], Dkhili and Dhiab [55], Mačaitytė and Virbašiūtė [56], and Odermatt [57].

Finally, empirical findings about the positive influence of renewable energy output on a country's food security were also proved by other scientists' and international organizations' reports, such as the International Renewable Energy Agency (IRENA) [75]. Namely, it is noted in the report that the increase of renewable energy has crucial importance because of several reasons:

- Electricity itself plays an important role in households' everyday life and agriculture business activity, while it is necessary for foodstuff production, storing, and distribution processes;
- Renewable energy and electricity allow the decrease of consumption of fossil fuels, both for private and business purposes;
- Substitution of traditional electricity production with renewable electricity production might help to solve some environmental problems, especially in terms of reduction of greenhouse gas emission;
- Renewable energy's prevalence in comparison with traditional energy sources is more fit for the Sustainable Development Goals, especially in terms of Goal #7: "Ensure access to affordable, reliable, sustainable and modern energy for all".

In terms of the practical implications of the empirical research results, they might become a background for the development of states' economic, social, and environmental policies in order to ensure countries' food security. Moreover, it also might be useful for the identification of the strategic and operational priorities of public policy.

In terms of further research perspectives, it might be noted that certain environmental determinants may be relevant to the general level of food security, but may not have a statistically significant effect on its components. Therefore, it is also important to identify specific environmental stimulants and inhibitors in terms of ensuring food availability, food access, food stability, and food utilization.

6. Conclusions

Thus, it can be concluded that this empirical research aimed at the identification of factors affecting countries' food security in short- and long-run perspectives allows us to confirm previous empirical research results and theoretical findings (especially about the influence of CO_2 emissions, sufficiency of arable lands, forest areas, and other natural resources, access to electricity, and use of fertilizers). On the other hand, results that were revealed allowed us to obtain empirical evidence and quantitatively clarify the kinds of relationships that were identified mostly on a theoretical level (about influence of access to clean fuels and technologies for cooking).

Therefore, taking into account the results obtained regarding the impact of environmental determinants on countries' food security in short- and long-run perspectives for the 28 former socialist countries, the following can be noted:

- The main operational target in terms of ensuring a country's food security might be an intensification of efforts in reducing greenhouse gas emissions (both methane and carbon dioxide), as well as the reorientation towards the production and consumption of electricity from renewable sources rather than traditional ones, which are more destructive to the ecosystem (in countries where the use of alternative energy sources is limited, a possible solution of the problem may be reducing the number of cogeneration and nuclear power plants in favor of hydroelectric power plants);
- Among the key vectors of mitigating the long-run risks of deterioration of a country's food security can be mentioned the following: Intensification of efforts to reduce carbon dioxide emissions not only in the agricultural sector, but also in the industrial sector; continuation of rural electrification and the provision of environmentally friendly fuels and electricity sources to the population, with the reorientation from traditional sources of energy production towards alternative ones; growth of arable land (or more effective usage of the existing ones) and increasing forest areas, while moving to intensive rather than extensive agricultural management (using fewer resources in order to ensure bigger yields).

Consideration of these proposals might become a basis for the development of state policies in the field of ensuring national food security.

Despite the fact that the obtained empirical results correlate with previous empirical findings, and that, on their basis, some practical recommendations that might be used by governmental authorities while ensuring country food security were developed, there are some limitations of this research, such as: (1) The sample of Countries consists of only 28 post-socialist countries, so expansion of the sample of countries might help to get more comprehensive, complex, and reliable results; (2) other than the expansion of the country sample, it might be valuable to realize cluster analysis and specify recommendations for certain clusters; (3) as the Global Food Security Index, which is considered as a unified proxy of countries' food security, covers the period starting from 2012, it is too small for reliable empirical results; thus, despite constructing our own index, a better option may be the use of the methodology of the Global Food Security Index in order to get more reliable assessments.

Moreover, this research was aimed at identification of specific environmental determinants that influence a country's food security in short- or long-run perspectives, but in order to develop efficient public policy in terms of ensuring country food security, lags of postponed impact of environmental determinants on the FSI might be specified.

Author Contributions: Conceptualization, A.V. and N.S.; methodology, N.S.; software, A.V. and G.M.; formal analysis, investigation, and resources, A.V. and G.M.; writing—original draft preparation, N.S. and T.O.; writing—review and editing, A.V.; visualization, N.S. and T.O.; supervision, A.V. All authors have read and agreed to the published version of the manuscript.

References

1. Sustainable Development Goals. Available online: https://sustainabledevelopment.un.org/?menu=1300 (accessed on 18 January 2020).

2.   Food Insecurity: When People Live with Hunger and Fear Starvation. The State of the Food Insecurity in the World 2000. Available online: http://www.fao.org/FOCUS/E/SOFI00/img/sofirep-e.pdf (accessed on 18 January 2020).

3.   The State of Food Security and Nutrition in the World 2019–Safeguarding Against Economic Slowdowns and Downturns. Available online: http://www.fao.org/3/ca5162en/ca5162en.pdf (accessed on 18 January 2020).

4.   Food and Agriculture Organization of the United Nations. Available online: http://fao.org (accessed on 18 January 2020).

5.   Musová, Z.; Musa, H.; Ludhova, L. Environmentally responsible purchasing in Slovakia. *Econ. Sociol.* **2018**, *11*, 289–305. [CrossRef]

6.   Dwikuncoro, R.A.; Ratajczak, S. Analysis of green activities impact on purchase intention. *Pol. J. Manag. Stud.* **2019**, *20*, 159–167. [CrossRef]

7.   Vasa, L. Economic coherences between food consumption and income conditions in the Hungarian households. *Ann. Agrar. Sci.* **2005**, *1*, 228–232.

8.   Jakubowska, D.; Radzymińska, M. Health and environmental attitudes and values in food choices: A comparative study for Poland and Czech Republic. *Oeconomia Copernic.* **2019**, *10*, 433–452. [CrossRef]

9.   Dabija, D.C.; Bejan, B.M.; Dinu, V. How sustainability oriented is generation z in retail? A literature review. *Transform. Bus. Econ.* **2019**, *18*, 140–156.

10. Gadeikienė, A.; Dovalienė, A.; Grase, A.; Banytė, J. Sustainable Consumption Behaviour Spill-Over from Workplace to Private Life: Conceptual Framework. *Pol. J. Manag. Stud.* **2019**, *19*, 142–154.

11. Arslan, Y. Exploring the Effects of Consumers' Nutritional Knowledge and Information Interest on the Acceptance of Artificial Sweetener Usage in Soft Drinks. *Mark. Manag. Innov.* **2019**, *3*, 33–44. [CrossRef]

12. Olasiuk, H.; Bhardwaj, U. An Exploration of Issues Affecting Consumer Purchase Decisions towards Eco-friendly Brands. *Mark. Manag. Innov.* **2019**, *2*, 173–184. [CrossRef]

13. Ahmad, J. Analyzing the Employee Satisfaction and Demand vs Fulfillment of the Food and Beverage Sector in Bangladesh. *Bus. Ethics Leadersh.* **2018**, *2*, 74–83. [CrossRef]

14. Morkūnas, M.; Volkov, A.; Pazienza, P. How Resistant is the Agricultural Sector? Economic Resilience Exploited. *Econ. Sociol.* **2018**, *11*, 321–332. [CrossRef]

15. Morkūnas, M.; Volkov, A.; Bilan, Y.; Raišienė, A.G. The role of government in forming agricultural policy: Economic resilience measuring index exploited. *Adm. Si Manag. Public* **2018**, *31*, 111–131. [CrossRef]

16. Tomchuk, O.; Lepetan, I.; Zdyrko, N.; Vasa, L. Environmental activities of agricultural enterprises: Accounting and analytical support. *Econ. Ann. XXI* **2018**, *169*, 77–83. [CrossRef]

17. Handayani, R.; Wahyudi, S.; Suharnomo, S. The effects of corporate social responsibility on manufacturing industry performance: The mediating role of social collaboration and green innovation. *Bus. Theory Pract.* **2017**, *18*, 152–159. [CrossRef]

18. Mikhaylova, A.; Mikhaylov, A.; Savchina, O.; Plotnikova, A. Innovation landscape of the Baltic region. *Adm. Manag. Public* **2019**, *33*, 165–180. [CrossRef]

19. Akhtar, P. Drivers of Green Supply Chain Initiatives and their Impact on Economic Performance of Firms: Evidence from Pakistan's Manufacturing Sector. *J. Compet.* **2019**, *11*, 5–18. [CrossRef]

20. Kheyfets, B.A.; Chernova, V.Y. Sustainable agriculture in Russia: Research on the dynamics of innovation activity and labor productivity. *Entrep. Sustain.* **2019**, *7*, 814–824. [CrossRef]

21. Stjepanović, S.; Tomić, D.; Škare, M. Green GDP: An analyses for developing and developed countries. *E A M: Ekon. A Manag.* **2019**, *22*, 4–17. [CrossRef]

22. Cismas, L.M.; Miculescu, A.; Negrut, L.; Negrut, V.; Otil, M.D.; Vadasan, I. Social Capital, Social Responsibility, Economic Behavior and Sustainable Economic Development—An Analysis of Romania's Situation. *Transform. Bus. Econ.* **2019**, *18*, 605–628.

23. Jayasundera, M. Economic development of Ceylon Tea Industry in Sri Lanka. *Financ. Mark. Inst. Risks* **2019**, *3*, 131–135. [CrossRef]

24. Harold, N.Y. Econometric analysis of long and short-run effects of exports on economic growth in Cameroon (1980–2016). *Financ. Mark. Inst. Risks* **2018**, *2*, 50–57. [CrossRef]

25. Haninun, H.; Lindrianasari, L.; Denziana, A. The effect of environmental performance and disclosure on financial performance. *Int. J. Trade Glob. Mark.* **2018**, *11*, 138. [CrossRef]

26. Ortikov, A.; Smutka, L.; Benešová, I. Competitiveness of Uzbek agrarian foreign trade–different regional trade blocs and the most significant trade partners. *J. Int. Stud.* **2019**, *12*, 177–194. [CrossRef] [PubMed]

27. Shuquan, H. The Impact of Trade on Environmental Quality: A Business Ethics Perspective and Evidence from China. *Bus. Ethics Leadersh.* **2019**, *3*, 43–48. [CrossRef]

28. Smutka, L.; Maitah, M.; Svatoš, M. Policy impacts on the EU-Russian trade performance: The case of agri-food trade. *J. Int. Stud.* **2019**, *12*, 82–98. [CrossRef] [PubMed]

29. Falkowski, K. Trade interdependence between Russia vs. the European Union and China within the context of the competitiveness of the Russian economy. *Equilib. Q. J. Econ. Econ. Policy* **2018**, *13*, 667–687. [CrossRef]

30. Kadochnikov, S.M.; Fedyunina, A.A. Explaining the performance of Russian export: What role does the soft and hard infrastructure play? *Int. J. Econ. Policy Emerg. Econ.* **2018**, *11*, 541–559. [CrossRef]

31. Krajnakova, E.; Navickas, V.; Kontautiene, R. Effect of macroeconomic business environment on the development of corporate social responsibility in Baltic Countries and Slovakia. *Oeconomia Copernic.* **2018**, *9*, 477–492. [CrossRef]

32. Grenčíková, A.; Bilan, Y.; Samusevych, Y.; Vysochyna, A. Drivers and Inhibitors of Entrepreneurship Development in Central and Eastern European Countries. In Proceedings of the 33rd International Business Information Management Association Conference, IBIMA 2019: Education Excellence and Innovation Management through Vision, Granada, Spain, 10–11 April 2019; Volume 2019, pp. 2536–2547.

33. Bilan, Y.; Lyeonov, S.; Vasylieva, T. and Samusevych, Y. Does tax competition for capital define entrepreneurship trends in Eastern Europe? *Online J. Model. New Eur.* **2018**, *27*, 34–66. [CrossRef]

34. Bilan, Y.; Vasilyeva, T.; Lyeonov, S.; Bagmet, K. Institutional complementarity for social and economic development. *Bus. Theory Pract.* **2019**, *20*, 103–115. [CrossRef]

35. Bilan, Y.; Raišienė, A.G.; Vasilyeva, T.; Lyulyov, O.; Pimonenko, T. Public Governance efficiency and macroeconomic stability: Examining convergence of social and political determinants. *Public Policy Adm.* **2019**, *18*, 241–255.

36. Bilan, Y.; Vasilyeva, T.; Lyulyov, O.; Pimonenko, T. EU vector of Ukraine development: Linking between macroeconomic stability and social progress. *Int. J. Bus. Soc.* **2019**, *20*, 433–450.

37. Lyulyov, O.; Pimonenko, T.; Stoyanets, N.; Letunovska, N. Sustainable development of agricultural sector: Democratic profile impact among developing countries. *Res. World Econ.* **2019**, *10*, 97–105. [CrossRef]

38. Akhmadeev, R.; Redkin, A.; Glubokova, N.; Bykanova, O.; Malakhova, L.; Rogov, A. Agro-industrial cluster: Supporting the food security of the developing market economy. *Entrep. Sustain. Issues* **2019**, *7*, 1149–1170. [CrossRef]

39. Bhandari, M.P. Sustainable Development: Is This Paradigm The Remedy of All Challenges? Does Its Goals Capture The Essence of Real Development and Sustainability? With Reference to Discourses, Creativeness, Boundaries and Institutional Architecture. *Socioecon. Chall.* **2019**, *3*, 97–128. [CrossRef]

40. Bello, H.S.; Galadima, I.S.; Jibrin, A.-M.A. Appraisal of the Salam Islamic Mode of Financing Agribusiness and Agriculture among Rural Farmers in Bauchi State of Nigeria. *Socioecon. Chall.* **2018**, *2*, 56–62. [CrossRef]

41. Sokolenko, L.F.; Tiutiunyk, I.V.; Leus, D.V. Ecological and economic security assessment in the system of regional environmental management: A case study of Ukraine. *Int. J. Ecol. Dev.* **2017**, *32*, 27–35.

42. Lizińska, W.; Marks-Bielska, R.; Babuchowska, K. Intervention on the agricultural land market in relation to the end of the transitional period for purchasing agricultural land by foreigners. *Equilib. Q. J. Econ. Econ. Policy* **2017**, *12*, 171–183. [CrossRef]

43. Vacca, A.; Onishi, H. Transparency and privacy in environmental matters. *Int. J. Econ. Policy Emerg. Econ.* **2018**, *11*, 333–343. [CrossRef]

44. Kostyuchenko, N.; Petrushenko, Y.; Smolennikov, D.; Danko, Y. Community-based approach to local development as a basis for sustainable agriculture: Experience from Ukraine. *Int. J. Agric. Resour. Gov. Ecol.* **2015**, *11*, 178–189. [CrossRef]

45. Popp, J.; Oláh, J.; Kiss, A.; Lakner, Z. Food security perspectives in Sub-Saharan Africa. *Amfiteatru Econ.* **2019**, *21*, 361–376. [CrossRef]

46. Vasylyeva, T.A.; Pryymenko, S.A. Environmental economic assessment of energy resources in the context of Ukraine's energy security. *Actual Probl. Econ.* **2014**, *160*, 252–260.

47. Mekhum, W. Sustainable development facets: Role of renewable energy production, consumption and research and development expenditure. *J. Secur. Sustain. Issues* **2020**, *9*, 252–263. [CrossRef]

48. Lu, Z.; Gozgor, G.; Lau, C.K.M.; Paramati, S.R. The dynamic impacts of renewable energy and tourism investments on international tourism: Evidence from the G20 countries. *J. Bus. Econ. Manag.* **2019**, *20*, 1102–1120. [CrossRef]

49. Androniceanu, A.; Popescu, C.R. An Inclusive Model for an Effective Development of the Renewable Energies Public Sector. *Adm. Manag. Public* **2017**, *28*, 81–96.

50. Lyeonov, S.; Pimonenko, T.; Bilan, Y.; Štreimikienė, D.; Mentel, G. Assessment of Green Investments' Impact on Sustainable Development: Linking Gross Domestic Product Per Capita, Greenhouse Gas Emissions and Renewable Energy. *Energies* **2019**, *12*, 3891. [CrossRef]

51. Abdimomynova, I.; Kolpak, E.; Doskaliyeva, B.; Stepanova, D.; Prasolov, V. Agricultural Diversification in Low- And Middle-Income Countries: Impact on Food Security. *Montenegrin J. Econ.* **2019**, *15*, 167–178. [CrossRef]

52. Mentel, G.; Vasilyeva, T.; Samusevych, Y.; Pryymenko, S. Regional differentiation of electricity prices: Social-equitable approach. *Int. J. Environ. Technol. Manag.* **2018**, *21*, 354–372. [CrossRef]

53. Aitkazina, M.A.; Nurmaganbet, E.; Syrlybekkyzy, S.; Koibakova, S.; Zhidebayeva, A.E.; Aubakirov, M.Z. Threats to sustainable development due to increase of greenhouse gas emissions in a key sector. *J. Secur. Sustain. Issues* **2019**, *9*, 227–240. [CrossRef]

54. Sibanda, M.; Ndlela, H. The link between carbon emissions, agricultural output and industrial output: Evidence from South Africa. *J. Bus. Econ. Manag.* **2020**, *21*, 301–316. [CrossRef]

55. Dkhili, H.; Dhiab, L.B. Management of Environmental Performance and Impact of the Carbon Dioxide Emissions (CO_2) on the Economic Growth in the GCC Countries. *Mark. Manag. Innov.* **2019**, *4*, 252–268. [CrossRef]

56. Mačaitytė, I.; Virbašiūtė, G. Volkswagen Emission Scandal and Corporate Social Responsibility—A Case Study. *Bus. Ethics Leadersh.* **2018**, *2*, 6–13. [CrossRef]

57. Odermatt, C.C. Clean coal project: Carbon certificate pricing. *Int. J. Trade Glob. Mark.* **2018**, *11*, 149–159. [CrossRef]

58. Vasylieva, N. Ukrainian Agricultural Contribution to the World Food Security: Economic Problems and Prospects. *Montenegrin J. Econ.* **2018**, *14*, 215–224. [CrossRef]

59. Aliyas, I.M.; Ismail, E.Y.; Alhadeedy, M.A.H. Evaluation of Applications of Sustainable Agricultural Development in Iraq. *Socioecon. Chall.* **2018**, *2*, 75–80. [CrossRef]

60. Bilan, Y.; Lyeonov, S.; Stoyanets, N.; Vysochyna, A. The impact of environmental determinants of sustainable agriculture on country food security. *Int. J. Environ. Technol. Manag.* **2018**, *21*, 289–305. [CrossRef]

61. The World Bank DataBank. Available online: http://databank.worldbank.org/data/home.aspx (accessed on 18 January 2020).

62. UNEP. *Environmental Data Explorer*; UNEP: Geneva, Switzerland, 2020; Available online: http://geodata.grid.unep.ch/results.php (accessed on 18 January 2020).

63. FAOSTAT. Available online: http://fao.org/faostat/en/ (accessed on 18 January 2020).

64. Beiragh, R.G.; Alizadeh, R.; Kaleibari, S.S.; Cavallaro, F.; Zolfani, S.H.; Bausys, R.; Mardani, A. An integrated multi-criteria decision-making model for sustainability performance assessment for insurance companies. *Sustainability* **2020**, *12*, 789. [CrossRef]

65. Pesaran, M.H.; Shin, Y.; Smith, R.P. Pooled mean group estimation of dynamic heterogeneous panels. *J. Am. Stat. Assoc.* **1999**, *94*, 621–634. [CrossRef]

66. Blackburne, E.F.; Frank, M.W. Estimation of nonstationary heterogeneous panels. *Stata J.* **2007**, *7*, 197–208. [CrossRef]

67. Summary for Policymakers. *Climate Change 2014: Mitigation of Climate Change. Contribution of Working Group III to the Fifth Assessment Report of the Intergovernmental Panel on Climate Change*; Edenhofer, O., Pichs-Madruga, R., Sokona, Y., Farahani, E., Kadner, S., Seyboth, K., Adler, A., Baum, I., Brunner, S., Eickemeier, P., et al., Eds.; Cambridge University Press: Cambridge, UK; New York, NY, USA, 2014; Available online: https://www.ipcc.ch/site/assets/uploads/2018/02/ipcc_wg3_ar5_full.pdf (accessed on 18 January 2020).

68. Sipos, G.; Urbányi, B.; Vasa, L.; Kriszt, B. Application of by-products of bioethanol production in feeding, environmental and feeding safety concerns of utilization. *Cereal Res. Commun.* **2007**, *35*, 1065–1068. [CrossRef]

69. Mesterházy, A.; Oláh, J.; Popp, J. Losses in the grain supply chain: Causes and solutions. *Sustainability* **2020**, *12*, 2342. [CrossRef]

70. Impacts of Bioenergy on Food Security–Guidance for Assessment and Response at National and Project Levels. Available online: http://www.fao.org/3/i2599e/i2599e00.pdf (accessed on 18 January 2020).

71. Sola, P.; Ochieng, C.; Yila, J.; Iiyama, M. Links between energy access and food security in sub Saharan Africa: An exploratory review. *Food Secur.* **2016**, *8*, 635–642. [CrossRef]

72. Energy-Smart Food at FAO: An Overview. Available online: http://www.fao.org/3/an913e/an913e.pdf (accessed on 18 January 2020).

73. Wambua, B.N.; Omoke, K.J.; Telesia, M.M. Effects of Socio-Economic Factors on Food Security Situation in Kenyan Dry lands Ecosystem. *Asian J. Agric. Food Sci.* **2014**, 2, 52–59.

74. Mbuthia, K.W.; Kioli, F.N.; Wanjala, K.B. Environmental Determinants to Household Food Security in Kyangwithya West Location of Kitui County. *J. Food Secur.* **2017**, 5, 129–133. [CrossRef]

75. IRENA. Renewable Energy in the Water, Energy & Food Nexus. Abu Dhabi, UAE, 2015. Available online: https://www.irena.org/documentdownloads/publications/irena_water_energy_food_nexus_2015.pdf (accessed on 18 January 2020).

15

Sustainable Value Creation in the Food Chain: A Consumer Perspective

József Tóth [1,2,*], Giuseppina Migliore [3], Giorgio Schifani [3] and Giuseppina Rizzo [3,*]

[1]   Department of Agricultural Economics and Rural Development, Corvinus University of Budapest, 1093 Budapest, Hungary

[2]   Faculty of Economics, Socio-Human Sciences and Engineering, Sapientia Hungarian University of Transylvania, 530104 Miercurea Ciuc, Piața Libertății nr. 1, Romania

[3]   Department of Agricultural, Food and Forest Sciences, University of Palermo, 90128 Palermo, Italy; giuseppina.migliore@unipa.it (G.M.); Giorgio.schifani@unipa.it (G.S.)

*   Correspondence: jozsef.toth@uni-corvinus.hu (J.T.); giuseppina.rizzo03@unipa.it (G.R.)

Abstract: The growth of diet-related diseases is becoming an important societal concern and a challenge for a more sustainable society. This has developed important trends in food consumption, including the increasing demand for food with a natural attribute and with health claims (e.g., enriched food). Consumers tend to evaluate these two attributes as superior ones and tend to pay a premium price for them. Accordingly, the value added by producers also will upturn if they take into consideration the consumers' preferences. However, to the best of our knowledge, consumer preference over the two types of products (natural and enriched) is not yet completely clear. The present study tries to contribute to reducing this gap by analyzing Hungarian consumer preferences for natural fruit juices over enriched ones and exploring the drivers which guide consumer choices for the two attributes. For this purpose, we analyze young consumers' willingness-to-pay (WTP) for natural and enriched fruit juices using a seemingly unrelated regression (SUR) to derive the two value-added activities. Our results show that the fruit juice with the natural attribute is preferred over the enriched one, and that there is a common feature behind the perception of the two attributes, namely the healthiness. Based on the natural fruit juice characteristic, these results open space for local production in gardens or in small-medium sized farms. This could have beneficial effects, both for sustainable development of rural areas and for the promotion of healthy food systems towards sustainability in food consumption.

Keywords: willingness to pay; enriched attribute; natural attribute; healthy attribute; seemingly unrelated regression (SUR); fruit juice; Hungary

1. Introduction

The growth of diet-related diseases is becoming an important societal concern and a challenge for a more sustainable society [1]. As a result, today, consumers are aware that their diet affects their health and so prefer to choose food that helps them to have a healthy lifestyle. [2,3]. This has contributed to the development of important trends in food consumption, which has seen, amongst others, the growing consumer interest towards foods with natural and health claims attributes [4]. The category of food with health claims includes food enriched with healthy components, such as polyphenols, vitamins, and other healthy components [2], while natural food is food without additives and human interventions, considered by consumers harmful for their health [5].

The Kampffmeyer Food Innovation Study [6] revealed that food naturalness is a decisive buying incentive and that the majority of the consumers perceived a strong connection between "natural" and "healthy." Furthermore, it has been demonstrated that consumers living in developed countries prefer natural foods over the conventional ones, as they are considered to have positive health effects [7]. Similarly, foods with health claims have registered a growing market success. According to the latest available data [8,9], 27% of global respondents, on average, are very willing to pay a premium price for health claims. This percentage is slightly higher in western countries and particularly in the U.S., where the majority of consumers believe that health claim foods give real benefits in improving and maintaining overall health, and nearly 30% indicate that they buy products with health claims on the labels [10].

The growing consumers' interest towards these product characteristics has pushed the food industry to provide healthier products [11,12]. The use of health as a selection criterion has already been offering new possibilities to the food market and continues to provide new challenges for producers [13]. One of these challenges for the food industry is to give consumers product options with a natural and healthy image.

From a consumers perspective, interest shown towards these two attributes (natural and health claims) is due to the common will of consumers to improve or maintain their health, although the two attributes have different exceptions [14]. Health motivations as factors for purchasing natural and health claim products have already been investigated in several studies [14,15]. However, to the best of our knowledge, consumer preference over the two types of attributes is not yet completely clear.

The present study tries to reduce this gap, by analyzing consumer preferences for the attributes natural and health claim, and exploring the drivers affecting consumer choices for both attributes. Knowing which attribute is more valued by consumers could give important indications to the food industry more oriented to provide products with a health image. Furthermore, understanding the drivers behind consumer preference could be useful for planning successful marketing strategies for those enterprises oriented to satisfying those consumers' needs.

Based on these premises, three objectives have been set in this study: 1. to investigate which attribute, between natural and health claims, is more appreciated by consumers; 2. to explore the drivers behind consumers' choice for both health attributes, and 3. to verify whether the price premium of two types of attributes is explained by common factors.

To answer these research questions, the present study used a laboratory experiment in order to derive the consumer's willingness to pay (WTP) for natural and enriched attributes. The remainder of the paper is organized as follows: after the introduction, Section 2 explains the theoretical framework and consumer choices on healthy products in Hungary; Section 3 shows the data collection and methods; Section 4 presents results and discussion and, finally, in Section 5, we conclude, and the study limitations are provided.

2. Theoretical Framework

Consumers consider the naturalness of foods as a highly valued quality characteristic [16]. They interpret the food naturalness as an indicator of the healthiness and quality of the product, derived from the 'integral integrity of the product' [17]. As a result, natural products are perceived as good for your health as they are free of additives and other synthetic substances, perceived to reduce the healthiness of food [18–21]. Furthermore, the idea of natural eating seems to generate a perception of physical and emotional well-being [22]. This is supported also by Rozin [18], according to whom natural food evokes a positive association in consumers' minds, following the idea that 'natural entities are inherently better than non-natural entities'. According to the literature, higher natural food consumption seems to be associated with the perception that processed food can cause high health risks [20,23].

Similarly, health claim products have registered a faster market growing in the last few years, responding to consumers concerns on health and providing messages about specific benefits of products that potentially increase perceived wellbeing [24]. Indeed, enriched foods communicate their health-related benefits with the help of claims that may contain a bulk of information [25].

The typical elements that health claims may be built from are the components that trigger the function by generating physiological and psychological benefits [1]. Enriched food looks similar to conventional food and is consumed as part of a regular diet and has been shown to have health benefits and/or reduce the risk of chronic diseases beyond the basic nutritional functions of food [26].

However, how consumers respond to enriched food varies from product to product. From a consumer's perspective, enriched foods are not a homogenous category of products, and consumers' attitudes seem to affect the purchasing intention for various enriched products differently [15]. Moreover, some enriched foods are perceived as less natural, since the beneficial components which trigger the function are derived from technology-based enhancement, and they may include foods with chemical additives and preservatives.

Thus, consumer perception is influenced by both the health element that has been added in food and the process by which this addition happened [2]. Acceptance of food products depends on the health image of the product category or the ingredients [27], on the production method [1] and how the product was enriched and 'tampered with' [28]. According to Lähteenmäki [1], the familiarity with the product greatly influences the perception of the consumer. In fact, familiarity is a key factor in acceptance of enriched food [27]. However, although it may be thought that natural and enriched foods are contradictory (for example due to the difference in health-related message), there are several studies that indicate a link between the two concepts, because both are chosen by consumers to improve or maintain their health [29]. For example, Caracciolo and colleagues [30] investigated consumer preferences for the two attributes and their empirical findings revealed that consumers evaluate both attributes, natural and enriched, similarly.

Among the products having a health image, fruit juices are among the most recognizable, thanks to their natural contents of vitamins and minerals [31]. The fruit juice market is one of the most innovative and competitive segments of the food sector [31]. Manufacturers striving to expand sales are focusing on product diversification, developing fruit juices that go beyond the taste of the product and providing general health benefits. Since fruits are the primary source of ascorbic acid in the diet, the enrichment of fruit juices has been concerned mainly with this vitamin [32]. In this context, vitamin enrichments are more accepted by consumers compared to other types of fruit juice enrichment, for example with calcium, since the latter is perceived an unnatural type of enrichment [28,33]. This would position vitamin-enriched fruit juices closer to natural food, creating a more "holistic health image" [28]. Similarly, consumers are increasingly preferring fruit juices with the natural attribute, containing 100% in fruit. These drinks are free from added sugars and artificial colors or flavors, and they represent an opportunity for those companies that want to create a competitive advantage in the fruit juice market. Furthermore, to create a strong sense of community and add value to the product, the local origin of the ingredients is often emphasized and well specified in natural fruit juices. As a result, there is a growing consumer interest in local products that position these fruit juices as healthier and more sustainable than their conventional counterparts [31]. However, to date, it is not clear which type of attribute (enriched or natural) is more preferred by consumers in fruit juices. This information could be very helpful for those companies operating in those market segments where healthy products are becoming of primary importance in consumers' buying behavior. For this reason, we carried out an experimental study in Győr, Hungary, because this country is among those where consumers are starting to pay more attention to their health [34,35]. Indeed, in Hungary, alongside economic growth, the healthy diet and lifestyle are becoming increasingly important for consumers [36]. Literature reveals that health issues represent the main reason for purchasing health food and that health attributes have become as important as sensory ones, during the buying decision-making process [37]. For example, Balázs [38] in his study showed that more than half of the respondents were willing to pay an extra 10% for healthy products. Moreover, Balázs's findings showed that consumers of healthy food generally have higher levels of education and higher incomes, while their age ranged between young and middle-aged. Furthermore, according to the literature young consumers seem to exceed all prior generational expenditure [39], making a large direct contribution to the economy [40] and an even larger indirect economic impact, by influencing the majority of family purchase decisions [41]. In addition, young consumers have significant current and future impact on the Western economies and are accordingly considered the most powerful consumer group in the marketplace [42]. For this reason, we have focused our attention on the university student generation in Hungary, in order to understand which health attribute, enriched or natural, is preferred in fruit juices. Knowing the theoretical framework is fundamental to developing this research, which may contribute to better marketing design strategies and, successively, contribute to creating a competitive market advantage for food companies.

3. Materials and Methods

3.1. Experimental Auctions

The experiments were conducted over a two-week period, in autumn 2018, at the "Széchenyi István" University (Hungary). The consumers of this study were students, who were recruited randomly and informed they were participating in a consumer preference research study for different types of fruit juice. Using the Vickrey auction methodology [43], an experimental evaluation process was chosen, which is identified in the fifth-price auction. Ten 25-min experimental sessions involving ten people each were organized. The choice of the fifth-price auction allows, at the same time, the number of participants in the auction and their degree of involvement to be increased. Lusk et al. [44] showed that bidders would generally be more involved if at least half of them could potentially win the product at auction. In addition, participants were told that only one round and one product would be binding, to avoid reductions in demand and effects on wealth in subsequent rounds [45]. Each participant in the auction received 2000 Hungarian forints (HUF) (approximately € 5.50) as a reward for his/her participation in the auction. All respondents rated the three fruit juices containing the same amount of information.

In the initial phase, participants were selected from among those who said they had been drinking fruit juice for the past two weeks. In the second phase, every individual received the monetary compensation and signed a consent form and a form committing him/her to buy the product in the case of a victory. In the third phase, the auction mechanism was explained, and in the fourth phase, a researcher described the three fruit juices' characteristics. The three products were (1) conventional fruit juice, used as a control product, compared to the other two types of fruit juices, (2) 100% natural fruit juice made from fruit straight from the garden, with no dilution and no concentrate and (3) fruit juice enriched with sea buckthorn to strengthen the immune system and with a high vitamin C content. The three fruit juices were packaged in three white and unbranded packs, to avoid the effects of the brand and the label. In the fifth phase, the participants wrote their sealed bids on anonymous tickets. Finally, in the last phase, everyone completed a questionnaire and one fruit juice and one price (market price) were randomly extracted. Those participants who bid more for the auctioned fruit juice compared to the market price won the fruit juice, paying the extracted price for it.

3.2. Questionnaire

The questionnaire included two main sections. The first section collected information on consumers' socio-demographic characteristics (age, gender, number of household members and monthly net income), on their consumption frequency of fruit juice and the characteristics that are sought in the product (good taste and smell, vitamin and mineral content, geographical origin, nice appearance, calories content, free from artificial materials, price and brand name). The second section included three psycho-attitudinal scales: natural product interest (NPI), general health interest (GHI), and reward from using functional food (RFF). These scales are widely used in the literature [46,47]. More precisely, GHI and NPI scales were developed by Roininen, Lähteenmäki and Tuorila [48]; the first consists of eight articles that reveal the consumers' attitude towards healthy eating, while the second scale includes six articles aimed at capturing the consumers' attitude towards the consumption of unprocessed food. The RFF scale was proposed by Lähteenmäki [49] and includes seven items that explain the declaration of gratitude deriving from the use of enriched foods. These validated GHI, NPI and RFF attitude scales were collected by means of 7-point Likert scales, where 1 corresponds to totally disagree and 7 to totally agree.

3.3. Statistical Analysis

The data collected were processed in four distinct phases, using the STATA 15.0 (Budapest, Hungary) integrated statistical software. In the first phase, the socio-demographic characteristics of the sample were defined, through descriptive analyses; in the second phase, the psycho-attitudinal scales were interpreted, checking their internal consistency (alpha-coefficient) and calculating the average of each item. In the third part, a description of the WTPs detected for the three types of fruit juices was made; in addition, by means of parametric (t-test) and non-parametric tests (Wilcoxon tests), it was verified whether the three WTPs were significantly different, and therefore, two deltas (premium prices) were calculated. The two premium prices were obtained, one at a time, by first calculating the difference between the WTP for natural and conventional fruit juices and then the difference between WTP for enriched and conventional fruit juice:

$$\Delta WTP_{NAT} = (WTP_{NAT} - WTP_{CONV})$$
$$\Delta WTP_{ENR} = (WTP_{ENR} - WTP_{CONV})$$

Later, the seemingly unrelated regressions (SUR) [50] were presented, together the Breusch-Pagan test of independence, to measure how the price premium of the two fruit juices can be influenced and, at the same time, to verify whether the price premium of the two types of juices is explained by common attributes.

This stochastic model may be expressed by the following relationship:

$$y = X\beta + u$$

where y and u are vectors with n elements, X is a matrix with n rows and $k + 1$ columns (with k the explanatory variables + 1 for the constant) and β is the vector containing $k + 1$ unknown coefficients.

4. Results and Discussion

4.1. Sample Description

The consumers participating in the experiment were 100 students of the "Széchenyi István" University of Győr (Hungary), including 29 males and 71 females, between 18 and 28 years of age (mean age = 22; S.D. = 2.23). The number of family members the students had ranged from 1 to 5, where 1 indicates that the student lives alone and 5 indicates that he/she lives with more than 4 people. The average number of members per family was 3 people. Finally, the monthly net income was in a range from "below 60 thousand" and "more than 350 thousand", with an average of about 120 thousand HUF (about €360). The socio-demographic characteristics of the participants are shown in Table 1.

Table 1. Socio-demographic characteristics of the sample.

Variables	Mean	Std. dev.	Min	Max
Gender [1]	0.29	0.50	0	1
Age [2]	21.65	2.23	18	28
Family Members [3]	2.77	1.12	1	5
Monthly Income [4]	1.81	0.64	1	3

[1]: dummy variable, 1 = male and 0 = female; [2]: continuous variable; [3]: categorical variable, 1 = single, 2 = two members, 3 = three members, 4 = four members, 5 = family with more than 4 members; [4]: categorical variable, 1 = <60 thousand HUF; 2 = 60–120 thousand HUF; 3 = 121–220 thousand HUF; 4 = 221–350 thousand HUF; 5 = >350,000.

4.2. Psycho-Attitudinal Scales

Attitudes have been shown to have a great effect during the consumers' decision-making process, and for this reason, they were used in the present study to explain consumers' food choices, through appropriate attitudinal scales [51]. In particular, the GHI scale was chosen because it is expected to correlate positively with attitudes towards enriched foods [48]; the NPI scale is hypothesized to have a positive correlation with natural product consumption [48], while the RFF is expected to have a positive correlation with the consumer's willingness to feed himself/herself with enriched foods in order to improve or maintain a state of health [52,53]. Furthermore, for those items with negative meaning, Likert scale scores were reversed to improve the attitude scales' readability.

The Cronbach's alpha value was 0.85 for natural product interest, 0.83 for general health interest and 0.89 for reward from using functional food, indicating a good internal reliability (Table 2).

Table 2. Internal reliability of the scales.

Scale	Cronbach's Alpha
GHI	0.83
NPI	0.85
RFF	0.89

The results show a high awareness of consumers about the health consequences of their food choices. Indeed, the higher GHI item scores were: "The healthiness of food has little impact on my food choices" (reversed mean = 5.00) and "I am very careful about the healthiness of food I eat" (mean = 4.51). Concerning NPI, the items with the highest scores were: "Foods containing artificial flavor enhancers are not harmful to health" (reversed mean = 5.98) and "Organically grown vegetables are no healthier than others" (reversed mean = 6.10). The items with the highest values for RFF were: "I get pleasure from eating functional foods" (mean = 5.77) and "The idea that I can take care of my health by eating functional foods gives me pleasure" (mean = 6.01).

Finally, correlation coefficients were computed and the presence of a positive and statistically significant correlation was found among all the three scales (Table 3).

Table 3. Correlation coefficients of the scales.

	GHI	NPI	RFF
GHI	1.0000		
NPI	0.5890	1.0000	
RFF	0.6601	0.4233	1.0000

The descriptive statistics of individual items composing the three scales are shown in Tables 4–6.

Table 4. Items' statistics of general health interest (GHI) scale.

	General Health Interest (GHI)	Mean	S.D.	Min	Max
GHI_1	The healthiness of food has little impact on my food choices.	5.00	1.6	1	7
GHI_2	I am very particular about the healthiness of food I eat.	4.51	1.39	1	7
GHI_3	I eat what I like, and I do not worry much about the healthiness of food.	4.06	1.67	1	7
GHI_4	It is important for me that my diet is low in fat.	3.53	1.40	1	7
GHI_5	I always follow a healthy and balanced diet.	3.85	1.54	1	7
GHI_6	It is important for me that my daily diet contains a lot of vitamins and minerals.	4.67	1.49	1	7
GHI_7	The healthiness of snacks makes no difference to me.	3.64	1.94	1	7
GHI_8	I do not avoid foods, even if they may raise my cholesterol.	3.78	1.66	1	7

Table 5. Items' statistics of natural product interest (NPI) scale.

Natural Product Interest (NPI)		Mean	S.D.	Min	Max
NPI_1	I try to eat foods that do not contain additives.	3.98	1.70	1	7
NPI_2	I do not care about additives in my daily diet.	4.33	1.72	1	7
NPI_3	I do not eat processed foods, because I do not know what they contain.	3.42	1.83	1	7
NPI_4	I would like to eat only organically grown vegetables.	5.59	1.56	1	7
NPI_5	In my opinion, artificially flavored foods are not harmful for my health.	5.98	1.16	1	7
NPI_6	In my opinion, organically grown foods are no better for my health than those grown conventionally.	6.10	1.37	1	7

Table 6. Items' statistics of reward from using functional food (RFF) scale.

Reward from using Functional Food (RFF)		Mean	S.D.	Min	Max
RFF_1	I get pleasure from eating functional foods.	5.77	1.35	1	7
RFF_2	The idea that I can take care of my health by eating functional foods gives me pleasure.	6.01	1.17	1	7
RFF_3	Functional foods make me feel more energetic.	5.12	1.43	1	7
RFF_4	Functional foods help to improve my mood.	4.69	1.53	1	7
RFF_5	My performance improves when I eat functional foods.	4.81	1.43	1	7
RFF_6	I actively seek out information about functional foods.	4.27	1.57	1	7
RFF_7	I willingly try even unfamiliar products if they are functional.	4.04	1.63	1	7

4.3. Willingness to Pay (WTP)

Consumer bids describe how much participants are willing to pay for conventional, natural and enriched fruit juice. The estimated average WTPs were the following: 646.76 HUF (about €1.93) for the conventional fruit juice, 794.09 HUF (about €2.37) for the enriched fruit juice, and 957.93 (about €2.86) for the natural fruit juice (Table 7).

Table 7. Consumers' willingness to pay.

	Mean	S. D.	Min	Max
WTP_{CON}	646.76	322.32	50	1800
WTP_{ENR}	794.09	369.98	100	2000
WTP_{NAT}	957.93	489.08	200	2500
ΔWTP_{ENR}	147.32	178.91	−500	1050
ΔWTP_{NAT}	311.17	287.97	−270	1400

By means of a t-test and Wilcoxon test, it was possible to verify that there are statistically significant differences between the two attributes and that the natural attribute was preferred to the enriched attribute. Indeed, ΔWTP_{NAT}, that is the differential value between the natural fruit juice and the conventional one, has an average value of 311.17 HUF (about 0.93€), while ΔWTP_{ENR}, which is the differential value between the enriched fruit juice and the conventional one, has an average value of 147.32 HUF (about €0.44) (Table 7).

4.4. Drivers Behind Consumers' WTP for Both Attribute

It is clear that consumer choice depends on many factors [48]. By performing a SUR between the two WTP for natural and enriched fruit juices and the other variables collected through the questionnaire, such as the consumer characteristics and psycho-attitudinal scales, it was possible to understand which are the drivers affecting consumer WTP for the two attributes. In Table 8 drivers behind consumers' WTP for both enriched and natural attributes, and the estimated coefficients as well as their statistical significance, are shown.

Table 8. Drivers behind consumers' willingness to pay.

Equation	Obs	Parms	"R-sq"	p
ΔWTP_{ENR}	95	4	0.1675	0.0007
ΔWTP_{NAT}	95	4	0.1093	0.0201
ΔWTP_{ENR}	Coef.	Std. Err.	z	$P > |z|$
GHI	−61.76	23.05	−2.68	0.007
NPI	−0.56	21.35	−0.03	0.979
RFF	70.81	19.62	3.61	0.000
INCOME	37.01	15.02	2.46	0.014
ΔWTP_{NAT}	Coef.	Std. Err.	z	$P > |z|$
GHI	−21.24	38.34	−0.55	0.580
NPI	−16.61	35.50	−0.47	0.640
RFF	72.22	32.64	2.21	0.027
INCOME	60.09	24.97	2.41	0.016

Breusch-Pagan test of independence: chi2(1) = 63.129, Pr = 0.0000.

For the section on consumers' WTP for vitamin-enriched fruit juice as a dependent variable, the results showed that the participants' preference is mainly affected by participants' attitude towards healthy eating (through their importance attributed to the items of the general interest scale for health (GHI)), towards the reward from using enriched foods (RFF) and consumers' monthly net income. Looking at this in more detail, RFF attitude and the monthly net income are positively correlated with the dependent variable; therefore, as the value of these independent variables increases, the average of the WTP for the enriched fruit juice tends to increase. On the contrary, the negative coefficient of GHI attitude suggests that as they increase, the dependent variable tends to decrease. This means that the attitude towards healthy eating negatively affects the preference for the enriched fruit juice. This research is in line with other studies that describe the choice for the two attributes to improve or maintain a state of health [14,54,55].

Relatively to the consumers' WTP for the natural fruit juice, the results show that, contrary to what was showed by Caracciolo and colleagues' study, the preference for natural fruit juice is not explained by the NPI attitude. According to the results, the WTP values for natural fruit juice seem to be also affected by RFF attitudes, which in this case is related to the rewards from using natural fruit juice rich in vitamins. This suggests the interest for both attributes (enriched and natural) seems to be affected by common drivers, that is, rewards from using fruit juices richer in vitamins compared to the conventional one. Differences in results, compared to other research findings, may depend on consumers familiarity with the product [1], suggesting, in line with Urala and Lähteenmäki's study [15], that effects on consumer choice have to be studied not as one homogenous group of product, but rather as separate products within the various food categories. Furthermore, monthly net income positively affects consumers' WTP for both products. This is in line with Bruchi and colleagues' study [50] showing how as the level of monthly net income increases, the WTP for natural and enriched fruit juice increases.

5. Conclusions

The present study had multiple objectives: to investigate the preferences for natural and enriched products and to understand which drivers affect their preference and if there are common drivers between the two WTP. To respond to these research questions, consumers' preferences for enriched and natural attributes were measured via an experimental auction on fruit juices. Outcomes point out that consumers prefer natural fruit juice more than the enriched ones, but the motivations underlying consumers' preferences for both products are the same (the perceived reward from consuming fruit juices richer in vitamins compared to the conventional one).

These results can help us to understand how much and how consumers accept innovations in the food market, and therefore, help companies put their products on the markets.

Our analysis also reveals the primacy of natural fruit juice against the enriched fruit juices at WTP level. Regarding the participants' cohort and the way the juice was produced (fresh apples direct from the garden/local farms, without burdening the environment) we can also conclude that producers along the food chain may create additional value if they consider the consumer preferences of the younger generation.

Furthermore, the preference for natural fruit juice opens space for local production in gardens or in small-medium sized farms. This could have beneficial effects, on one hand for sustainable development of the rural area due to the resulting lower CO_2 emissions from short-distance transportation and the recirculation of financial capital in rural areas. On the other hand, the preference for the natural attribute could be a leverage for the promotion of healthy and sustainable food systems more oriented towards sustainability in food consumption. This direction is very much in line with the sustainability requirements of the globe. The study carries significant implications for consumer research on the preference of sustainable fruit juices, as well as practical management implications. Regarding the former, our study is one of the first to analyze consumer behavior towards fruit juices with health attributes, thus enriching extant literature on the willingness to pay a premium price for health attributes, and reinforcing business literature, which supports that consumers have a positive attitude towards sustainable products. In addition, our results corroborate the importance of consumer attitudinal characteristics in explaining the purchasing decision process for products with sustainable characteristics. As for the managerial perspective, our results offer entrepreneurs suggestions to differentiate their product offerings. In fact, considering the growing awareness, among consumers, of the importance of healthy food consumption, the Hungarian fruit juices industry is called upon to develop effective marketing strategies that will help consumers identify and distinguish fruit juices on the market. From this point of view, the ability of industries to develop innovations in this direction, which could boost the competitive performance of companies, is particularly important.

Although the study offers much food for thought, it has some limitations, such as having used a non-representative sample; thus, the generalizability of the results is limited. Furthermore, the RFF attitude in explaining the preference for both attributes in fruit juices opens space for further analysis in order to validate the results of the present study or overcome its limitations. Therefore, further studies have to take into account statistically representative samples in order to capture a full picture of consumers' preferences for healthy fruit juices. Moreover, further research could be repeated in different markets also for comparison. This would offer cross-cultural insights and help adapt marketing strategies to the individual and/or global perspectives.

Author Contributions: Conceptualization, J.T. and G.R.; methodology, J.T. and G.S.; formal analysis, investigation and writing—original draft preparation, G.R.; writing—review and editing, J.T. and G.M. All authors have read and agree to the published version of the manuscript.

References

1. Lähteenmäki, L. Claiming health in food products: Ninth Pangborn Sensory Science Symposium. *Food Qual. Prefer.* **2013**, *27*, 196–201. [CrossRef]

2. Grunert, K.G. *Consumer Trends and New Product Opportunities in the Food Sector*; Academic Publishers: Wageningen, The Netherlands, 2017.

3. Goetzke, B.; Nitzko, S.; Spiller, A. Consumption of organic and functional food. A matter of well-being and health? *Appetite* **2014**, *77*, 96–105. [CrossRef] [PubMed]

4. Willer, H.; Schaack, D.; Lernoud, J. Organic farming and market development in Europe and the European Union. In *The World of Organic Agriculture. Statistics and Emerging Trends*; Research Institute of Organic Agriculture FiBL and IFOAM-Organics International: Bonn, Germany, 2019; pp. 217–254. Available online: https://shop.fibl.org/chde/2020-organic-world-2019.html (accessed on 10 November 2019).

5. Asioli, D.; Aschemann-Witzel, J.; Caputo, V.; Vecchio, R.; Annunziata, A.; Næs, T.; Varela, P. Making sense of the "clean label" trends: A review of consumer food choice behavior and discussion of industry implications. *Food Res. Int.* **2017**, *99*, 58–71. [CrossRef] [PubMed]

6. GoodMills Innovation. Available online: http://goodmillsinnovation.com/sites/kfi.kampffmeyer.faktor3server.de/files/attachments/1_pi_kfi_cleanlabelstudy_english_final.pdf/ (accessed on 15 October 2016).

7. Roman, S.; Sánchez-Siles, L.M.; Siegrist, M. The importance of food naturalness for consumers: Results of a systematic review. *Trends Food Sci. Technol.* **2017**, *67*, 44–57. [CrossRef]

8. Nielsen Company. Available online: https://www.nielsen.com/wp-content/uploads/sites/3/2019/04/health-wellness-report-feb-2017.pdf (accessed on 7 December 2019).

9. Nielsen Company. Available online: https://www.nielsen.com/wpcontent/uploads/sites/3/2019/04/Nielsen20Global20Health20and20Wellness20Report20-20January202015-1.pdf (accessed on 12 October 2018).

10. Sloan, E.; Adams Hutt, C. Beverage trends in 2012 and beyond. *Agro. FOOD Ind. Hi Tech.* **2012**, *23*, 8–12.

11. Nguyen, H.V.; Nguyen, N.; Nguyen, B.K.; Lobo, A.; Vu, P.A. Organic Food Purchases in an Emerging Market.: The Influence of Consumers' Personal Factors and Green Marketing Practices of Food Stores. *Int. J. Environ. Res. Public Health* **2019**, *16*, 1037. [CrossRef]

12. Seiders, K.; Ross, D.P. Obesity and the Role of Food Marketing: A Policy Analysis of Issues and Remedies. *J. Public Policy Mark.* **2004**, *23*, 153–169. [CrossRef]

13. Casey, D.K. *Three Puzzles of Private Governance: Global GAP and the Regulation of Food Safety and Quality*. UCD Working Papers in Law, Criminology & Socio-Legal Studies. United Kingdom. 2009. Available online: http://dx.doi.org/10.2139/ssrn.1515702 (accessed on 1 November 2019).

14. Aschemann-Witzel, J.; Maroscheck, N.; Hamm, U. Are organic consumers preferring or avoiding foods with nutrition and health claims? *Food Qual. Prefer.* **2013**, *30*, 68–76. [CrossRef]

15. Urala, N.; Lähteenmäki, L. Reasons behind consumers' functional food choices. *Nutr. Food Sci.* **2003**, *33*, 148–158. [CrossRef]

16. Rozin, P.; Spranca, M.; Krieger, Z.; Neuhaus, R.; Surillo, D.; Swerdlin, A.; Wood, K. Preference for natural: Instrumental and ideational/moral motivations, and the contrast between foods and medicines. *Appetite* **2004**, *43*, 147–154. [CrossRef]

17. Kahl, J.; Załęcka, A.; Ploeger, A.; Bügel, S.; Huber, M. Functional food and organic food are competing rather than supporting concepts in Europe. *Agriculture* **2012**, *2*, 316–324. [CrossRef]

18. Rozin, P. The meaning of "natural" process more important than content. *Psychol. Sci.* **2005**, *16*, 652–658. [CrossRef]

19. Migliore, G.; Borrello, M.; Lombardi, A.; Schifani, G. Consumers' willingness to pay for natural food: Evidence from an artefactual field experiment. *Agric. Food Econ.* **2018**, *6*, 21. [CrossRef]

20. Dickson-Spillmann, M.; Siegrist, M.; Keller, C. Attitudes toward chemicals are associated with preference for natural food. *Food Qual. Prefer.* **2011**, *22*, 149–156. [CrossRef]

21. Evans, G.; de Challemaison, B.; Cox, D.N. Consumers' ratings of the natural and unnatural qualities of foods. *Appetite* **2010**, *54*, 557–563. [CrossRef] [PubMed]

22. Moscato, E.M.; Machin, J.E. Mother natural: Motivations and associations for consuming natural foods. *Appetite* **2018**, *121*, 18–28. [CrossRef] [PubMed]

23. Lockie, S.; Lyons, K.; Lawrence, G.; Grice, J. Choosing organics: A path analysis of factors underlying the selection of organic food among Australian consumers. *Appetite* **2004**, *43*, 135–146. [CrossRef]

24. Nocella, G.; Kennedy, O. Food health claims–What consumers understand. *Food Policy* **2012**, *37*, 571–580. [CrossRef]

25. Annunziata, A.; Vecchio, R. Functional foods development in the European market: A consumer perspective. *J. Funct. Foods* **2011**, *3*, 223–228. [CrossRef]

26. Falguera, V.; Aliguer, N.; Falguera, M. An integrated approach to current trends in food consumption: Moving toward functional and organic products? *Food Control* **2012**, *26*, 274–281. [CrossRef]

27. Lähteenmäki, L.; Lampila, P.; Grunert, K.; Boztug, Y.; Ueland, Ø.; Åström, A.; Martinsdóttir, E. Impact of health-related claims on the perception of other product attributes. *Food Policy* **2010**, *35*, 230–239. [CrossRef]

28. Siro, I.; Kápolna, E.; Kápolna, B.; Lugasi, A. Functional food. Product development, marketing and consumer acceptance—A review. *Appetite* **2008**, *55*, 456–467. [CrossRef] [PubMed]

29. Verbeke, W. Consumer acceptance of functional foods: Socio-demographic, cognitive and attitudinal determinants. *Food Qual. Prefer.* **2005**, *16*, 45–57. [CrossRef]

30. Caracciolo, F.; Vecchio, R.; Lerro, M.; Migliore, G.; Schifani, G.; Cembalo, L. Natural versus enriched food: Evidence from a laboratory experiment with chewing gum. *Food Res. Int.* **2019**, *122*, 87–95. [CrossRef]

31. Priyadarshini, A.; Priyadarshini, A. Market dimensions of the fruit juice industry. In *Fruit Juices Fruit Juices, Extraction, Composition, Quality and Analysis*; Rajauria, G., Tiwari, B., Eds.; Academic Press; Elsevier: Dublin, Ireland, 2018; Volume 19, pp. 15–32.

32. Bunnell, R.H. Enrichment of fruit products and fruit juices. *J. Agric. Food Chem.* **1968**, *16*, 177–183. [CrossRef]

33. Verbeke, W.; Scholderer, J.; Lähteenmäki, L. Consumer appeal of nutrition and health claims in three existing product concepts. *Appetite* **2009**, *52*, 684–692. [CrossRef]

34. Oláh, J.; Virglerova, Z.; Popp, J.; Kliestikova, J.; Kovács, S. The Assessment of Non-Financial Risk Sources of SMES in the V4 Countries and Serbia. *Sustainability* **2019**, *11*, 4806. [CrossRef]

35. Lakner, Z.; Kiss, A.; Merlet, I.; Oláh, J.; Máté, D.; Grabara, J.; Popp, J. Building Coalitions for a Diversified and Sustainable Tourism: Two Case Studies from Hungary. *Sustainability* **2018**, *10*, 1090. [CrossRef]

36. Nagy-Pércsi, K.; Fogarassy, C. Important Influencing and Decision Factors in Organic Food Purchasing in Hungary. *Sustainability* **2019**, *11*, 6075. [CrossRef]

37. Lee, H.J.; Yun, Z.S. Consumers' perceptions of organic food attributes and cognitive and affective attitudes as determinants of their purchase intentions toward organic food. *Food Qual. Prefer.* **2015**, *39*, 259–267. [CrossRef]

38. Balázs, B. Local Food System Development in Hungary. *Int. J. Sociol. Agric. Food* **2012**, *19*, 403–421.

39. O'Donnell, J. Gen Y sits on top of consumer food chain; they're savvy shoppers with money and influence. *USA Today* **2006**, *11*, 322–358.

40. Yang, Z.; Lu, W.; Long, Y.; Bao, X.; Yang, Q. Assessment of heavy metals contamination in urban topsoil from Changchun City, China. *J. Geochem. Explor.* **2011**, *108*, 27–38. [CrossRef]

41. Bucic, T.; Harris, J.; Arli, D. Ethical consumers among the millennials: A cross-national study. *J. Bus. Ethics* **2012**, *110*, 113–131. [CrossRef]

42. Ferris, G.R.; Berkson, H.M.; Harris, M.M. The recruitment interview process: Persuasion and organization reputation promotion in competitive labor markets. *Hum. Resour. Manag. Rev.* **2002**, *12*, 359–375. [CrossRef]

43. Vickrey, W. Counter speculation, auctions, and competitive sealed tenders. *J. Financ.* **1961**, *16*, 8–37. [CrossRef]

44. Lusk, J.L.; Alexander, C.; Rousu, M.C. Designing experimental auctions for marketing research: The effect of values, distributions, and mechanisms on incentives for truthful bidding. *Rev. Mark. Sci.* **2007**, *5*. [CrossRef]

45. Shogren, J.F.; Shin, S.Y.; Hayes, D.J.; Kliebenstein, J.B. Resolving differences in willingness to pay and willingness to accept. *Am. Econ. Rev.* **1994**, *30*, 255–270.

46. Onwezen, M.C.; Reinders, M.J.; van der Lans, I.A.; Sijtsema, S.J.; Jasiulewicz, A.; Guardia, M.D.; Guerrero, L. A cross-national consumer segmentation based on food benefits: The link with consumption situations and food perceptions. *Food Qual. Prefer.* **2012**, *24*, 276–286. [CrossRef]

47. Zandstra, E.H.; De Graaf, C.; Van Staveren, W.A. Influence of health and taste attitudes on consumption of low- and high-fat foods. *Food Qual. Prefer.* **2001**, *12*, 75–82. [CrossRef]

48. Roininen, K.; Lähteenmäki, L.; Tuorila, H. Quantification of consumer attitudes to health and hedonic characteristics of foods. *Appetite* **1999**, *33*, 71–88. [CrossRef] [PubMed]

49. Lähteenmäki, L.; Lyly, M.; Urala, N. Consumer attitudes towards functional foods. *Underst. Consum. Food Prod.* **2004**, 412–427. [CrossRef]

50. Cameron, A.C.; Trivedi, P.K. *Microeconometrics, Methods and Applications*; Cambridge University Press: New York, NY, USA, 2005; ISBN 13 978-0-521-84805-3.

51. Tuorila, H. Attitudes as determinants of food consumption. In *Encyclopedia of Human Biology*, 2nd ed.; Academic Press: San Diego, CA, USA, 1997; Volume 1.

52. Bruschi, V.; Teuber, R.; Dolgopolova, I. Acceptance and willingness to pay for health-enhancing bakery products–Empirical evidence for young urban Russian consumers. *Food Qual. Prefer.* **2015**, *46*, 79–91. [CrossRef]

53. Rizzo, G.; Borrello, M.; Dara Guccione, G.; Schifani, G.; Cembalo, L. Organic Food Consumption: The Relevance of the Health Attribute. *Sustainability* **2020**, *12*, 595. [CrossRef]

54. Apaolaza, V.; Hartmann, P.; D'Souza, C.; López, C.M. Eat organic–Feel good? The relationship between organic food consumption, health concern and subjective wellbeing. *Food Qual. Prefer.* **2018**, *63*, 51–62. [CrossRef]

55. Barauskaite, D.; Gineikiene, J.; Fennis, B.M.; Auruskeviciene, V.; Yamaguchi, M.; Kondo, N. Eating healthy to impress: How conspicuous consumption, perceived self-control motivation, and descriptive normative influence determine functional food choices. *Appetite* **2018**, *131*, 59–67. [CrossRef]

Permissions

The contributors of this book come from diverse backgrounds, making this book a truly international effort. This book will bring forth new frontiers with its revolutionizing research information and detailed analysis of the nascent developments around the world.

We would like to thank all the contributing authors for lending their expertise to make the book truly unique. They have played a crucial role in the development of this book. Without their invaluable contributions this book wouldn't have been possible. They have made vital efforts to compile up to date information on the varied aspects of this subject to make this book a valuable addition to the collection of many professionals and students.

This book was conceptualized with the vision of imparting up-to-date information and advanced data in this field. To ensure the same, a matchless editorial board was set up. Every individual on the board went through rigorous rounds of assessment to prove their worth. After which they invested a large part of their time researching and compiling the most relevant data for our readers.

The editorial board has been involved in producing this book since its inception. They have spent rigorous hours researching and exploring the diverse topics which have resulted in the successful publishing of this book. They have passed on their knowledge of decades through this book. To expedite this challenging task, the publisher supported the team at every step. A small team of assistant editors was also appointed to further simplify the editing procedure and attain best results for the readers.

Apart from the editorial board, the designing team has also invested a significant amount of their time in understanding the subject and creating the most relevant covers. They scrutinized every image to scout for the most suitable representation of the subject and create an appropriate cover for the book.

The publishing team has been an ardent support to the editorial, designing and production team. Their endless efforts to recruit the best for this project, has resulted in the accomplishment of this book. They are a veteran in the field of academics and their pool of knowledge is as vast as their experience in printing. Their expertise and guidance has proved useful at every step. Their uncompromising quality standards have made this book an exceptional effort. Their encouragement from time to time has been an inspiration for everyone.

The publisher and the editorial board hope that this book will prove to be a valuable piece of knowledge for researchers, students, practitioners and scholars across the globe.

List of Contributors

Deepak Kumar and Prasanta Kalita
ADM Institute for the Prevention of Postharvest Loss, University of Illinois at Urbana-Champaign, Urbana, IL 61801, USA

Tshudufhadzo Mphaphuli
Department of Horticulture, Tshwane University of Technology, Pretoria West 0001, South Africa
Phytochemical Food Network Group, Department of Crop Sciences Tshwane University of Technology, Pretoria West 0001, South Africa

Vimbainashe E. Manhivi and Dharini Sivakumar
Phytochemical Food Network Group, Department of Crop Sciences. Tshwane University of Technology, Pretoria West 0001, South Africa

Retha Slabbert
Department of Horticulture, Tshwane University of Technology, Pretoria West 0001, South Africa

Yasmina Sultanbawa
Australian Research Council Industrial Transformation Training Centre for Uniquely Australian Foods, Queensland Alliance for Agriculture and Food Innovation, Center for Food Science and Nutrition, The University of Queensland, St Lucia, QLD 4069, Australia

Anna Walaszczyk and Barbara Galińska
Faculty of Management and Production Engineering, Lodz University of Technology, 90-924 Lodz, Poland

Zofia Utri and Dominika Głąbska
Department of Dietetics, Institute of Human Nutrition Sciences, Warsaw University of Life Sciences (SGGW-WULS), 159c Nowoursynowska Street, 02-776 Warsaw, Poland

Kevin Serrem, Anna Dunay and Csaba Bálint Illés
Institute of Business Economics, Leadership and Management, SzentIstván University, 2100 Gödöllő, Hungary

Charlotte Serrem
Department of Consumer Sciences, School of Agriculture and Biotechnology, University of Eldoret, Eldoret 1125-30100, Kenya

Bridget Atubukha
Faculty of Bioscience Engineering, Katholieke Universitiet Leuven, 3001 Leuven, Belgium

Judit Oláh
Institute of Applied Informatics and Logistics, Faculty of Economics and Business, University of Debrecen, 4032 Debrecen, Hungary
TRADE Research Entity, Faculty of Economic and Management Sciences, North-West University, Vanderbijlpark 1900, South Africa

Gyula Kasza, Annamária Dorkó and Atilla Kunszabó
Risk Management Directorate, National Food Chain Safety Office, 1024 Budapest, Hungary

Ákos Mesterházy
Cereal Research Non-Profit Ltd., 6701 Szeged, Hungary

József Popp
Faculty of Economic and Management Sciences, TRADE Research Entity, North-West University, Vanderbijlpark 1900, South Africa
Faculty of Economics and Social Sciences, Szent István University, 2100 Gödölő, Hungary

Dan Costin Niţescu and Valentin Murgu
Bucharest University of Economic Studies, Faculty of Finance and Banking, Department of Money and Banking, 010961 Bucharest, Romania

Jana Majerova, Anna Krizanova, Lubica Gajanova and Margareta Nadanyiova
Department of Economics, Faculty of Operation and Economics of Transport and Communications, University of Zilina, Zilina 026 01, Slovakia

Wlodzimierz Sroka
Management Department, WSB University, Dąbrowa Gornicza 41-300, Poland
North-West University, Potchefstroom 2351, South Africa

George Lazaroiu
Department of Economic Sciences, Spiru Haret University, Bucharest 030045, Romania

Chun-Chieh Ma
Department of Public Administration and Management, National University of Tainan, No.33, Sec. 2, Shu-Lin St., Tainan 70005, Taiwan

Han-Shen Chen and Hsiao-Ping Chang
Department of Health Diet and Industry Management, Chung Shan Medical University, No.110, Sec. 1, Jianguo N. Rd., Taichung City 40201, Taiwan
Department of Medical Management, Chung Shan Medical University Hospital, No. 110, Sec. 1, Jianguo N. Rd., Taichung City 40201, Taiwan

Dániel Fróna and János Szenderák
Karoly Ihrig Doctoral School of Management and Business, University of Debrecen, 4032 Debrecen, Hungary
Institute of Sectoral Economics and Methodology, University of Debrecen, 4032 Debrecen, Hungary

Mónika Harangi-Rákos
Institute of Sectoral Economics and Methodology, University of Debrecen, 4032 Debrecen, Hungary

Faheem Bukhari
Faculty of Business Administration, IQRA University, Defense view Shaheed-e-Millet Road (Ext), Karachi 75500, Pakistan

Saima Hussain
Faculty of Management Sciences, Shaheed Zulfiqar Ali Bhutto Institute of Science and Technology, SZABIST 90, Block 5, Clifton Karachi 75600, Pakistan

Rizwan Raheem Ahmed
Faculty of Management Sciences, Indus University, Karachi 75300, Pakistan

Dalia Streimikiene
Institute of Sport Science and Innovations, Lithuanian Sports University, Kaunas 44221, Lithuania

Riaz Hussain Soomro
Institute of Health Management, Dow University of Health Sciences, Karachi 74200, Pakistan

Zahid Ali Channar
Department of Business Administration, Sindh Madressatul Islam University, Karachi 74000, Pakistan

Dávid Szakos, László Ózsvári and Gyula Kasza
Department of Veterinary Forensics an Economics, University of Veterinary Medicine Budapest, 1078 Budapest, Hungary

Alina Vysochyna
Department of Accounting and Taxation, Sumy State University, 40000 Sumy, Ukraine

Natalia Stoyanets
Faculty of Economics and Management, Sumy National Agrarian University, 40000 Sumy, Ukraine

Grzegorz Mentel
Department of Quantitative Methods, The Faculty of Management, Rzeszow University of Technology, 35-959 Rzeszow, Poland

Tadeusz Olejarz
Department of Management Systems and Logistics, The Faculty of Management, Rzeszow University of Technology, 35-959 Rzeszów, Poland

József Tóth
Department of Agricultural Economics and Rural Development, Corvinus University of Budapest, 1093 Budapest, Hungary
Faculty of Economics, Socio-Human Sciences and Engineering, Sapientia Hungarian University of Transylvania, 530104 Miercurea Ciuc, Piața Libertății nr. 1, Romania

Giuseppina Migliore, Giorgio Schifani and Giuseppina Rizzo
Department of Agricultural, Food and Forest Sciences, University of Palermo, 90128 Palermo, Italy

Index

9 781641 168410